OWLS

OF THE WORLD

OWLS
OF THE WORLD

a photographic guide

Heimo Mikkola

FIREFLY BOOKS

A FIREFLY BOOK

Published by Firefly Books Ltd. 2012

Text copyright © 2012 Heimo Mikkola
Photographs copyright © 2012 With the individual
photographers – see credits on pages 505–506

First printing

Publisher Cataloging-in-Publication Data (U.S.)

A CIP record for this title is available from the Library of Congress

Library and Archives Canada Cataloguing in Publication

Mikkola, Heimo
     Owls of the world : a photographic guide/Heimo Mikkola.
– 1st ed.
Includes index
ISBN 978-1-77085-136-8
     1. Owls. 2. Owls –Pictorial works. I. Title.
QL696.S8M54 2012   598.9'7   C2012-900630-0

Published in the United States by
Firefly Books (U.S.) Inc.
P.O. Box 1338, Ellicott Station
Buffalo, New York 14205

Published in Canada by
Firefly Books Ltd.
66 Leek Crescent, Richmond Hill
Ontario L4B 1H1

Design: Julie Dando, Fluke Art
Printed in China by C&C Offset Printing Co Ltd.

First published in the UK by
Christopher Helm,
an imprint of Bloomsbury Publishing Plc,
50 Bedford Square,
London WC1B 3DP

CONTENTS

49. SUNDA SCOPS OWL	*Otus lempiji*	162
50. NICOBAR SCOPS OWL	*Otus alius*	163
51. SIMEULUE SCOPS OWL	*Otus umbra*	164
52. ENGGANO SCOPS OWL	*Otus enganensis*	165
53. MENTAWAI SCOPS OWL	*Otus mentawi*	166
54. RAJAH SCOPS OWL	*Otus brookii*	168
55. SINGAPORE SCOPS OWL	*Otus cnephaeus*	169
56. PHILIPPINE SCOPS OWL	*Otus megalotis*	170
57. PALAWAN SCOPS OWL	*Otus fuliginosus*	172
58. WHITE-FRONTED SCOPS OWL	*Otus sagittatus*	173
59. REDDISH SCOPS OWL	*Otus rufescens*	174
60. SERENDIB SCOPS OWL	*Otus thilohoffmanni*	176
61. ANDAMAN SCOPS OWL	*Otus balli*	177
62. JAVAN SCOPS OWL	*Otus angelinae*	178
63. WALLACE'S SCOPS OWL	*Otus silvicola*	179
64. FLORES SCOPS OWL	*Otus alfredi*	180
65. MINDANAO SCOPS OWL	*Otus mirus*	181
66. LUZON SCOPS OWL	*Otus longicornis*	182
67. MINDORO SCOPS OWL	*Otus mindorensis*	183
68. MOLUCCAN SCOPS OWL	*Otus magicus*	184
69. WETAR SCOPS OWL	*Otus tempestatis*	186
70. SULA SCOPS OWL	*Otus sulaensis*	188
71. BIAK SCOPS OWL	*Otus beccarii*	189
72. SULAWESI SCOPS OWL	*Otus manadensis*	190
73. SANGIHE SCOPS OWL	*Otus collari*	192
74. MANTANANI SCOPS OWL	*Otus mantananensis*	193
75. FLAMMULATED OWL	*Philoscops flammeolus*	196
76. WESTERN SCREECH OWL	*Megascops kennicottii*	198
77. EASTERN SCREECH OWL	*Megascops asio*	201
78. PACIFIC SCREECH OWL	*Megascops cooperi*	204
79. OAXACA SCREECH OWL	*Megascops lambi*	205
80. WHISKERED SCREECH OWL	*Megascops trichopsis*	206
81. BEARDED SCREECH OWL	*Megascops barbarus*	208
82. BALSAS SCREECH OWL	*Megascops seductus*	210
83. BARE-SHANKED SCREECH OWL	*Megascops clarkii*	211
84. TROPICAL SCREECH OWL	*Megascops choliba*	212
85. MARIA KOEPCKE'S SCREECH OWL	*Megascops koepckeae*	216
86. PERUVIAN SCREECH OWL	*Megascops roboratus*	218
87. TUMBES SCREECH OWL	*Megascops pacificus*	220
88. MONTANE FOREST SCREECH OWL	*Megascops hoyi*	222
89. RUFESCENT SCREECH OWL	*Megascops ingens*	223
90. SANTA MARTA SCREECH OWL	*Megascops* sp.	224
91. COLOMBIAN SCREECH OWL	*Megascops colombianus*	225
92. CINNAMON SCREECH OWL	*Megascops petersoni*	226

ACKNOWLEDGEMENTS

Many people have assisted me in the preparation of this book. The photographers who forwarded hundreds of their amazing images make the book truly a work of art.

Special thanks go to Freidhelm Adam, Yves Adams, Roger Ahlman, Deborah Allen, Desmond Allen, Michael Anton, Christian Artuso, Robin Arundale, David Ascanio, Nick Athanas, Aurélien Audevard, Nick Baldwin, Matt Bango, Paul Bannick, Glenn Bartley, Eyal Bartow, Bill Baston, David Behrens, Bonnie Block, Amir Ben Dov, Nik Borrow, Adrian Boyle, Neil Bowman, Chris Brignell, Dušan M. Brinkhuizen, Jim and Deva Burns, Thomas M. Butynski, John Carlyon, Pei-Wen Chang, HY Cheng, Robin Chittenden, Rohan Clarke, Erwin Collaerts, Jesus Contreras, Murray Cooper, Mike Danzenbaker, Abhiskek Das, Subharghya Das, Mario Dávalos P., Santiago David-R., Kleber de Burgos, Roy de Haas, Gehan de Silva Wijeyeratne, Matthias Dehling, Bram Demeulemeester, Arpit Deomurari, Kathleen Deuel, Eric Didner, K.-D.B. Dijkstra, Lee Dingain, Andrés Miguel Domingues, Guy Dutson, James Eaton, Knut Eisermann, Willy Ekariyono, Anne Elliott, Stuart Elsom, Jacques Erard, Hanne and Jens Eriksen, Mandy Etpison, John Evesen, Augusto Faustino, Ian Fisher, Tim Fitzharris, Dick Forsman, Christian Fosserat, Errol Fuller, Guilherme Gallo-Ortiz, Nick Gardner, Tom and Pam Gardner, Hans Germeraad, Steve Gettle, Martin Goodey, Michael Gore, Martin Gottschling, Arthur Grosset, Jon Groves, Roberto Güller, Stefan Hage, Martin Hale, Trevor Hardaker, Karen Hargreave, Hugh Harrop, Sebastian K. Herzog, Uditha Hettige, John Hicks, Ron Hoff, Stefan Hohnwald, David Hollands, John and Jemi Holmes, Marcel Holyoak, Dieter Hopf, Jon Hornbuckle, David Hosking, Steve Huggins, Rob Hutchinson, David Jirovsky, Donald M. Jones, Jayesh Joshi, Arto Juvonen, Adam Scott Kennedy, Vicki Louise Kennedy, Kazuyasu Kisaichi, Seig Kopinitz, Evgeny Kotelevsky, Rolf Kunz, René-Marie Lafontaine, Markus Lagerqvist, Sander Lagerveld, Frank Lambert, Hennie Lammers, Martjan Lammertink, Danny Laredo, Ch'ien C. Lee, Wil Leurs, Tasso Leventis, Jerry Ligon, Markus Lilje, Lucas Limonta, Kevin Lin, Dario Lins, Scott Linstead, Dan Lockshaw, James Lowen, Ram Mallya, Thomas Mangelsen, Charles Marsh, Ralph Martin, Daniel Martínez-A., Jonathan Martinez, S. & D. & K. Maslowski, Marco Mastorilli, Bence Mate, Marko Matesic, Juan Matute, András Mazula, Luiz Gabriel Mazzoni, Rob McKay, Phil McLean, Ross McLeod, Ian Merrill, Jérome Micheletta, Dominic Mitchell, John Mittermeier, David Monticelli, Lee Mott, Werner Müller, José Carlos Motta-Junior, Rebecca Nason, David W. Nelson, Jonathan Newman, Paul Noakes, Rolf Nussbaumer, Daniele Occhiato, János Oláh, Fabio Olmos, Scott Olmstead, Atte Ivar Olsen, Erica Olsen, Alain Pascua, Jari Peltomäki, Vincenzo Penteriani, Niall Perrins, Winnie Poon, Richard Porter, Hira Punjabi, Mathias Putze, Esko Rajala, Yves-Jacques Rey-Millet, Adam Riley, Amano Samarpan, Niranjan Sant, Ali Sadr, Ran Schols, Yeray Seminario, David Shackleford, Dubi Shapiro, Tadao Shimba, Jussi Sihvo, Oliver Smart, Paul Smith, Lars Soerink, Eric Sohn Joo Tan, Franz Steinhauser, Ulf Ståhle, Matti Suopajärvi, Harri Taavetti, Kenji Takehara, Toyonari Tanaka, Stan Tekiela, Gary Thoburn, Russell Thorstrom, Chris Townend, Gaku Tozuka, Keith Valentine, Rick van der Weijde, Erik van der Werf, Peter van der Wolf, Menno van Duijn, Lesley van Loo, Fred van Olphen, Chris van Rijswijk, Alex Vargas, Filip Verbelen, Rollin Verlinde, Jan Vermeer, Tom Vezo, S.P. Vijayakumar, Thomas Vinke, Choi Wai Mun, Dave Watts, Doug Wechsler, Roger Wilmshurst, Martin B. Withers, Michelle and Peter Wong and Simon Woolley.

Published and unpublished taxonomic data, measurements, biological and distributional information, or other useful materials and data were provided by numerous people, including Christian Artuso, Roseanna Avento, Paula L. Enríquez, Jon Fjelså, John Gray, Monika Kirk, Niels Kaare Krabbe, Rolf G. Krahe, Mauricio Ugarte-Lewis, Ossi V. Lindqvist, Karl Mayer, Theodor Mebs, Anita Mikkola, Kariuki Ndanganga, Darcy Ogada, Juan Freile Ortiz, Wolfgang Scherzinger, Jevgeni Shergalin, Friedhelm Weick, Michael Wink and Tamás Zalai.

At Bloomsbury my editor, Jim Martin, was invaluable in sourcing from his many contacts a considerable number of the images that I would otherwise have never found; he also provided much useful guidance and stimulating 24-hour email discussions throughout the entire process. Nigel Redman helped find several obscure taxa and took photos of specimens at the Natural History Museum (BMNH) at Tring, England. Ernest Garcia and Nigel Redman read the proofs and improved the text, maps and photo captions by pointing out inconsistencies and errors. Special thanks go to Julie Dando at Fluke Art: as project manager and designer, she sifted through literally thousands of photos, carefully selecting the images finally chosen and exercising much-needed quality control. She also sourced many images herself and spent countless hours re-working layouts as new images were found. The final result is a tribute to her skills.

◀ Luzon Scops Owl (*Bram Demeulemeester*).

INTRODUCTION

In late 2009, I was approached by Jim Martin from Bloomsbury Publishing with a view to writing this new book on the world's owls, then more than 25 years after the publication of my own *Owls of Europe*, which is still available (as an e-book) from the same publisher. Since 1983, much has been published on owls and owl biology, including an excellent survey *Owls of the Northern Hemisphere* by Karel Voous (1988) and *Owls of the World* by James Duncan (2003), as well as the second edition of a comprehensive book on the taxonomy, distribution and identification of the owls of the world (König, Weick and Becking 2008). For the present work, all of these books have been widely consulted and the distribution maps are largely based on those in the last-mentioned book, with the kind permission of the publisher.

The majority of the world's owls, 68 per cent, live in the Southern Hemisphere, the remaining 32 per cent occurring in the Northern Hemisphere. Most are forest-adapted birds, and thus feeling the pressures of worldwide deforestation trends. An estimated 75 per cent of the nearly 250 surviving species of owl are associated with dense and undisturbed forests, the very habitat that is nowadays being destroyed at ever-increasing rates. Deforestation, usually in order to make way for agriculture, has been underway for decades, Brazil and Indonesia being the hotspots. It was once widely believed that the voices of owls emanating from the dark forests were an omen of impending ill fortune, if not death. Perhaps we should now realise that it is, in fact, the increasing absence of owl voices that should be taken as a sign of impending ill fortune for the human species. In logged and silent forests the future will be a stressful one for the owls, as well as for all of us human beings. Although a majority of the world's people now live in cities, we are dependent more than ever on forests, in a way that few of us understand. We should recognise the great value of trees and forests in helping us to deal with the excess carbon that we are generating. The burning of forests not only ends their ability to absorb carbon, but it also produces an immediate flow of carbon back into the atmosphere, making it one of the leading sources of greenhouse-gas emissions. Trees themselves could become victims on a massive scale if climate change eventually causes widespread forest death in such places as Amazonia. One potentially promising plan calls for wealthy countries to help those in the tropics to halt the destruction of their immense forests for agriculture and timber. It is hoped that this book will help us all to understand more about how owls relate to their environment, and how important it is for us to use that environment wisely.

In the main, internet and literature searches for this book have not extended beyond the end of November 2011. It is expected that many new discoveries about owl distribution and biology will be made in the near future. There will certainly also be a major revision of some of the ideas about owl taxonomy presented in these pages, as they do, inevitably, raise more unanswered questions than give valid answers. It would be a great pleasure to receive readers' comments and criticisms on any matters connected with the owls of the world. Any photographs of new or less well-known species or subspecies would be similarly welcomed for future editions of the book. All communications should be addressed to the author c/o Bloomsbury Publishing or e-mail Jim.Martin@bloomsbury.com.

◀ Magellanic Horned Owl (*Rob Hutchinson*).

WHAT MAKES AN OWL?

With very few exceptions, owls look like nothing other than owls. They are soft-plumaged, short-tailed, big-headed birds, with large eyes surrounded, usually, by a broad facial disc. Owls probably have the most frontally situated eyes of all birds. This, together with their ability to blink with the upper eyelids, gives them a semi-human appearance, in which surely lies much of their appeal to man.

Owls exhibit a number of adaptations which enable them to operate with outstanding efficiency as nocturnal predators.

Vision

First of all, owls have undergone the most successful visual adaptation to nocturnal living by having eyes well suited for hunting their prey in poor light, although in absolute darkness they cannot see at all. The owl eye has become elongated through the development of an enormously enlarged cornea, pupil and lens. Owl eyes are 2.2 times larger than the average for birds of the same weight. The eyeballs of some large owls are, in fact, larger than the human eye. Even in the medium-sized Tawny Owl the overall length of the eye exceeds that of the human one. The huge eyes are shaped like tapering cylinders so as to provide the largest possible expanse of retina, and a notably thickened cornea acts as an additional lens. By these means, in addition, the amount of light entering the eye is increased, so that a brighter image falls on the retina.

Owls have forward-facing eyes, and the effect is heightened by the fact that their bills, unlike those of other birds of prey, are deflected more or less downwards to avoid obstructing their field of vision. They have a rather narrow visual field of 110°, of which 60° to 70° is overlapping. Frontally situated eyes give a considerable degree of binocular vision, but the eyes themselves are nearly immobile owing to the fusing of the sclerotic ring with the skull. For the Tawny Owl, the area of binocularity is estimated to be only 48°. The owls must turn the whole head to look sideways, but they have an exceptional ability to rotate the head; that of the Long-eared Owl is capable of turning through at least 270°. Like man, owls employ their binocular vision effectively, looking at an object with both eyes in order to judge its position accurately. This is known as the parallax method, and becomes more effective the farther apart the eyes are placed. Larger owls have their eyes well spaced, and it seems that the smaller species, by having flatter skulls, have developed as far as is physically reasonable in order to achieve wide interocular spacing. Owls can further improve this three-dimensional vision by constantly moving and bobbing the head. The head-bobbing movements of many owls may be amusing, but this

▲ This Southern White-faced Owl's *Ptiliopsis granti* head shows well the huge size of the eyes (*Mark Bridger*).

peculiar behaviour is totally under visual control. Movement of the visual world alone will cause the owl to bob its head, as it is merely assessing as fully as possible what it sees before deciding what action to take.

Anatomical studies of the Burrowing Owl have revealed that binocular integration of visual inputs from a pair of eyes occurs at a higher level of the brain in a specific structure called the visual 'wulst'. This appears to be very similar to the striate cortex, which mediates stereo vision in mammals. The wulst varies in size and this is directly proportional to the extent of the binocular visual field. Electrophysiological studies have shown that the wulst of the Barn Owl has a large population of neurons selective for binocular disparity, a prerequisite for stereopsis. Scientific tests have proven that a Barn Owl's visual sensitivity is at least 35 times better than that of a human being.

The number of light-sensitive elements (mostly rods) in the retina is very high, and this increases both the owls' visual acuity at low light levels and their sensitivity to light. Not surprisingly, all owls have excellent powers of adaptation to the dark, but this is particularly the case with nocturnal species such as the Tawny Owl, the

▼ Oriental Bay Owl *Phodilus badius* showing the pale blue transparent fold of skin, the nictitating membrane, which can be drawn across the eye (*James Eaton*).

▼ Owls have extraordinarily mobile necks to allow views from a variety of angles. Little Owl *Athene noctua* (*Chris Brignell*).

retina of which has about 56,000 light-sensitive rods per square millimetre. These decidedly nocturnal owls are seldom seen during the day and sometimes appear to be blinded by very strong sunlight. On the other hand, many owls, such as the Eurasian Pygmy Owl and the Burrowing Owl, are able to hunt in daylight, and none is really helpless during the day. The Eurasian Eagle Owl has slightly better daytime vision than humans. This is made possible by an exceptional range of aperture sizes (pupil size) controlled by the iris. It is always difficult to prove whether an animal can perceive colour. As owls have some cones in their retina, it is likely that they can do so when the light is good. Experiments on the crepuscular Little Owl have demonstrated that this species can see at least yellow, green and blue, but it confused red and the darkest grey. More diurnal owls such as the Eurasian Pygmy Owl are able to distinguish colours.

Recent research has assessed the claims that owls can see in the ultra-violet part of the spectrum as can many diurnal raptors, such as Common Kestrels *Falco tinnunculus*, which are known to use ultraviolet vision and the ultraviolet-reflecting scent markings (urine and faeces) of small mammals as a cue to the location of areas of food abundance. Examination of the eyes of the Tawny Owl suggests that owls lack the ultraviolet-sensitive/violet-sensitive cone class associated with ultraviolet vision. Laboratory experiments also confirmed that nocturnal Tengmalm's Owls are not able to use UV light to find suitable hunting areas. New experiments with diurnal pygmy owls, however, are showing that these can detect near-ultraviolet and can use UV to gain information about prey in the same way as do diurnal raptors.

Hearing

The owl's sense of hearing is no less remarkable than its exceptional sight. This is made possible by adaptations in ear structure, which have modified the skull as profoundly as have those of the eyes. The most obvious of these adaptations is the sheer size of the ear openings. Instead of fairly small round openings as in most birds, owls have surprisingly large, half-moon-shaped vertical slits, nearly as deep as the head itself. The inner ear of an owl is very large, and the auditory region of the brain is provided with many more nerve cells than in other birds of comparable size. Laboratory tests have shown that the Barn Owl's sense of hearing is some ten times better than that of humans. Aural abilities are aided by a wide outer ear tube, and in some species by the presence of large conchae, surrounded by the feathers of the facial disc, which can be erected at will. The exact amount of muscular control that the owl has over the facial disc and the skin flaps in the ear openings is not fully known. This aural dexterity, however, enables owls to scan different parts of their environment for sounds in the same way as many mammals can do by moving their external ears.

In some species, a striking asymmetry in the shape and relative position of the external part of the ear, including the bones surrounding the tympanic region and the operculum, has been described. Asymmetrical ear openings are known to help the owls to locate the source of sound with precision. The owl can discriminate changes in the location of sound sources as small as 3°, and can direct its head to within 2° of a source.

In laboratory experiments, Barn Owls have been found to be capable of locating and striking an unseen living prey in complete darkness, by using only their acute sense of hearing. The high-frequency squeaks of small prey and the rustle of dry leaves contain all the information that the owl needs to pinpoint the exact position of its intended prey.

Similarly in the wild, many owls rely only on their remarkable auditory powers, at least when hunting rodents and shrews during the winter, when these mammals often move unseen under the snow cover. The American Barn Owl, Great Grey Owl and Long-eared Owl are able to localise invisible small mammals from the air, and to catch them blindly by pouncing

▲ Australian Barn Owl *Tyto delicatula* showing the huge ear opening – normally hidden by skin flaps and facial disc feathers, which can be erected at will (*Rohan Clarke*).

on the correct spot below the snow surface. Those owl species not having the ears so highly developed can always combine their acute visual and aural abilities, because there is never total darkness in places where an owl is likely to hunt.

Recently, it has been noted that the owl, in order to guide its aural accuracy, uses a mental map tuned to the location of sounds. This auditory map cannot maintain its accuracy by sound alone, but is modified by visual information so that it continually updates its preciseness and accuracy. There is a 'gate' in the brain which, when opened, allows the auditory map to receive the visual information that it needs. So, it can be concluded that an owl is using both its visual and its aural abilities to pinpoint the scurrying of a mouse or other small animal, before it swoops out of the night sky and dispatches its prey with lethal accuracy.

It has been stated many times, and very strongly, that owls are not able to use echolocation when moving in the dark, just as firmly as it was claimed that owls cannot use UV light as a cue to find prey. American Barn Owls and related species emit metallic clicking sounds when flying, suggesting possible echolocation for orientation, and further investigation is perhaps needed in order to avoid similar incorrect generalisations to those involving UV light.

Significance of ear-tufts

About 44 per cent of the world's 249 species of owl have on their heads tufts of feathers commonly called 'ears' or 'horns'. The ear-tufts which so many of the species possess do not, in fact, have anything to do with the sense of hearing. The adaptive value of ear-tufts has been construed in three ways. First, ear-tufts do no more than express mood and act as night-time recognition signals, i.e. the presence or absence of tufts may help to distinguish species at short range: in other words, tufts provide a silhouette which, when combined with voice, facilitates species recognition; owls can see well enough at night to distinguish head shapes, and tufted and untufted species are often sympatric. Second, ear-tufts mimic the ears of mammals, and thus, during threat displays, make an owl's face resemble that of a mammal: owl nests are often visited by mammalian predators such as lynx, fox and pine marten, and in such confrontations the owl's mammal-like face, complete with ears, might make the predator withdraw. Third, ear-tufts also aid camouflage by breaking up the outline of the owl's head when the owl stretches upwards, ear-tufts erect, and so it looks more like a broken-off stub: this effect is possible because owls sit upright and most are coloured in grey-browns and greys. Even some untufted species,

▼ Long-eared Owls *Asio otus* have prominent ear-tufts, as their name suggests (*Steve Huggins*).

▼ In flight even the prominent ear-tufts of a Long-eared Owl *Asio otus* may not be visible (*Jussi Sihvo*).

▼ Madagascar Scops Owl *Otus rutilus* has very small ear-tufts (*Roy de Haas*).

▼ Barred Eagle Owl *Bubo sumatranu*s has prominent sideways-pointing ear-tufts (*HY Cheng*).

▼ A female Tengmalm's Owl *Aegolius funereus* showing the false 'ear-tufts' well (*Matti Suopajärvi*).

▼ The Oriental Bay Owl *Phodilus badius* has perhaps the most impressive false 'ear-tufts' (*Roy de Haas*).

such as Tengmalm's Owl and the Northern Saw-whet Owl, assume 'tufted' postures when alarmed during diurnal roosting. The change in body shape, from a rounded appearance to a narrow vertical oblong with outer crown feathers erected (as ear-tufts), certainly makes the owl less conspicuous.

Silent flight

Most owls have relatively large, rounded wings, which are shorter in those species which hunt in cover, and much longer in those which hunt in open country or are highly migratory. There are ten primary feathers, not including the rudimentary first primary concealed beneath the coverts, and twelve rectrices. The remiges and rectrices are relatively soft, particularly at their outer edges, which form a softened fringe. This feature was thought to assist the owl in flying silently, but the experimental removal of this fringe seems to make no difference to the wing noise of the Tawny Owl. Nevertheless, owls do fly silently, which is a necessary asset for a group of birds that hunt their prey, by sound as well as by sight, close to the ground.

Owls' wings are broad, with a large area in comparison with the weight of the bird. In aeronautical terms, many species, but not all, have a low wing-loading, especially if compared with that of some other birds (Table 1).

Table 1. Wing-loading (as total weight in g per cm^2 of wing area) of owls and some other birds.

Bird species	Wing-loading g/cm²
Great Horned Owl	0.80
Eurasian Eagle Owl	0.71
Tawny Owl	0.40
Eastern Screech Owl	0.37
Short-eared Owl	0.36
Great Grey Owl	0.35
Ural Owl	0.34
American Barn Owl	0.32
Long-eared Owl	0.29
Tengmalm's Owl	0.29
Eurasian Pygmy Owl	0.26
Barn Owl	0.21
Carrion/Hooded Crow	0.42
Peregrine Falcon	0.63
Golden Eagle	0.65
Black Grouse	1.34

The higher the wing-loading, the more effort is required for the bird to support itself in the air, and the more noise the hard-working wings are likely to make. Owls, with their low wing-loading, fly very buoyantly and effortlessly, without too much flapping and loss of energy. This characteristic enables them to glide easily and to fly slowly for long periods at a time.

◄ Long-eared Owl *Asio otus* primary, with a softened fringe at the outer edge that may have a role in silent flight (*Marco Mastrorilli*).

► Part of a Barn Owl's wing, showing the soft feather fringes that assist the owl in flying silently (*Lars Soerink*).

► Barn Owl *Tyto alba* has the lowest wing-loading of all studied owls (*Menno van Duijn*).

► Great Grey Owl *Strix nebulosa* has huge wings and a wide tail, that may put Great Greys ahead of the barn owls in aeronautical terms (*Harri Taavetti*).

Bill and claws

The bill and claws of owls are often clearly suited to their predatory way of life. The bill is hooked, usually short and not conspicuously strong, and it is directed downwards, a modification to reduce obstruction of the already limited frontal visual field. As with the diurnal birds of prey, the nostrils are placed in a soft cere at the base of the bill, which is partly hidden by the feathers of the facial disc. The feet are always four-toed, with the fourth toe reversible. The legs and toes are often feathered, perhaps as a protection against possible bites from animal prey. In the case of the Snowy Owl, the feet are well covered in feathers as a protection against the freezing ground. The fish-eating owls have bare legs and toes with rough spiny soles, and many insect-eating owls have bare legs and bristled toes. All owls have sharp, markedly curved raptorial claws, which they employ for striking and gripping prey.

◄ Short-eared Owl *Asio flammeus* has a strong raptorial bill; this allows it to switch to eating larger prey such as seabirds when small mammals are in short supply (*Jari Peltomäki*).

◄ Ural Owl *Strix uralensis* has a very strong bill, which can kill large prey like squirrels and gamebirds (*Matti Suopajärvi*).

▶ Eurasian Eagle Owl *Bubo bubo* has the strongest bill; this owl can kill prey as large as foxes, marmots and young deer (*Vincenzo Penteriani*).

▶ This leucistic Long-eared Owl *Asio otus* demonstrates the reversible outer toe characteristic of owls (*Chris van Rijswijk*).

▶ Great Grey Owl *Strix nebulosa* has fully feathered legs – protection against both frost and prey-bites. Note the very long, slightly curved claws, which can cover a large area when grasping small rodents from snow (*Hugh Harrop*).

THE NATURE OF OWLS

Shape and size

Owls are mostly birds of medium size, with some species fairly large or very small. The largest is the Eurasian Eagle Owl, reaching 75cm in length, and the smallest the Elf Owl, which is only 12–14cm long. Like most other birds, owls have dense, soft plumage which makes them look much bigger than they really are and helps to keep them warm during long periods of inactivity between hunting forays. In 92 per cent of 156 species studied, the female is bigger than the male, and only in six species is the male larger and heavier; in five species, the sexes exhibit no size difference (Table 2). See box on Sexual size dimorphism.

Table 2. Sexual size differences in the world's owls.

Larger sex	Female	Male	Sexes equal in size	Size of sexes not known
Number of species	144	7	5	93
Per cent of total	58	2	2	38
Per cent of studied species	92	5	3	

Sexual size dimorphism

As with many birds of prey, the female is often larger than the male. Incorrectly, this is called in human terms 'reversed sexual size dimorphism', although females are normally the larger sex for the vast majority of all living animals.

Over the years, numerous explanations have been offered to explain why female owls are often bigger than males of the same species. Factors that could favour the large female include egg production, incubation and nest defence. A larger female can produce more eggs, generate more heat energy for incubation, and defend her eggs and young better against predators. On the other hand, smaller males are more agile and faster fliers, using less energy to hunt for the family.

► Eurasian Eagle Owl *Bubo bubo* is the largest living owl species (*Vincenzo Penteriani*).

◄ Elf Owl *Micrathene whitneyi* is the smallest of all owls (*Jim and Deva Burns*).

Calls

Owls are heard more often than they are seen, and several species of owl sing, but songbirds they are not. Unlike most songbirds, owls do not open their bill visibly when singing. They merely inflate the throat into a small ball, which is often white and shines in the dark. In contrast to many songbirds, there is little correlation between the frequency range of an owl's hearing and the range of sounds produced by these same birds. Whereas, among songbirds, the dominant vocal frequency range is usually above the zone of maximum hearing sensitivity, for those owls so far tested the dominant frequencies of vocalisations are lower than those of their best hearing range. This would, of course, suggest that the hearing capacity of owls has evolved largely in association with adaptive prey-localisation functions, whereas their vocalisations have obviously been adaptively adjusted to provide for efficient intraspecific information-transfer under varying ecological conditions.

In temperate regions it is the owls which, in late winter or early spring, fill the night with musical hoots, while in the tropics owls are just part of a formidable chorus of animal songs and calls. Throughout the world, they have an extraordinary repertoire of caterwauling, shrieks, screams, screeches, squawks and whistles, in a range of frequencies that carry far on the night air. These vocalisations announce their presence and the existence of occupied territories. Calls are completely diagnostic of species, and owls are as likely to recognize other individuals by voice as by sight during their travels in the dark. Individual Tawny Owls can be identified from sonograms of their hooting. Similarly, recordings of Great Horned Owls reveal that each one has its own personal call, even though the song of that owl has no more than two notes, a long *woooo* and a short *woo*. These subtle differences will be enough for individuals to identify one another, to signal territory ownership or to defend the hunting grounds. Should one owl stop hooting, his territory will often be absorbed by the rival males.

There is ongoing debate over whether owls are more active and vocal on moonlit nights. It has been noted that the vocal displays of Eurasian Eagle Owls are strongly related to the phase of the moon, silence being more frequently associated with darker nights compared with brighter nights. The reason is that this species uses moonlight to increase the conspicuousness of its white throat patch, which is visible only during the vocal

► Aggressive calling of a Great Horned Owl *Bubo virginianus,* showing the balloon-like white throat, with tail and ear-tufts raised (*Paul Bannick*).

◄ Eurasian Eagle Owl *Bubo bubo* call displays relate strongly to the phases of the moon (*Vincenzo Penteriani*).

display. The Mexican subspecies *lucida* of the Spotted Owl calls more during the last quarter and new moon phases. While this is contrary to previous findings, it should be noted that Mexican Spotted Owls do not display any white plumage while calling. Moreover, it would not be advantageous for them to call more during moonlit nights, because the combination of moonlight and calling could increase the risk of predation of this small owl by the much bigger Great Horned Owl.

An excellent study of the vocal behaviour of Tawny Owls in Denmark found that the calls can be heard throughout the year, although they are not of equal intensity at all seasons. In Denmark, calling reached maximum intensity from mid-February to early May, followed by a minimum level in June–July, then another maximum from August to October, and a secondary minimum in December–January. These fluctuations reflect the reproduction cycle, moulting, territorial disputes, etc. Weather had a distinct influence on the owls' calling activity. Strong winds had an adverse effect, and wind and cold together brought a particularly pronounced reduction in calling. Owls also called less frequently in rainy weather. Tawny Owls called less when the moon was up than when the night was cloudy and overcast. I would suggest that small mammals and even some small birds are more active on moonlit nights, with the result that owls then hunt more and call less.

Colour variation and ageing

The colour of the plumage is often cryptic, which makes the owl less conspicuous when it is resting during the day. Colour variation appears to be clinal in nature, changing gradually over distance and habitat. Northern species that survive in snowbound winter conditions tend to be pale and grey, whereas those living in more southern areas are darker and brown. In Siberia, many birds tend to have an arctic form, in which the white is highly developed. Large and white Siberian Eagle Owls are well known, but Short-eared Owls and Ural Owls, too, have a pale morph in Siberia. Species having large geographical ranges that span totally different habitat types, such as the Great Horned Owl, vary strikingly in colour from pale whitish to dark brown and to burnt-orange tones, and this seems to relate to habitat type. Woodland owls tend to be brown or grey in basic coloration, whereas owls living in open habitats are typically paler, and those inhabiting desert areas, such as Hume's Owl and the Lilith Owl, are distinctly sandy-coloured.

In many species markedly different colour morphs exist, and the frequency of these is sometimes climate-related. Rufous-morph Eastern Screech Owls are relatively scarce in northern areas, because it appears that they suffer greater mortality than grey ones do during severe winter cold snaps. It has also been noted that

▼ Grey morph of Tawny Owl *Strix aluco* (*Peter Krejzl*).

▼ Reddish-brown morph of Tawny Owl (*Andrés Miguel Domínguez*).

rufous females are likely to survive cold spells better than rufous-coloured males. The most common British owl, the Tawny Owl, is typically brown, but it also has a rarer grey morph. In Finland, numbers of brown and grey morphs are also climate-related, as the grey ones survive the harsh winters better. Over the last three decades, the warming of the winter climate has produced a micro-evolutionary response in morph frequency: the brown morph has increased in number at the expense of the grey. The two morphs differ also in several life-history aspects related to immune defence against parasites and somatic maintenance costs, possibly explaining their different sensitivity to winter conditions.

Distinct sexual colour dimorphism is usually lacking among owls, but there are a few exceptions. For example, the female of the Snowy Owl is barred, whereas the male is nearly pure white. Many species, however, exhibit slight colour differences between female and male, but sex-related plumage dimorphism is generally not well studied. Many species tend to become paler or more whitish in old age, the Short-eared Owl being one such example.

The plumage of an owl chick on hatching is downy and mainly whitish in coloration. This natal plumage may be termed protoptile. However, a few groups of birds, such as penguins and owls, produce a second set of down and semiplumes, and this second nestling plumage is known as mesoptile. On fledging, the juvenile plumage is retained until the adult plumage is acquired, usually after only a few months. Differences between juvenile and adult plumages are often slight or absent, but there are some notable exceptions; in some owl species the mesoptile plumage, or parts of it, remains even years after leaving the nest, making the identification of such owls very difficult if not seen with the parents. For example, young Tengmalm's Owls are mainly chocolate-brown but a plumage rather closely, or exactly, resembling that of the adults is gained in October or November of the first year of life. The young Snowy Owl, however, which is very heavily barred, does not lose this plumage until considerably later. In South America, the Spectacled Owl has juveniles that are often known as 'white owls', as if they were of a different species. In captivity one young took five years to attain adult plumage, but a shorter time may be normal in the wild.

Below are some examples of juvenile owls that differ a lot from their respective adults.

◀ Black-and-white Owl juvenile is dirty whitish overall but has the typical black face of an adult (*Jonathan Martinez*).

▶ Spectacled Owl juvenile also has a black mask, but no white eyebrows or lores like the adults. Partly whitish-headed birds may be found more than four years after fledging.

◀ Great Horned Owl juvenile has a yellowish appearance and is fairly different from the adult. It is densely barred buff and grey; the dark ruff is already visible as in the adult (*Matt Knoth*).

▶ Barred Owl juvenile still has downy feathers on its crown and belly, but no concentric rings on the facial disc (*Paul S. Wolf*).

Abnormalities in owl plumage

Much text here has been allocated to colour mutations in an effort to correct some common misunderstandings concerning albinism. Colour mutations can also lead easily to erroneous identification of species, and it has recently been speculated, with some possible justification, that albinism has become more common in recent times as a result of climate change and ozone-layer depletion. There is already evidence that rapid human-induced climate changes have led to the observed evolutionary changes in the Tawny Owl's polymorphism. So, we should obviously pay more attention to colour mutations.

Abnormalities such as albinism, leucism and melanism may be due to changes in the amount and distribution of the pigments normally present, chemical changes in the pigments producing abnormal colours, changes in feather patterning, or changes in the structure of feathers. Abnormal pigmentation is the most frequently occurring type of plumage abnormality, at times affecting most of the commoner pigments, the melanins. Eumelanin is seen as black and grey colour, phaeomelanin as brown or buff, and erythromelanin as chestnut-red. Lipochrome pigments are responsible for the red and yellow colours. Although extreme colour abnormalities are often referred to as chance mutations, they are rather genetically controlled and usually recessive, and with captive individuals they can be produced by subsequent controlled breeding. In albinism, the complete loss of pigments makes the entire plumage white, resulting also in red eyes and pale pink legs and bill. Partial loss of pigments causes leucism and makes the plumage appear paler or white, but the normal body colours are retained. In the non-eumelanic form of schizochroic plumage, only the black is lacking and the owl appears a pale buffish-brown, with white markings where black alone was present. This is controlled by a sex-linked gene and in the wild is usually found only in females. In melanism, the black and/or brown melanin present increases, and may spread to parts of the plumage which normally lack melanin. The owl usually appears entirely black or very dark brown, or a mixture of the two.

Albinism in owls is extremely rare. So far, it has been recorded in only five species:

Eastern Screech Owl Long Island, New York, USA, 1982–87
Brown Wood Owl Sri Lanka, 1970s
Tawny Owl Italy, 1996; UK, 1996; Germany, 2005
Barred Owl North America, 1976
Short-eared Owl The Netherlands, 1967

Adult Snowy Owls are predominantly white, but their feather colour is derived from a schemocrome feather structure, which possesses little or no pigment. Light reflects within the feather structure and produces the white coloration.

Melanism Among owls, black melanism seems to be almost as rare as albinism. Only ten black owls have been listed: a Little Owl from Turkey; a Spotted Little Owl from Saswad, in India; two Tawny Owls, from Croatia and Switzerland, respectively; a Great Grey Owl from Usinskaya, Russia; a Eurasian Eagle Owl in captivity in Europe; a Great Horned Owl from Canada; and three Barn Owls from Britain, all in captivity. In the UK, there is an old report of grey and brown Short-eared Owls, but no further details are known.

Leucism is the most common abnormality among owls. Thus far I have collected the following cases:

Barn Owl One pure white male, but with normal eyes, in a small private owl collection in Norfolk, UK.
Lesser Sooty Owl One chick fledged in pure white plumage, with brown eyes, in an Australian Raptor Wildlife Centre in 2009.
Great Horned Owl Two leucistic individuals known from USA; one, named 'Casper', lives in captivity in Missouri.
Spectacled Owl Antwerp Zoo acquired a pair of leucistic owls from Central America which first bred in 1976, since when the pair has produced at least 14 leucistic progeny, although it is understood that some young have been normally coloured. Consequently, leucistic Spectacled Owls could now occur almost anywhere within the captive population.
Indian Scops Owl One pair produced one leucistic young every year between 1994 and 1996 in a collection in the UK.
Western Screech Owl One adult and one young leucistic individuals in Washington State, USA.
Eastern Screech Owl Lincoln Children's Zoo, Nebraska, USA had a pure white bird with a few tan feathers on the breast, and the eyes were not pink.
Barred Owl One leucistic owl seen often in Montana during 2010–11, and a good photograph of a very white individual from central Illinois is shown on the internet.

Great Grey Owl Five individuals with some abnormal white feathers were seen during the handling of more than 300 live and some 80 dead adult owls in Canada up to the mid-1980s. Since that time, a further two leucistic owls have been reported from Canada. In USA, the first leucistic individual in the Targhee National Forest, Idaho, in 1980, and between 1990 and 1992 there were several observations of a more strikingly white owl some 112km from Targhee. This male occupied the same breeding area over three seasons, and fathered three normally plumaged grey owlets in two out of the three breeding seasons with a normally coloured female. Two or three leucistics have also been seen in the Yellowstone National Park and one leucistic Great Grey was again seen in 2008 in Bozeman, not too far from Idaho.

In the 1990s, it was noted that one leucistic Great Grey Owl in captivity in Germany was becoming whiter from year to year after the moult. In Finland, a pale-coloured and large Great Grey Owl was followed over nine years, its abnormal colour making it almost as easy to find as it would have been if radio-tagged. See the box headed 'Linda' on page 32.

In Sweden, another light-coloured Great Grey Owl was seen in 1994. She was given the name of 'Husky' as she had light blue eyes like those of that well-known dog from Siberia, Alaska and northern Canada. Husky was not so pale as Linda, but it was speculated that she could have originated from Siberia, where Eurasian Eagle, Ural and Short-eared Owls are known to be unusually light-coloured.

Tawny Owl In Germany, there have been a number of white mutations in Tawny Owls kept in captivity. All of these white Tawnies had normal dark brown eyes, but the bill and toes were yellowish. In 1996 alone, five white Tawny Owls hatched in Germany, and thus far tens of leucistic owls have been produced.

Little Owl Old observations from Italy, in 1902, of leucistic individual with dark eyes. There is, or was, a local population of white Little Owls in Jerez, Spain, some of which were exhibited in the local zoo; although uniformly white, they all had normal eye colour. One leucistic wild-born young Little Owl was discovered in 2006 in an army aircraft hangar in Gloucestershire, UK.

Burrowing Owl One leucistic owl reported from the United States before 1912, and another more recently, in 1996, from Brazil.

Hawk Owl One leucistic caught in 1996 in Duluth, Minnesota, USA, had a number of white feathers in both wings. In the following year, another leucistic owl was caught in North America.

◄ A leucistic Burrowing Owl *Athene cunicularia* – completely white but with normal yellowish eyes (*José Carlos Motta-Junior*).

Pearl-spotted Owl A leucistic skin at British Museum, Tring, was collected in 1902.

Long-eared Owl One leucistic owl was shot in Minsk region, Belorussia, in 1998; it was all white, but with natural yellow eyes. In the next year, a completely white Long-eared Owl was photographed in Netherlands; this owl seems also to have had some normal colour in the eyes.

Short-eared Owl One perfectly white-breasted owl, with much white on the face and back, was seen in Minnesota, USA, in 1980, and another leucistic individual, killed in Connecticut before 1983, was suffused with white throughout its plumage.

Northern Saw-whet Owl Between 1977 and 1983, only three partially leucistic individuals were captured during the ringing of more than 14,000 birds of prey in Minnesota, USA.

Seventeen owl species have been listed above as having occured in leucistic plumage, and with several specimens of some species (more than 15 Spectacled, Great Grey and Tawny Owls). In previous publications, however, many of my colleagues and I have erroneously named these owls as being incomplete, imperfect or partial albinos, but albinism is an 'all-or-nothing' matter. Total loss of pigmentation always affects the retina of the eye, and can lead to impaired sight in bright light. That is why albinos usually do not live for very long. Leucistic owls, on the other hand, have normal eye colour and normal eyesight, which explains why they often live for many years in the wild.

Linda

In Finland, a very white and large Great Grey Owl was observed in March–April and again in November 1994 in Vesanto, in the central part of the country. Initially this owl was regarded as a partial albino, but it was later considered leucistic or even as having atypical pale buffish-brown pigmentation as a non-eumelanic form of schizochroic plumage. This shows only how difficult it is to name abnormal pigmentation in wild birds. The crown, nape, mantle, scapulars, back, breast and flanks of that 'blond' owl were almost white, or at least yellowish. The ruff and facial disc lacked the normal black barring and the typical concentric circles of the Great Grey, although the face and edge of the ruff had some light brown markings. The eyes, however, were yellow, edged on the inside with a touch of blackish-brown; two large outward-facing 'commas' were white, as is usual. The bill was normal, but surrounded by a brown 'beard' instead of the normal black one. The owl had white 'moustaches' and fairly prominent white patches in the middle of the foreneck, as is usual. The flight and tail feathers were also very light, but with some faint brown markings, giving the owl a somewhat yellowish look. Also, the toes, including the talons, and tarsus were much paler than normal.

In 1995, a similar white owl was seen in March near Kajaani, about 170km north of Vesanto. Good photos of both owls proved that this was the same individual. During the next two years the 'Great White' was not seen in Finland, but in March 1998 it was found again, this time some 200km NW from Kajaani, at a coastal village called Liminka. Here, the owl was caught and ringed, and was found to be at least four years old, confirming that it was the same individual all the time. It was given the nickname 'Linda' when it was seen to be consorting with a normally coloured male. Unfortunately, no nest was found, so it is possible that the pair did not breed. After spring 1998, Linda disappeared once more.

Two years later she was found again, 200km east in Puolanka, not far from the Russian border. Linda still bore the ring and it was checked. The nest was found, but because of the poor vole population Linda reared only one chick, which had perfectly normal colouring like the male.

After 2000 Linda seemingly disappeared, but she was found again in February–March 2003 not far from Sweden, in Ii, some 200km NW from Puolanka. In the latter half of March she was not seen in that small pasture area, as she had obviously gone to nest in deep forest, like all of her fellow Great Greys.

Linda lived for over nine years, proving that leucistic owls can survive in the wild despite being so conspicuous. Linda has further demonstrated that at least the female Great Grey Owls may migrate, according to the food situation, over long distances from south to north and from east to west, and vice versa.

Moult

Owl feathers wear out through everyday use, as the feathering is soft and loose. Owls do not replace each feather every year, and this partial moult facilitates age determination of some species. Unfortunately, the moult pattern of many owl species remains completely unknown. Moult of the primaries by European owls is summarised in Table 3.

Table 3. Moult of primaries in European owls.

Owl species	Annual moult	Definitive / Adult (basic plumage)	Time of moult
Barn Owl	partial	3 years	May–October
Eurasian Eagle Owl	partial	4 to 5 years	not known
Snowy Owl	partial	4 years	shortly after laying
Northern Hawk Owl	almost complete	2 years	shortly after laying
Eurasian Pygmy Owl	complete	1 year	after breeding
Little Owl	complete	1 year	after breeding
Tawny Owl	partial	3 to 4 years	after breeding
Ural Owl	partial	4 years	after breeding
Great Grey Owl	partial	4 years	after breeding
Long-eared Owl	complete	2 years	May–October
Short-eared Owl	complete	1 year	May–October
Tengmalm's Owl	partial	2 years	May–October

In all owls, the moult of the primaries occurs slowly enough that their effectiveness for flight is not noticeably impaired. For instance, the moult period of Barn Owls last for about three months, and the wing-moult pattern differs from that of many owls (which shed their primaries in sequence from the inner ones outwards) in that it begins in the middle of the primaries and proceeds in both directions. The tail moult of this species is unusually slow and is irregular in sequence. In comparison, some smaller owls, such as the Little Owl, change all tail feathers almost simultaneously. In the case of the South American Spectacled Owl, full adult plumage requires up to five years, moulting through a complicated series of intermediate plumages in the meantime. Moult is a lengthier process also for the Great Horned Owl, taking three to six years, but the replacement sequence is relatively predictable, and it is believed possible to assess the age of individuals of many owls accurately up to the third and possibly fourth year. It can, however, be difficult to distinguish subtle differences between third-generation and fourth-generation feathers by looking only for contrast in wear and colour. The use of ultraviolet light, which causes porphyrin pigments in feathers to fluoresce, provides a new, effective means of distinguishing multiple generations of flight feathers in owls. This permits a simple and more accurate age classification of adult owls.

Food and hunting

It has been fairly easy to analyse in detail what owls are eating during different seasons. This is because the smaller prey are generally swallowed whole, and indigestible matter such as fur, feathers, bones and chitin is regurgitated some hours later in the form of large pellets. Owls are opportunistic hunters by nature, and many species take a great variety of prey animals. Small mammals such as rodents and shrews predominate as food items for the majority of larger owls, and insects and other arthropods for the smaller ones. Many species supplement their diet by eating other birds, as well as reptiles, amphibians, fish, crabs and earthworms. Of the world's 249 owl species, 52 per cent are mainly insectivorous, 34 per cent are carnivorous and 3 per cent are piscivorous. The diet and hunting behaviour of the remaining 11 per cent are totally unknown.

Hunting methods vary according to the prey. A few species hunt actively on the wing, taking bats, moths and other small creatures in flight. Semi-aquatic fish-eating owls use their talons to seize fish from the surface of rivers, or they hunt crabs on shores and in river shallows. The majority of owl species, however, quarter the ground in silent flight, or scan it from a convenient perch, waiting and intently listening for ground-dwelling insects and small mammals. They consume food mainly in its freshly caught state; only occasionally do they eat carrion.

Consumers of voles and rats are becoming more and more popular as biological control elements in forest management and in the biodiesel industry. Thousands of nestboxes are introduced every year by the forest-owners in Finland in order to encourage owls and thereby reduce forest damage done by rodents. Rats have caused serious damage to oil-palm plantations, which are becoming increasingly important in large-scale biodiesel production. Barn Owls were first introduced in the 1980s in Malaysia as a method of controlling the rats, and this is now becoming a common practice in many Asian oil-palm plantations. Nestbox-occupancy rates have been around 50–60 per cent, and oil-palm damage has dropped considerably after the introduction of owls.

◀ A female Snowy Owl *Bubo scandiacus* about to vomit a pellet – excellent material for food studies (*Chris van Rijswick*).

◀ Northern Hawk Owl *Surnia ulula* dropping a pellet. Owls often have favourite trees, under which a lot of pellets can be found (*Harri Taavetti*).

▶ Many owls leave pellets and food remains at the bottom of the nest. These can be collected to study diet composition (*Rohan Clarke*).

▶ Pellet contents from a Barn Owl *Tyto alba*: mouse and vole skulls and bones (*Yves Adams*).

▶ 52 per cent of owls, such as this Elf Owl *Micrathene whitneyi* in Arizona, eat insects and other small creatures (*Jim and Deva Burns*).

◀ The mainly insectivorous Lilith Owl *Athene lilith* catching a large carabid bettle (*Danny Laredo*).

◀ Little Owl *Athene noctua* will eat earthworms (*Christian Fosserat*).

► Little Owl *Athene noctua* will also eat mice (*Chris van Rijswijk*).

► Burrowing Owl *Athene cunicularia* with a lizard (*Jim and Deva Burns*).

◀ Northern Saw-whet Owl *Aegolius acadicus* mainly eats small mammals (*Matt Bango*).

◀ Barn Owl *Tyto alba* bringing a juvenile bird to the nest (*Andrés Miguel Domínguez*).

◀ Ural Owl *Strix uralensis* has a varied diet, with small rodents, like this wood mouse, being among their favourite prey (*Gaku Tozuka*).

► Only eagle owls eat larger mammals. Here a male Eurasian Eagle Owl *Bubo bubo* has killed a female Mongoose *Herpestes ichneumon*; the prey will weigh around 7.5kg – three times the weight of the owl (*Vincenzo Penteriani*).

► A handful of owls are fish-eaters, like this Blakiston's Fish Owl *Bubo blakistoni* from Japan. This is the second-largest of all owl species (*Hans Germeraad*).

Habitat

As a group, owls are able to occupy all kinds of habitats, ranging from tundra, deserts and grasslands to marshes, swamps, woods and luxuriant rainforests, and from lowland areas to mountains and islands, but the majority of the species, 76 per cent, live in woodlands or forest edges. Approximately one fifth of all owl species live in semi-open or open-country habitats with scattered trees and bushes, and a further 20 per cent have adapted to life in or near human settlements. Only a few species, such as the arctic Snowy Owl and the desert-living species of North America, such as the Elf Owl, prefer habitats where trees cannot grow. Some species are terrestrial and live in flat country or among rocks. Long-legged terrestrial species are known both in the Tytonidae, with the African Grass Owl in the Afrotropics and the Eastern Grass Owl from India to Australia, and in the Strigidae, with the Burrowing Owl of North and South America. The recent decline of many owls with arboreal habits can be associated with the destruction of forest habitat that is now occurring all over the world.

◀ Eurasian Eagle Owl *Bubo bubo* is generally a bird of wilderness, far from human settlements (*Vincenzo Penteriani*).

◀ Eurasian Eagle Owl *Bubo bubo* has also adapted to breeding in cities – Helsinki has at least seven pairs (*Vincenzo Penteriani*).

► Little Owls *Athene noctua* favour the presence of livestock, as they eat dung beetles (*Erica Olsen*).

Behaviour

Owls usually hide away by day in holes, or in dark places in thick foliage, taking advantage mainly of their highly protective plumage coloration. At least 69 per cent of the species are nocturnal and only 3 per cent are diurnal; some 22 per cent are partly diurnal or crepuscular, and the remaining 6 per cent are so little studied that their habits are unknown. The borderline between diurnal and nocturnal is sometimes difficult to draw, as in northern latitudes Snowy Owls and Northern Hawk Owls hunt during the light nights of the Arctic summer and in winter during the short hours of daylight.

Most owls seem to be highly territorial, but nomadic species, such as the Snowy and Short-eared Owls, sometimes form loose colonies. Highly territorial owls, of which the Eurasian Eagle Owl and Tawny Owl are examples, are very aggressive towards other birds of prey, especially during the breeding season. Indeed, birds of prey and smaller owls often make up as much as 3–5 per cent of the total food of the Eurasian Eagle Owl.

Studies of the Little Owl in Germany demonstrated that males defended their territories at all seasons. The largest territories, of 28ha, were defended in March/April, during the courtship season. During the breeding season, in May–June, all males reduced the size of territories to an average of 13ha. The minimum territory size, of 1.6ha, was reached in the summer months of July and August, when the fledglings were still being fed by the parents. In September–October the defended area increased again, to 10ha, when the young of the year started to disperse. A further increase in territory size (to 20ha) was observed in winter, but males were aggressive only during warm weather. Tawny Owls have shown that there is a limit to the owl density in a given habitat, this being determined by territorial behaviour. In northern England, male Tawny Owls had much larger home ranges than females, 167ha as against the females' 44ha. Female home ranges overlap less than do those of males, and male ranges included those of several females. In the south of England, Tawny Owl pairs had much smaller territories of about 13ha in closed woodland and 20ha in mixed woodland and open ground. Fluctuations in food resources did not lead to changes in the number of adult owls in the woods. Failure to breed, the laying of fewer eggs than was thought possible, failure to hatch laid eggs, and mortality of the young in autumn and winter were the main factors keeping the numbers virtually constant from one year to the next. Quite the opposite population dynamic, however, applies to the nomadic species. For instance, the Short-eared Owl alters the size of its territory from month to month in according with the abundance of its main prey, *Microtus* voles. If food is scarce, the owl becomes nomadic and seeks new breeding and hunting areas.

An owl's entire life history can be studied by using data gained from ringing programmes. Annual checks and recapturing of owls that exhibit nest-site fidelity provide information not only on the number of young, but also on the adults' 'marital problems'. Recoveries from Finland reveal that 80–90 per cent of Tawny Owls remain in their territories throughout their life, and 90 per cent of their nestlings bred less than 50km from their birthplace. As many as 98–100 per cent of male Ural Owls and 90–95 per cent of the females are faithful to their previous breeding sites. The most interesting ringing results, however, are those which provide intimate details on several species' breeding ecology.

Being mobbed is one of the strangest aspects of an owl's life. If an owl appears in the open during the daytime, it is quickly surrounded and attacked by numbers of other birds. Attacking birds in Europe include finches, tits, buntings, warblers and blackbirds, to name but a few. A conspicuous stationary owl is a major target for mobbing. The small birds gather around and confront the owl, approaching remarkably closely. This mobbing behaviour does not occur when the owl is actively hunting. The question remains of why, everywhere in the world, daytime birds suffer from a lifelong fear of owls and they identify them as a threat. By using stuffed owls and even wooden dummies, researchers have shown that a big head, a short tail, solid contours, brownish or greyish colouring, a patterned surface with spots and streaks, a visible beak and frontally directed eyes are the important properties to stimulate mobbing. But it is also possible to attract some mobbing birds simply by imitating the hooting of the local owls. In this context, the eye-spots, or 'false eyes', on the nape of many small owls, especially pygmy owls, have an important ecological function. In experiments when eye-like patterns were present on the nape of the model owl, mobbers shifted away from the model, providing the first empirical evidence for a link between eye-spots and avian mobbing.

Especially large eagle owls are recognised by other birds as being serious predators. As a result, they are subjected to constant attacks by corvids. This is why eagle-owl dummies or stuffed specimens are commonly used when seeking out crows and ravens.

One important behaviour of most, if not all, owls is mutual preening. Allopreening is one of the strongest behaviour patterns evident during the pair-bonding of Great Grey Owls. Mutual preening by this species probably serves to reduce aggressiveness between individuals and may provide for sexual recognition and pair-bond maintenance. As even young owls often perform this behaviour, however, it is not necessarily a kind of courtship ritual. In the Barn Owl, allopreening occurs regularly between female and male partners throughout winter. It is often performed by the female, which approaches the male while uttering squeaks or soft whistles and preening him around the face and the back of the head. The preened bird emits twittering noises and chirrups through apparent pleasure.

Owls are very poor nest-builders, but they are excellent nest-defenders. If any intruder comes too close to an occupied nest or to young just out of the nest, the parent owl often performs a dramatic defensive display, injury-feigning or carries out a fierce attack. The defensive display consists of ruffling all the feathers, spreading the wings wide and then turning them in such a way that their upper surface faces the intruder. This makes the owl appear huge. The head may sway from side to side, and the tail is often spread and sometimes raised. Large, brightly coloured eyes, bill-snapping and hissing noises add to the intimidation.

The injury-feigning distraction display consists of a parent owl flapping on the ground around the nest as if it is badly injured and therefore easy prey. In this way it draws the attention of the intruder away from the nest. If the intruder tries to catch the bird, the apparently vulnerable adult owl suddenly flies away to safety, and often the helpless chicks in the nest will be overlooked.

At times a full attack may be delivered, in which the parent owl, usually the female, swoops low over the intruder's head and tries to slash it with its sharp claws. Nest defence increases significantly throughout the breeding season, because the older chicks are defended more strongly than younger chicks and eggs. Ural and Great Grey Owls in the north and Tawny, Long-eared and Eurasian Eagle Owls farther south are notorious for these forceful attacks near the nest. In Europe alone, the author knows half a dozen people who have lost an eye to an owl in just such an encounter. Recently, in Finland, more and more Great Grey and Eurasian Eagle Owl attacks on dogs have been recorded when the dog has been walking with its owner, and this has happened even outside the breeding season and not too close to the owl's nesting territory. In Africa, a female Spotted Eagle Owl attacked the author's Rhodesian ridgeback if it went too near the owl's nest in a water tower in Malawi.

When a human or other terrestrial predator approaches, an owl may also hide by slowly stretching upwards while compressing its plumage, assuming a 'tall-thin' posture as a concealment display. Some owls augment this display by erecting the ear-tufts and closing the eyes to the point that they become mere slits, but all the while keeping a peeping eye on the intruder.

▶ A Barn Owl *Tyto alba* pair allopreening – this is outside the breeding season (*Andrés Miguel Domínguez*).

▶ Juvenile Eastern Screech Owls *Megascops asio* performing mutual preening; this is not just a courtship behaviour (*Christian Artuso*).

▶ A juvenile Great Horned Owl *Bubo virginianus* preening its tail feathers (*Anne Elliott*).

◀ Mottled Wood Owl *Strix ocellata* demonstrating almost perfect camouflage in its nest tree (*Ron Hoff*).

◀ Rock Eagle Owl *Bubo benghalensis* performing an injury-feigning broken-wing display (*Niranjan Sant*).

◀ A Eurasian Pygmy Owl *Glaucidium passerinum* uses its fully spread wings to keep it near the surface after a snow-plunge. After capturing a vole the owl nervously glances around for possible predators (*Harri Taavetti*).

► Eurasian Pygmy Owls *Glaucidium passerinum* often flick their tails like Wrens *Troglodytes troglodytes*. Tail-flicking behaviour is also known in Forest Spotted Owl *Heteroglaux blewitti,* maybe suggesting a taxonomic relationship between these genera (*Lesley van Loo*).

▼ Northern White-faced Owl *Ptilopsis leucotis* in an alert position, with thin body and erect ear-tufts (*Ron Hoff*).

▼► A pair of Short-eared Owls *Asio flammeus* playing in the air, as part of the courtship (*Lee Mott*).

◄ Little Owls *Athene noctua* mating in daylight (*Christian Fosserat*).

◄ Wing and leg-stretching by a Burrowing Owl *Athene cunicularia* (*Lesley van Loo*).

◄ Burrowing Owl *Athene cunicularia* stretching both wings over its head (*José Carlos Motta-Junior*).

▶ A threat or defensive display makes this Long-eared Owl *Asio otus* seem larger than it really is (*Ian Fisher*).

▶ A recently fledged Long-eared Owl *Asio otus* chick performs an inverted wing-display, making it appear much larger and more formidable to would-be assailants (*Evgeny Kotelevsky*).

▶ The 'tall-thin' posture of a male Great Grey Owl *Strix nebulosa* in Finland. This posture is adopted should a human or other terrestrial predator approach the owl (*Matti Suopajärvi*).

◀ Northern Pygmy Owl *Glaucidium californicum* with head turned showing false eye spots (*Donald M. Jones*).

◀ Little Owl also has an occipital 'face', but with less clear 'false eyes' than in many pygmy owls. It has only two white longitudinal spots above a broken white nuchal band (*Eyal Bartov*).

◀ Blue Jay *Cyanocitta cristata* mobbing an Eastern Screech Owl *Megascops asio*, which is hiding in its nest hole (*Christian Artuso*).

► Peruvian Pygmy Owl *Glaucidium peruanum* being mobbed by a hummingbird (*Christian Artuso*).

► Prior to the taking of this photo, six different species were mobbing this Ferruginous Pygmy Owl *Glaucidium brasilianum*. The owl eventually caught one of its harassers, a Fork-tailed Flycatcher *Tyrannus savanna*; mobbing is not always advantageous to small birds (*José Carlos Motta-Junior*).

► If a Eurasian Eagle Owl *Bubo bubo* flies in the open during the day it will often get a crow escort (*Atle Ivar Olsen*).

Interspecific aggression

Until very recently, interspecific aggression has been largely overlooked in owl conservation, mainly because of our limited knowledge of the subject. The existence of interspecific aggression makes it pointless to attract 'weaker' owl species into territories of 'stronger' species, and vice versa. After collecting all known records of interspecific killings among owls, we have calculated an interspecific-aggression index for the European owls (Table 4).

Table 4. Interspecific aggression index for the European owls. This index is positive (maximum +1) if the owl is aggressive towards other owls (i.e. if it has killed more owl species than are known to prey on it), and negative (maximum -1) if it is killed by more owl species than it has been able itself to kill.

Owl species	Interspecific-aggression Index
Eurasian Eagle Owl	+ 1.0
Tawny Owl	+ 0.5
Ural Owl	+ 0.4
Great Grey Owl	+ 0.3
Short-eared Owl	+ 0.2
Long-eared Owl	+ 0.1
Snowy Owl	0
Barn Owl	-0.3
Northern Hawk Owl	-0.3
Tengmalm's Owl	-0.8
Hume's Owl	-1.0
Little Owl	-1.0
Common Scops Owl	-1.0
Eurasian Pygmy Owl	-1.0

The three most aggressive owls, the European Eagle, Tawny and Ural Owls, are highly territorial, sedentary species with a varied diet. Less aggressive owls are often nomadic species, such as the Great Grey, Snowy and Northern Hawk Owls, which feed primarily on small mammals. The smallest owls will inevitably have an index of -1, because of their inability to kill species larger than themselves. In Finland, the increase in Ural Owls has caused a decline in numbers of Tengmalm's Owls by 2 per cent annually, showing the importance of this inter-specific aggressiveness. The competition between Barred and Spotted Owls in the United States has attracted more publicity. The latter species' population has continued to shrink by about 3 per cent owing to the invasion into the Spotted Owls' territories by the Barred Owls. As a drastic solution, it has been proposed that an experimental killing of Barred Owls be carried out in order to ascertain if that would save the Spotted Owl.

◄ Eurasian Eagle Owl *Bubo bubo* will attack and kill all smaller owls. Here a Little Owl *Athene noctua* provides lunch for the 80-day-old young (*Vincenzo Penteriani*).

► ▲ Tengmalm's Owl *Aegolius funereus* will occasionally prey on the smaller Eurasian Pygmy Owl *Glaucidium passerinum* (*Matti Suopajärvi*).

Breeding strategies

Weather influences the start of breeding activities, and late snowfalls can cause even advanced broods to be abandoned. Many owls make an assessment of rodent abundance, their commencement of breeding activities being influenced mainly by the quantity of available food. During good vole years the owls lay eggs much earlier than they do during poor years. An abundant food supply and warm temperatures can encourage Barn Owls and Short-eared Owls to breed even in winter.

Owls exhibit a wide range of breeding strategies. For instance, the Barn Owl breeds when it is only one year old, lays large clutches of up to as many as 14 eggs, and can breed more than once in a breeding season. Other owls, among them Eurasian Eagle and Ural Owls, mature more slowly over several years and have lower productivity, and they may forego breeding in years when prey animals are scarce. With very few exceptions, owls make hardly any nest themselves; instead, they use old nests of other birds, such as the abandoned nest of a raptor or corvid, and holes in trees or rocks and a great variety of other places, including human habitations. Eurasian Eagle Owls sometimes dig their nest cavities into uninhabited anthills. Owls living in the taiga and tundra zones often nest on open ground or in low vegetation; they make a shallow scrape and even add some lining material to their nest, typical examples being the Snowy and Short-eared Owls. Desert species tend to live underground, taking over abandoned rodent burrows to escape the heat of the sun.

Owl eggs are chalky-white and roundish, the number in a clutch varying from one to 14. Clutch size is dependent on the food supply available at the time, the differences from one season to another being most notable in species which feed on rodents that are subject to cyclic population fluctuations. Hence, in years of vole abundance, the Snowy Owl may have clutches of 10–14 eggs, whereas in years of food scarcity it may have clutches of 2–4 or even not breed at all. Owls lay their eggs often several days apart, and the incubation starts with the first egg laid, resulting in marked differences in the size of the young in the nest. In good vole years all the young may survive, but in poor years the oldest progeny compete with their siblings for scarce food, and the clutch then produces just one well-fed fledgling instead of three or four starved weaklings. Such flexibility maximises success in good years, while minimising the risk of total failure in rodent-poor years.

Usually only the female incubates, while the male forages and brings food to her. Both sexes care for the young. The incubation period is long, 26–28 days in the case of the Long-eared Owl, 32–34 in the Barn Owl, and 34–36 in the Eurasian Eagle Owl. The young are nidicolous after hatching, with the ears and eyes closed and the body lacking any independent means of temperature regulation. After the natal down, the young acquire the so-called mesoptile feathers, which are followed by true feathers, these appearing in the same feather papillae. Some owls become capable of breeding at about the age of one year, but larger owls often breed for the first time only when they are two to four years old.

◀ During good vole-years, the male Tengmalm's Owl *Aegolius funereus* will bring surplus food to the nest, even when the chicks under the female are too tiny to consume all of it (*Esko Rajala*).

◀ Short-eared Owl *Asio flammeus* almost always breeds on the ground, with asynchronous hatching of the eggs (*Christian Artuso*).

Longevity records

Owls have the reputation of reaching a great age but there are amazingly few reliable data on their longevity. Keeping a hatchling in captivity until it dies is one way of determining how long that species may live, but captive birds tend to live longer than birds in the wild. For estimating population dynamics, we need to know the longevity of owls living under natural conditions which requires the capturing, marking and subsequent recovery of wild owls.

Table 5. Longevity records of owls in the wild and in captivity in years (yrs) and months (mo).

Owl species	Weight (in g)	Longevity in the wild	Longevity in captivity
Milky Eagle Owl	1588–3115		>30 yrs
Malay Fish Owl	1028–2100		>30 yrs
Tawny-browed Owl	481 (1♀)		30 yrs
Blakiston's Fish Owl	3400–4500		30 yrs
Barn Owl	254–612	29 yrs 2 mo	34 yrs
Long-eared Owl	200–435	27 yrs 9 mo	
Great Horned Owl	900–2503	27 yrs 9 mo	>28 yrs
Eurasian Eagle Owl	1500–4600	27 yrs 4 mo	53 and 68 yrs?
Ural Owl	500–1300	23 yrs 10 mo	30 yrs
Tawny Owl	325–800	22 yrs 5 mo	27 yrs
Spotted Owl	520–760	21 yrs	25 yrs
Morepork	150–216		27 yrs
Short-eared Owl	206–500	20 yrs 9 mo	
Eastern Screech Owl	125–250	20 yrs 8 mo	
Barred Owl	468–1051	18 yrs 2 mo	
American Barn Owl	311–700	17 yrs 10 mo	
Northern Hawk Owl	215–450	16 yrs 2 mo	10 yrs
Great Grey Owl	568–1900	15 yrs 11 mo	27 yrs
Little Owl	105–260	15 yrs 10 mo	18 yrs
Tengmalm's Owl	90–215	15 yrs	
Flammulated Owl	45–63	13 yrs	14 yrs
Western Screech Owl	90–250	12 yrs 11 mo	
Snowy Owl	710–2950	11 yrs 7 mo	35 yrs
Indian Scops Owl	125–152		22 yrs
Rufous Fishing Owl	743–834		>21 yrs
Burrowing Owl	147–240	11 yrs	
Spotted Eagle Owl	550–850		15 yrs
Northern Saw-whet Owl	54–124	10 yrs 4 mo	
Common Scops Owl	60–135	6 yrs 10 mo	>12 yrs
Eurasian Pygmy Owl	47–100	6 yrs	7 yrs
Elf Owl	36–48	4 yrs 11 mo	
Collared Scops Owl	108–170	4 yrs 4 mo	

The larger the owl is, the longer it tends to live, both in the wild and, more so, in captivity, but many people have questioned the top ages of captive European Eagle Owls recorded in Russia (Table 5). In Britain, no captive eagle owl older than 35 years has ever been known. The Barn Owl has the greatest reported longevity in the wild, but its body mass is just a fraction of that of several owl species with shorter longevity records. Similarly, the Flammulated Owl has a greater longevity compared with many larger owls. These anomalies may simply reflect the fickle nature of longevity records in the wild, especially when, in global terms, relatively few owls have been ringed and recaptured.

◀ Ringing of young owls (this is a Barn Owl *Tyto alba*) is the best way to study the longevity and movements of owls (*Christian Fosserat*).

▶ Traffic is a serious killer of Barn Owls *Tyto alba* (*Rebecca Nason*).

▼ Barbed-wire fences kill many open field-hunting owls, like this American Barn Owl *Tyto furcata* (*José Carlos Motta-Junior*).

▼ ◀ The long lives of Eurasian Eagle Owls *Bubo bubo* often end in electrocution by power lines (*Jesus Contreras*).

Movements

Owls are often fairly sedentary, and regular and long migrations are known for only a relatively small number of species. Among European owls, most of the Common Scops Owl populations migrate regularly to tropical Africa, while the northern populations of the Pallid Scops Owl migrate to north-east Egypt, the Indus Valley and the Bombay region of India. Within eastern Asia, Oriental Scops Owl populations from East Siberia and Japan migrate to winter in India, China, Malaysia and Sumatra. The Brown Hawk Owl also migrates in Asia between temperate and tropical regions, but its movements are not well studied; nevertheless, one dead owl of this species was found as far south as in Australia. Two further migratory owls in Eurasia are some northern populations of the Long-eared and Short-eared Owls, which are partially migratory also in North America.

Long-eared Owl studies have revealed that, in Europe, the northern populations migrate through all countries bordering the North Sea and the species is not reluctant to cross such large areas of water, most migrant recoveries in Britain having been of northern birds moving south. Some of the owls move eastwards into Russia. Long-eared Owls reared in Switzerland have been found wintering in Spain, France, Sardinia and Italy, and German Long-eared Owls have been found in France and Portugal at distances of, respectively, 1200km and 2140km from where they were first ringed. A number of Short-eared Owls ringed in Finland have dispersed across Europe. Recoveries in Britain extend down the length of the east coast, from Scotland to the south-eastern corner of England. On record are the recoveries of birds that had travelled more than 3000km. The Short-eared Owl is known to migrate across the Sahara Desert, this species and the Common Scops Owl being the only western Palearctic owl species recorded as doing so.

Several species, among them the Snowy, Northern Hawk and Great Grey Owls, make nomadic winter movements, or even irruptions, more or less cyclically, these being triggered by fluctuations in rodent population levels. A Snowy Owl chick ringed in Norway was found in the following year, having travelled 1380km, while an adult ringed in Finland was recovered two years later 810km away in Norway. One Snowy Owl ringed on Fair Isle, Shetland, was seen three years later 310km away in the Outer Hebrides. A Swedish adult female Great Grey Owl was found seven years later in Russia, about 900km from the ringing site. Two Finnish-born Northern Hawk Owls were found 2659km and 2795km to the east towards the Ural Mountains, and three others moved 1200–1400km in the opposite direction, to southern Norway. On rare occasions, wandering Northern Hawk Owls have also been observed in central and western Europe.

Comparable irruptive migrations take place in the Canadian and Alaskan Northern Hawk Owl populations, which move south towards the northern states of the USA. Frequently, some or many of the individuals involved in these irruptions settle and breed, before once again returning north. In Canada, a radio-tagged Great Grey Owl was relocated 800km north of where it had been living two years earlier.

Tengmalm's Owl has evolved a strategy of partial migration, the adult males being resident and the females and young being migratory. The periodic food scarcity favours the migration of females and young, and the importance of guarding nest holes of good quality favours residence by the adult males. The same is not the case with the Northern Saw-whet Owl. A male ringed at a nest was found dead three years later more than 900km NNW.

For the past decade it has been recognised that the continuing future trend of global warming may drive species to shift permanently in order to remain in an ideal climate or habitat. There is already clear evidence that some southern species are now found much farther north, such as Common Scops Owls in Sweden and Finland, and that northern species, such as the Northern Hawk Owl, are moving away from south and central Finland. Even more so, global warming drives many owl species to move or to stay higher up in the mountains, or it facilitates such a change.

◄ Short-eared Owl *Asio flammeus* taking a free ride on a North Sea vessel – this type of travel may explain some amazing oceanic crossings by owls (*Martin Gottschling*).

EVOLUTION OF OWLS

The oldest known bird fossil dates back some 225 million years, but the oldest owl fossils are from 56 to 65 million years ago. From that time, it took 33–42 million years before real strigid owls were first recorded; these were *Strix brevis* from North America and *Bubo poirrieri* from France. In the Pleistocene, giant barn owls *Ornimegalonyx* existed in the Caribbean and Mediterranean areas between 10,000 and 30,000 years ago. These large owls were more than 100cm tall and perhaps twice the weight of modern eagle owls. They must have preyed on giant rodents such as the capybara *Hydrochoerus hydrochaeris*.

The history of the family Tytonidae is, therefore, an old one, and it is not possible to say whether barn owls originated in the Eastern or Western Hemisphere. Specimens of six genera of Tytonidae have been described from the Paleocene–Oligocene deposits of the Quercy caves, in southern France. This illustrates well the early differentiation of barn owls before the present era of mainly 'true owls' (Strigidae).

As no fossil bones belonging to the Common Scops Owl have been found in deposits in Hungary and France older than the Upper and Middle Pleistocene, and as the majority of scops owls occur in South-east Asia, the European and African scops owls probably have a common Asian origin. Fossil fragments of a kind of screech owl, resembling Western/Eastern Screech Owls, from the Upper Pliocene of Kansas, USA, would support the hypothesis that these screech owls have a tropical North American origin.

The Northern Hawk Owl is the only member of its genus, *Surnia*, and no fossil relatives are known. Fossil records exist for this species from the Late Pleistocene in Tennessee, USA, France, Switzerland, Austria and Hungary.

Fossil records of pygmy owls in the Americas are from Pleistocene deposits only, namely from California, Mexico and Brazil. Fossils of burrowing owls named as *Speotyto megalopeza*, somewhat larger and more robust than the present-day species *Athene cunicularia*, have been found in Upper Pliocene deposits in the states of Idaho and Kansas, USA.

In Europe, *Strix intermedia* was found in Middle Pleistocene fossils in the former Czechoslovakia and Hungary. As the name suggests, this owl was intermediate in size and structure between the today's Tawny and Ural Owls. A representative of the genus *Aegolius* was present in the Upper Pliocene of Hungary, and remains of Tengmalm's Owl itself have been reported from there, too, from the Upper Pleistocene onwards. 'Boreal Owls' and Northern Saw-whet Owls have also been reported from the Late Pleistocene of Tennessee, USA.

Many owl species that are nowadays rare in the area are also recorded from the Late Pleistocene of Britain. The Snowy and European Eagle Owls are two such examples.

DISTRIBUTION AND BIOGEOGRAPHY

On the whole, owls are successful birds, which have dispersed to all continents and even to several remote oceanic islands. Recently, a Barn Owl was found in Antarctica, which used to be the only continent without any owls. The great majority of the species occur in the tropics and subtropics: 18 per cent in the Afrotropical zoogeographical zone, 27 in the Neotropical, 17 in the Australasian, and 20 in the Oriental zone.

North America and the Palearctic zone of the Old World are inhabited by 18 per cent of the world's owl species. Of these, only six species occur in both North America and Eurasia.

TAXONOMY AND DNA-SEQUENCING

Although owls comprise a distinct and easily recognised group of birds, similarities in plumage and morphology, coupled with a general lack of knowledge of the ecology and behaviour of many species, have led to considerable uncertainty regarding species and even generic limits. The internal taxonomy of owls, which make up the order Strigiformes, may be in a greater state of turmoil than that of any other family of non-passerine birds.

The meaning of the term 'species' has gone through many changes, driven onwards by new methods, the differing priorities of each scientific age and the varied field of biological research. Four basic species definitions are presented in the box 'Species definitions'.

Owls have the lowest hybridisation rate among bird groups so far studied, the rate being only about 1 per cent, whereas the corresponding figure for gamebirds is more than 20 per cent and that for the swans, geese and ducks group is in excess of 40 per cent. The Biological Species Concept (BSC) therefore still serves quite well with owls. All species definitions, however, have been shown to have their limitations. The BSC encapsulates the idea that species are the real and fundamental units of evolution, while higher taxonomic categories such as genera, families and orders are more artificial groupings made for convenience, though loosely reflecting evolutionary relationships. The main problem with the Morphological Species Concept is the question of how different two groups have to be before they can be called separate species. The Evolutionary Species Concept is very appealing, but discovering the precise evolutionary history of organisms is practically impossible. The discovery of the DNA code revolutionised taxonomy,

> ### Species definitions
> **Biological Species Concept** – a group of actually or potentially interbreeding populations which are reproductively isolated from other such groups.
> **Morphological Species Concept** – a species is defined by a given set of common morphological features not shared by other groups.
> **Evolutionary Species Concept** – a species is defined by its shared evolutionary history and descent from a common ancestor.
> **Genotypic Cluster Species Concept** – a recently introduced definition, which is essentially a genetic version of the morphological concept. Genetic rather than morphological gaps identify the distinctions between species.

but the problem is that variability in DNA is often not correlated with variability in morphology or reproductive compatibility. It is obviously unrealistic to assume that we can impose and apply any single definition on a natural world made restless by evolutionary change.

It is no wonder, then, that owl taxonomy is currently in a state of flux and the number of 'acceptable' species varies between 150 and 250. One good example comes from Africa, where the Albertine *Taenioglaux albertina*, Chestnut *T. castanea*, Etchécopar's *T. etchecopari* and Sjöstedt's Owlets *T. sjostedti* could all be seen as subspecies of the African Barred Owlet *T. capensis*, mainly on the basis of their fairly similar vocalisations. So, one species or five species? What is not known is the degree of intergradation among these five populations. What happens in areas of contact? It seems clear from the distribution maps that some of these populations are not in contact with others. For instance, the owls that occur in Liberia and Ivory Coast (*etchecopari*) must be isolated from other populations. This kind of taxonomic problem is frequent, and different species concepts often lead to different answers. In the present book all possible new species have been presented. Should new studies of DNA, behaviour, vocalisations and biology either corroborate or reject the validity of any of these suggested 'new' species, it would be greatly appreciated if the author be advised of these findings.

Many owls are so rare that it has not been possible to obtain blood samples to examine nucleotide sequences in the cytochrome-b gene. At present, molecular data exist only for some 150 species, and 100 or so species therefore await 'official' confirmation when new material for DNA-testing becomes available.

One may ask where all the newly discovered owl species come from. To a large extent they have been known to the scientific community as subspecies, erroneously declared as such. To a much smaller degree there are some completely unknown owl species still being identified in the tropical forests. Even so, only five totally new owl species have been described since 2000, as shown in Table 6. These five are Pernambuco Pygmy Owl (2002), Little Sumba Hawk Owl (2002), Serendib Scops Owl (2004), Togian Hawk Owl (2004) and Sick's Pygmy Owl (2005). Four other owls are presented as new species for the first time in this book. They are Hume's Hawk Owl, which was only separated from Brown Hawk Owl in 2005; Northern Little Owl and Grey-bellied Little Owl (1988), which were first named as long ago as 1870 and 1988 respectively but which have since been overlooked as variants of the Little Owl; and the Santa Marta Screech Owl, which has yet to be officially described.

In this book we have listed 433 described and nine undescribed subspecies, which are the engines of evolution, and some may become the species of tomorrow's world. Already, 37 subspecies are thought possibly to be full species. We should, therefore, regard subspecies as part of the planet's biodiversity and realise that they are no less important than species. This means that the extinction of a subspecies should not be any less serious than that of a species. To gather DNA data and to undertake bioacoustic research on all owl species and subspecies may be a difficult task, but both species and subspecies are equally valuable and worthy of protection. If we use the same criteria for defining the conservation needs for owl subspecies as the one used for owl species, it would almost double the number of owls giving cause for concern.

Table 6. Timing of new owl descriptions.

Period	Number of owl species described
up to 1800	23
1801–1900	167
1901–2000	57
2001–2010	5

▼ Philippine Hawk Owl is one of those owls having many subspecies. The race *spilonota* is likely to be a future split. This male, from Cebu, is reddish-brown below and not clearly barred as in the birds from the island of Camiguin. It is possible that the forms of *spilonota* on Sibuyan, Cebu and Camiguin islands differ enough to be split further as new subspecies or even full species (*Bram Demeulemeester*).

OWLS AND HUMANS

One purpose of this book is to point the way towards a better understanding of how owls relate to their environment and how important it is for us, humans, to use that environment more wisely. Success in the conservation of various kinds of living creatures, including owls, however, depends not only on environmental issues, but also on social and cultural matters. The value of people's participation in resolving complex conservation problems has been rediscovered only lately.

It is a paradox that, in reality, owls are one of the most beneficial groups of birds but at the same time one of the least understood. Few other birds or, indeed, animals have attracted so many different and contradictory beliefs about them. Owls have been both feared and venerated, despised and admired, considered both wise and foolish, been associated with witchcraft and medicine, the weather, births and deaths – and have even found their way into *haute cuisine*.

Folklore has it that owls are birds of ill omen and that deception is one of their favourite ploys. As a counterbalance, it has to be said that the owl has been widely admired through the ages by deities, scholars, poets and animal-lovers in general. In France owls have been looked upon with great esteem, with several named as dukes, for example the Eurasian Eagle Owl as *Hibou Grand-Duc* and the Long-eared Owl as *Hibou Moyen-Duc*. Blakiston's Fish Owl is one of the most important gods of the native Ainu people of Hokkaido, in Japan. It is called *kotan kor kamuy*, meaning 'god of the village' or 'god who defends the village'.

It is highly likely that owls were among the first birds to be noticed by ancient man, probably because their vocalisations in the dead of night would cause havoc in the superstitious mind. Because human beings undoubtedly evolved in Africa, it must be there that the paths of humans and owls first crossed. My straightforward argument from this African-origin context was that beliefs about owls, since they are virtually universal (i.e. the same from Africa to North and South America), originated in Africa and moved outwards with the initial human migrations. Later, however, I have been made to wonder if it is not perhaps more the case that all pre-industrial societies shared the same beliefs because there is something intrinsic in owls – nocturnal behaviour, inhuman and sometimes almost preternatural calls and noises, large piercing all-seeing eyes, ghostly silence in movements – that results in each culture giving them the same meaning in the human world.

The Mesopotamian goddess Lilith was the goddess of death, and she was depicted on a Sumerian tablet of 2300–2000 BC as having a headdress of horns and taloned feet and being flanked by owls. The Barn Owl is extremely common as an Egyptian hieroglyph, but it is only rarely presented in art. A well-cut and faithfully painted Barn Owl is carved on a block from the 18th-dynasty temple of Tuthmosis III at Deir el-Bahari; the 18th dynasty (Dynasty XVIII) ruled between 1550 and 1307 BC. In addition, an 'eared' owl frequently appears as an Egyptian hieroglyph; this hieroglyph is clearly fashioned after some species of 'eared' owl, either the Long-eared Owl or the Pharaoh Eagle Owl, or both. The only positive symbol of an owl in Africa that I have found occurred also in ancient Egypt, where it was used for the letter 'M' in hieroglyphic inscriptions, some of which date back to 3000 BC.

In pre-Columbian America, the owl motif is widespread, and often highly stylised in carvings, totems, masks, fetishes, pottery and other objects. For example, from the Mochica people of northern Peru (c. 300 BC to AD 800) a jar was found featuring an owl with a face mask resembling a Barn Owl clutching a rodent in its bill, perhaps indicating very early recognition of this owl's beneficial association. Mayan hieroglyphics also include owl symbols, probably representing deities such as Ah Puch, the god of death, also linked with the god of war and human sacrifice. This association of death is found among the Zapotec of southern

▼ An eared owl in an ancient Egyptian papyrus painting - this is based either on Long-eared *Asio otus* or Pharaoh Eagle Owl *Bubo ascalaphus*, or both. Note the barn owl in the writing (as letter 'M') (*Heimo Mikkola*).

OWL.

Mexico, who believe that the owl is commissioned to give notice when a man is about to die and then to go and fetch the soul. In other New World indigenous cultures, owls may symbolise courage and strength and may play a symbolic role in warfare, or represent wisdom or animal trickery. These few examples demonstrate a universality in how the appearance and behaviour of owls have been interpreted in both the Old and the New Worlds well before the arrival of African slaves and European migrants.

Despite the possibility of a common origin, there are also striking contradictions between some owl associations. It is easy to see why owls, given their presumed gift of foresight, have become symbols of wisdom and intelligence, especially in Europe. More difficult to grasp, however, is why they have also become synonymous with stupidity. In Finland, the word *pöllö* means both an ignorant person and an owl. The ancient Romans also believed that some owls were so foolish that they could be induced to twist their head around and around until they throttled themselves. In India, too, the owl is synonymous with stupidity and pomposity. *Ullu ki tarah* ('in the manner of an owl: acting foolishly') and *ullu banana* ('to make a fool of someone') are commonly used expressions.

The hunting of Snowy Owls for food seems to have been an old tradition in Europe and is one that still continues to some extent among Eskimos, who describe Snowy Owls as 'lumps of fat'. The Burrowing Owl must have played a similar role in Uruguay, where Indians served the flesh of the owl as a delicacy to those convalescing from illness in the belief that it produces appetite for other food. For centuries, many rural people have held the common belief that the owls can also be used as medicine for healing some diseases. This was obviously based on the traditional precepts of sympathetic medicine, whereby the eating of an animal or parts of it enables the patient not only to benefit from the meat itself, but also to absorb the physical characteristics of that creature's good night vision, wisdom and the like, as shown in Table 7.

Table 7. Uses of owls in traditional medicine.

Disease or problem	Owl parts needed	Preparation of medicine	Where used
Eye complaints	Owl eggs or entire owl	Must be charred and powdered	UK, India
Earache	Owlet's brain or liver	Mix with oil and inject into ear	Italy
Whooping cough	Entire body of an owl	Eat as a soup	UK
Asthma	Body of an owl	Cures it since owls eat coffee beans!	Puerto Rico
Influenza	Magical owl hooting	Strain to hear; cures worst symptoms	Europe
Gout	Owl body without feathers	Mummify in oven, mash, mix with pig fat and apply to affected site	Europe
Haemorrhage	Screech Owl = Barn Owl	Boil in oil. Add ewe-milk butter and honey	Italy
Infection of sinews	Long-eared or Eagle Owl's head	Take ashes with lily root and honeyed wine	Italy
Rheumatism	Owl feathers	Burn over charcoal	Poland
Grey hair	Owl eggs	Use to darken hair	Europe
Hair-loss prevention	Owl eggs	A good cure for thinning hair	Europe
Stop child crying or to sleep well	Owl feathers	Put under pillow. Works for children or adults	Europe
Against epilepsy	Owl eggs	Prepare a soup when moon waning	Europe
Against snakebite	Owl feet	Burn with herb *Plumbago*	Europe
Hangover cure	Owl eggs	Cook eggs three days in wine	Europe
Give dislike of wine	Owl egg	Eating one gives lifelong aversion to wine	Europe
Absolute abstention from alcohol	Owl egg	Eat one and a child will never be a drunk	Europe

▼ In the superstitious minds of humans, owls are often connected with cemeteries, death and bad luck. This Little Owl *Athene noctua* brings down some impressive lightning (*Bence Mate*).

It is important to note that the effectiveness of any of the medicines from owl parts mentioned in Table 7 has not been scientifically proven, nor even studied, and their potency in all cases may be more than questionable. They are presented in this book only as a curiosity, and one which may explain the thinking behind some of the unnecessary and unfortunate killing of owls.

In Africa, owls are still used commonly for witchcraft and for killing people through 'juju', and by traditional medicine men. A recent comparative interview-based study (1,506 interviews) in the New and Old Worlds demonstrated that there are still many more superstitions relating to owls in Africa than in Middle and South America, Central Asia and Europe. In Africa 58 per cent of 661 respondents connected the owls with bad omen, witchcraft and death, 64 per cent of the females and 54 of the males. An additional 19 per cent regarded owls as frightening. Women tend to see owls as directly linked with witchcraft more often than the men do. In Central and South America 4 per cent of 477 respondents linked owls directly with witches and wizards, but in Europe (251) and in Central Asia (118) nobody did so. On the other hand, 75 per cent of the people in Europe, 49 per cent in Central Asia and 6 per cent in Central and South America saw owls as symbols of wisdom. In Africa, only one person gave that kind of positive label to owls.

Strong negative superstitions about owls are providing reasons for killing these birds in Africa, because people are thus protecting themselves against the witches and wizards. But owls are also killed elsewhere. In South America and Central Asia killing still takes place for traditional medicine: 11 per cent and 12 per cent respectively of interviewees knew people in these areas who use body parts of owls in medicine. In Asia, owl killings are common and are carried out in order to obtain a 'potent' aphrodisiac. Both in Asia and in South America, owl feathers are believed to bring good luck and to help in one's love life.

A more in-depth study of the general public's knowledge of owls was undertaken in Finland and Mexico in 1997–98 in order to determine the relationship of European and Middle American humans with owls. A general-knowledge interview with 251 Finnish and 210 Mexican people included: owl species known, common names, habitats, food, calls, owl killings, and superstitions relating to the owls. Public knowledge of owl species in Finland is fairly good, mainly as a result of school teaching, and is much better than that in Mexico.

A common local name in Finland, *kissapöllö* (literally, 'catowl'), means owl in general and Tawny or Tengmalm's Owl in particular. The name surely comes from the cat-like face of many owls. Interestingly, in Costa Rica the second most common name for an owl is *cara de gato* ('cat face'). In Mexico, common owl names often resemble the voice of the species in question, or they refer to the look of a particular species. *Tecolote* is the most common name for the owls and is derived from the 'nahautl' language, in which *teco* stands for 'stone' and *lotl* for 'bird'. A typical onomatopoeic name is *Búho*, which is also commonly used in Mexico for owls in general.

Persons interviewed in Finland very much agreed that the favoured habitat of the owls is the forest. Marshes, hillsides, fields and riverbanks were also selected as usual habitats. In Mexico, respondents listed mountains and forests as the two most common habitats for owls, but also mentioned cities, pastures, caves etc. Some people did not know where owls might live. Almost all respondents answered that mice and voles, in that order, are the most important food items of the Finnish owls. In Mexico, mice were similarly listed as the main food of owls, but fruits and seeds were incorrectly listed as the second most important food source of almost entirely carnivorous owls! One in five females had no idea what owls might eat.

▶ Many people around the world see owls as having a 'cat-face' – this picture of an Eastern Screech Owl *Megascops asio* helps explain the association (*Christian Artuso*).

▼ This white flyer in complete darkness may explain many ghost stories from old buildings, which are good breeding sites for Barn Owl *Tyto alba* (*Rollin Verlinde*).

In Finland more than 80 per cent of the respondents had heard owls calling or singing most often in the forest, but also near the houses. The corresponding figure for Mexico was 64 per cent.

Killing of owls continues in both of these countries, as shown in the box headed 'Owl killings'. At least witchcraft, use in traditional medicine and reasons of bad omen are no longer cited as justification for the killing of owls in Finland.

The interviewed people were asked how much superstition there still is with regard to owls. In Finland, half of the respondents answered that there is none at all. Common answers were also that 'I do not have any superstitions, but others do.': children have more superstitions than do adults, or elderly people have more than young ones (in other countries, yes, but not in Finland), superstitions are more common in rural areas than in cities, and in the past more than nowadays, and in Lapland more than in the south. These evasive ways of answering seem to indicate that people have inherited some superstitions, but try to forget or conceal them.

As a conclusion, the respondents were asked to classify the owls according to their knowledge and beliefs (Table 8).

Table 8. Owl classification in Finland and Mexico.

Owl classification	Finland female	Finland male	Mexico female	Mexico male
Symbol of wisdom	82	66	6	7
Beneficial	55	75	14	18
Harmful	0	1.9	1	5
Cause fear	15	11	1	0
Superstition	3.2	3.8	35	42
Bad omen	0.7	1.9	9	14

Interestingly, both in Finland and in Mexico men seemed to be more superstitious about owls.

This comparative study showed that there are perhaps a few more surviving superstitions in Latin America than in Europe. On the other hand, Mexican people seem to be less frightened of owls than Finnish people

Owl killings

Killing of owls has been very common in Finland in the past. At the beginning of the last century about 1,000 Eurasian Eagle Owls alone were killed annually and until the 1950s people were paid a reward for killing these owls. The species was given full legal protection only in 1983. Only 7 per cent of the women and 18 per cent of men questioned had witnessed some owl killings themselves, but most of the respondents (206) gave several reasons why owls are killed. Stupidity was listed as the principal reason. A number of people questioned whether anybody nowadays kills owls, and 28 per cent could not find any reason for killing these birds.

In Mexico, 18 per cent of the women and 43 per cent of men knew somebody who had killed an owl, but the reasons for killing were slightly different from those in Finland. In Mexico, owls were killed for their plumage/feathers (17 per cent), to mount for decoration (16 per cent), out of fear (8 per cent), to be eaten as food (8 per cent), because of superstitions and prejudice (7 per cent), because of the hunting instinct (6 per cent), out of ignorance (5 per cent), for purposes of witchcraft/magic (4 per cent), because owls bring bad luck (4 per cent), and purely for fun, or for medicinal purposes, or because of disturbance caused by the owl's voice, or because they do not like owls, or because owls know when we are about to die (each 3 per cent); and 8 per cent did not know why! So, it seems that little mercy is given to owls in Mexico, because the birds are still strongly associated with traditional medicine and magic.

are. So, one could speculate that the old superstitions are still there below the surface in Finland, too, but that people would rather admit to being afraid than to being superstitious.

Yet, it seems clear that superstitions and beliefs die hard, and all too often they give rise to behaviour quite disproportionate to the alleged 'offences'. The fear of owls is deep-seated and obviously dates back countless centuries. Thus, further education of the general public is needed, both in Europe and in South America, to demonstrate how beneficial owls are, and that the superstitious beliefs and myths about them are groundless.

All these interview results suggest that it is very important to consider how people's attitudes towards wildlife affect their action towards the conservation of species and ecosystems. Understanding both environmental/biological problems and the influence of human behaviour is indispensable if any success in the conservation of any owl populations is to be achieved. Only by informing people, through books, schools and television, about the roles which owls play in nature will we be able to do away with the incorrect and superstitious beliefs relating to them. A counteracting educational programme should be an integral part of any owl conservation work worldwide.

▶ The greek goddess Athene, goddess of wisdom and patron of Athens, with the sacred owl of Athens, the Little Owl *Athene noctua*.

CONSERVATION

Six of the owl species treated in this book have been listed as Critically Endangered and a further 26 as Endangered. In addition, 43 species are Vulnerable or Near-threatened. The reason is nearly always the same: loss of habitat. As stated before, two-thirds of the owls require forest, and forests are being destroyed all around the globe. For all 75 Endangered and Vulnerable owl species, therefore, the long-term future looks bleak as the devastation of forests continues throughout the world. Moreover, it is now not a matter just of the logging and clearing of forests for agriculture, but also of forest loss accelerated by global climate changes. The great *Euphoria* trees of southern Africa are succumbing to heat and water stress. So, too, are the cedars of north Algeria. Fires fed by hot, dry weather are killing enormous stretches of Siberian forest. Eucalypt trees are succumbing on a large scale to a heat blast in Australia, and the Amazon recently suffered two serious droughts just five years apart, resulting in the death of many large trees. In the south-west United States, namely in Arizona, New Mexico and Texas, the countryside landscape has become so dry that, in 2011, huge, explosive fires consumed millions of acres of vegetation and thousands of homes and other buildings. Doubtless thousands of owls also became homeless during these fires.

The second most important threat is that pesticides and rodenticides are being widely used to reduce the numbers of either insects or rodents, which most owls require as part of their diet. Further, as mentioned before, in a few underdeveloped regions there are still many serious superstitious beliefs about owls that justify the killing of these birds as evil spirits and messengers of witches and wizards.

There are many conservation breeding projects in place for endangered mammals, in many cases with the support of hi-tech medical care, and with enormously high costs in feeding and housing. It is surprising, therefore, that there is no conservation breeding project in place for any of the globally threatened owl species as defined by BirdLife International. Most of the owls are easy and cost-effective to keep and feed; they rarely develop health problems under human care and they live a long life in captivity.

Happily, some zoos and private individuals have started, or have plans in place to start, breeding projects for some endangered owls, such as the Christmas Hawk Owl, Spectacled Owl and two white-faced scops owls. Unfortunately, nobody prevented the mixing of the last two species, allowing hybridisation of the Northern White-faced Scops with the Southern White-faced Scops. The Society for the Conservation and Research of Owls (S.C.R.O) has a promising project with the Government of the Dominican Republic to maintain a conservation breeding project with the four owl species endemic on the island of Hispaniola. At least the Ashy-faced Owl and the Hispaniola race of Burrowing Owl have been bred under human care.

Luckily, the great majority of owls are still listed in the conservation strategy as being of Least Concern. A number of them seem to have nearly unassail-

▲ Many owls, such as Tawny and Ural Owls, are aggressive at the nest. Eye and head protection are advisable when ringing the young. Finland (*Harri Taavetti*).

able populations. For example, the Great Horned Owl has an estimated world population of more than 5.3 million individuals, and the Common and American Barn Owls together almost match this with 4.9 million. The owls seem to have a special, nocturnal solution to the problem of survival, and thus have always been a globally successful family of birds, and with figures like those above it would seem likely that they are here to stay for ever.

◄ Tengmalm's Owl *Aegolius funereus,* nesting in a man-made box in Sweden (*Rebecca Nason*).

Managing a habitat to suit owls is seldom a realistic option, given the considerable spatial requirements and the diverse prey needs of these birds. On the other hand, the providing of nest sites for owls is often a productive habitat-management option. For example, the provision of large numbers of nestboxes in Sweden and Finland has brought about a noticeable increase in the numbers of Ural Owls.

Erecting boxes for Barn Owls is frequently successful, and indicates that the occurrence of these owls is often limited by a lack of suitable nest sites. In Africa, artificial boxes work also for Spotted Eagle Owls, African Wood Owls, Pearl-spotted Owls and African Scops Owls. In Europe, nestboxes are readily utilised by Tawny, Ural, Northern Hawk, Tengmalm's and Little Owls, provided that they are tailored to imitate the natural situations preferred by these species. If any of these species occur in the area where owl encouragement is being considered, the act of putting up appropriate nextboxes could be rewarding.

◄ Long-eared Owl *Asio otus* will accept a nest-box if half of the front is left open (*Fred van Olphen*).

The existence of interspecific aggression, however, makes it prudent to ascertain which owls are resident in a particular area before nestboxes or other artificial nests are introduced. It is usually counter-productive to put a nestbox suitable for a Ural or a Tawny Owl within the territory of a Eurasian Pygmy or Tengmalm's Owl. Ural and Tawny Owl boxes should not be sited within 2km or less of Little, Tengmalm's and Eurasian Pygmy Owl nests. Nestboxes for any smaller hole-nesting owls should not be introduced in a European Eagle Owl territory, because the latter species will not tolerate the presence of other, competing birds of prey. The lack of natural holes suitable for breeding for the smaller owls is so severe that these will attempt to breed despite the proximity of eagle owls if they are attracted to the area by the provision of nestboxes.

Artificial nests other than nestboxes can be helpful in certain cases. Most pairs of Great Grey Owls will readily occupy old nests of raptors. In Canada, Finland, Sweden and Russia, there are several reports of Great Grey Owls breeding in man-made nests of branches and sticks or large baskets on trees. Smaller flat baskets are good breeding sites also for Long-eared Owls. In the USA, specially modelled plastic nests have been installed successfully at a depth of 1m in the ground for Burrowing Owls. These consist of a plastic box with two flexible tubes running from the box to ground level, and there is no bottom to the box, thus giving a natural soil flooring.

▲ Long-eared Owl *Asio otus* and Great Grey Owl *Strix nebulosa* like artificial twig-nests. These are young Long-eared Owls (*Fred van Olphen*).

◄ Great Grey Owl *Strix nebulosa* will nest in dog baskets and artificial twig nests but so far, none has accepted any of my huge nest-boxes (*Jari Peltomäki*).

EXTINCT OWLS

Many of the owls confined to islands became extinct soon after human settlement, mainly as a result of hunting, deforestation and predation by human commensals such as cats and rats. For instance, the Andros Island Barn Owl *Tyto pollens* became extinct when the island, in the Bahamas, was colonised by Europeans and their slaves in the 16th century. This 1m long, flightless owl lived in the old-growth forest, the destruction of which may well have led to its extinction. Réunion Island, in the north-west Indian Ocean, was the home until 1700 of the Réunion Owl *Mascarenotus grucheti*, which resembled the Long-eared Owl in size and appearance, but with nearly bare legs. It has been assumed that it preyed mainly on roosting songbirds. Similarly, Mauritius had two owls, *Mascarenotus murivorus* and *M. sauzieri*, both of which became extinct much later than the Réunion Owl; the former, the Rodrigues Owl, had certainly gone by 1761, but the larger (up to 60cm long) Mauritius Owl lasted until 1859. The genus *Mascarenotus* was most likely a distinct evolutionary lineage, related to the genus *Ninox*, that evolved convergently with *Otus* or *Asio* owls.

Ornithologists have been expecting that some living species would reach Mauritius and Réunion. There are now a few field records of an unidentified scops owl on Réunion. This was first thought to be a vagrant Madagascar Scops Owl *Otus rutilus*, but it could also be a new species of scops. This as yet undescribed owl has been omitted from this book, for the time being.

The following three owl species, which have not been seen for a long time, have also been omitted, as it is believed that they have been lost during recent times:

Stresemann's Scops Owl *Otus stresemanni* Holotype collected in 1914 from Mount Kerinci, Sumatra, and never seen since.

Siau Scops Owl *Otus siaoensi* Holotype collected in 1866 from Siau Island, off north-east Minahassa, Sulawesi, and no observations after that.

Laughing Owl *Sceloglaux albifacies* No records since the 1930s of this endemic New Zealand owl.

Nine recently extinct subspecies have been mentioned in the species accounts. The author will be most happy to remove the extinct label from any of these species and subspecies if they are rediscovered in the future.

▶ Photograph of a mounted specimen of a Laughing Owl *Sceloglaux albifacies* held in the Museum of New Zealand, Wellington (*Paddy Ryan*).

▼ One of the few known photographs of a live Laughing Owl, taken in captivity in 1892.

'OWLAHOLICS'

People who collect anything with an owl emblem or picture on it are referred to by the invented term 'owlaholics'. Thousands of such collectors and enthusiasts live in Europe and the United States, although many owlaholics reside elsewhere in the world. Large numbers of people with an obsession for collecting owl artifacts also live in Australia, and even such far-flung places as Japan and Fiji can claim a few. One New Zealander has a huge postage-stamp collection containing more than 1,200 different owl stamps from 192 countries across the globe.

The box on owlaholics gives some common reasons for why people started to collect every form of owl. Figurines, carvings, statues and paintings are just a few popular items. Owlaholics admit to sleeping on owl sheets, cooking with owl pots, and wearing owl jewellery. They also have owl key-rings, owl paperweights, owl oven-gloves, owl bottle-openers, owl money-boxes, owl ashtrays, owl playing cards, and so on; the list is almost endless.

Some reasons for becoming owlaholics

Preferring to work the graveyard shift or stay up at night. People whose biological clocks are set wrongly and who are at their mental best between midnight and 06.00 hours are called 'night owls'. This connection often gets them interested in collecting owls. For example, a group of nurses used to call themselves 'the Owl Team' and two became collectors.

Ex-Brownies in the Girl Guides movement (the female equivalent of the Boy Scouts). The adult leader of a Brownie Pack was called a 'Brown Owl', and she was sometimes assisted by individuals known as 'Tawny Owls' or 'Snowy Owls'. This long association with owls made many of them owlaholics.

Inheriting an owl-artefact collection. Once you have been given a collection, particularly by someone of whom you were fond, it is difficult not to add to it. This also covers people who started collecting when a family member or a friend passed on some duplicates in his or her collection.

Being involved with real owls. Like the author and many of you reading this, so we all know how that goes!

By accident. Some people suddenly realise that they are collecting when they put several owl items together, and then start thinking things such as 'Hmm, I'll buy that. It will go well on my owl shelf or owl wall'.

Women are more likely to collect owl artifacts, but men, once hooked, are more likely to become obsessed, and it seems to be men who own the really big collections (5,000 or more items). Since owlaholics are social, active people, this addiction will probably spread even further.

▼ Typical owl figurines collected by 'owlaholics'.

OWL ASSOCIATIONS AND GLOBAL RESEARCH ORGANISATIONS

A large number of associations and internet-groups have been formed to provide local, regional or worldwide forums for all those interested in owls – whether it be research and conservation or just in general. There are far too many to list here, but a few are mentioned as examples.

Arbeitgemeinschaft Eulenschutz (German Organisation for Owl Protection)
Publishes annual *Kauzbrief* magazine. **www.eulenwelt.de**

Global Owl Project
A worldwide project to resolve fundamental aspects of owl taxonomy (including DNA sample collection) and conservation of the world's owls. The team also has regional representatives. **www.globalowlproject.com**

Society for the Conservation and Research of Owls (SCRO)
Publishes *S.C.R.O.* magazine. info@scro.org or montowl.it@alice.it

The International Owl Society
Publishes a quarterly newsletter *Tyto News Brief* and the annual *Tyto* magazine. **www.international-owl-society.com**

The Raptor Research Foundation
Publishes quarterly *The Journal Of Raptor Research* which is an international scientific journal dedicated entirely the dissemination of information about diurnal and nocturnal birds of prey. **www.raptorresearch-foundation.org**

World Working Group on Birds of Prey and Owls
Publishes a raptor and owl research and conservation proceedings after each world conference, every fourth year since the 1970s. Operates raptor-conservation@yahoogroups.com for discussion and news reporting on worldwide *Falconiformes* and *Strigiformes* conservation matters. **www.raptors-international.org**

Owls on the Internet

Nowadays, the fastest way to search for any information or new photos is the internet. Owls are very well represented in this forum. If you want to know general facts on them, you can go to www.ageulen.de or www.eulenmanie.de (in German) and www.owlpages.com or www.owls.org (in English). Recently described new species can be viewed in www.ornitaxa.com and taxonomy debates in www.birdforum.com

One way is to enroll onto an internet discussion group. For instance, the newly-created buhos-neotropicales@yahoogroups.com deals with the current knowledge, distribution, natural history, ecology and conservation of owls in tropical America.

SPECIES ACCOUNTS

By and large the taxonomy, names and sequence follow that of *Owls of the World*, 2nd Edition (König *et al.* 2008). However, Morris (2009) gave valid reasons to reject two poorly defined small-island 'species'. In addition, four totally new species are included and three that are considered to be extinct are excluded, so the total number of species treated is 249. It is highly likely that genetic, behavioural and vocal studies will reveal even more species in future, mainly by the elevation of existing subspecies into full species, and where this possibility is currently suspected it is mentioned in the text. In any case, all accepted subspecies are described and many of them are illustrated. Professor Desmond Morris also introduced the streamlining of some confusing English names such as Long-whiskered Owl instead of Long-whiskered Owlet *Xenoglaux loweryi* and Forest Spotted Owl instead of Forest Owlet *Heteroglaux blewitti*. Thus, in this book the name 'owlet' is applied only to species in the genus *Taenioglaux;* all the rest are simply 'owls'. Former names are included in 'Other names' to avoid any confusion.

All species are numbered from 1 to 249. This numbering system is also used throughout the species accounts for cross-referencing purposes. Species accounts have a strong emphasis on identification and vocalisations, but other aspects of ecology are also covered. Some important biological aspects such as behaviour and breeding strategies of several species are virtually unknown, which is why current knowledge of these aspects is mainly presented in the introductory sections of the book.

Symbols and size classes

L = Total length, **Wt** = Weight, **W** = Wing length, **WS** = Wingspan (details for each given as and when available)

General size (total length) is described as follows:

tiny	12–16cm
very small	17–20cm
small	21–25cm
small to medium-sized	26–34cm
medium-sized	35–45cm
fairly large	46–56cm
large	57–63cm
very large	64–75cm

Each species account is subdivided as follows:

Names English and scientific names are given for every species.

Measurements and weight This includes, if available, total length (in centimetres), weight (in grams), wing length (in millimetres) and wingspan (in centimetres).

Other name(s) Some commonly encountered alternative names are given if relevant.

Identification Each species account begins with a summary of the key plumage features and such structural differences that will enable separation from similar species. All known colour morphs are presented. Separately in this section, *Juvenile* and *In flight* descriptions are given if recorded. Field identification can largely be based on the text but should be used in conjunction with the photographs which comprise the greatest part of the book.

Call Descriptions of the voice are given together with transcriptions of calls where known. In the latter, a sequence *huu huu* written without hyphenation indicates an interval of more than one second and *huu-huu* means a shorter interval of around half a second. *Huuhuu* indicates a call without noticeable breaks.

Food and hunting The main food items are given if recorded. Hunting methods are commented on, although these are often poorly known.

Habitat Habitat preferences are given, including favoured biotopes if known. Altitudinal ranges are also noted if available.

Status and distribution Distributional information is given briefly in the text but this should be interpreted in conjunction with the maps. For several species information on status and conservation needs is very scanty. Most species are not protected and many are under threat from habitat destruction. The most important research requirements are often highlighted in this section.

Geographical variation A species with no races or variation geographically is described simply as 'monotypic'. For polytypic species, the aim of this section is to describe briefly all accepted races, giving the main differences between them. The ranges of individual subspecies are also given here. The taxonomic situation of many species, and especially of numerous subspecies, is not fully resolved, and future research will undoubtedly result in many changes. Potential taxonomic changes are indicated in these sections.

Similar species Similar sympatric species are described, giving the best criteria for their separation. Sometimes allopatric species are also compared if they could be confused.

Photographs

Sourcing the photographs was a major part of the work. Key photographic agencies were consulted to find missing owl images. Internet bird groups, international bird magazines and personal communications were all used extensively to request photographs of owls. Some 14 species have apparently not been photographed at all, but for five of these we have included photographs of skins held in the Natural History Museum at Tring, England.

Thus, more than 850 photographs illustrate almost all known species of owls, showing their different colour morphs, and sometimes their juvenile plumage and birds in flight. Many of the more distinct subspecies are also illustrated, resulting in unrivalled coverage of the world's owl taxa. The photo captions give quick comments for each species, with information on range or diagnostic features. The captions also include the locality and date (month only) of the photograph, together with the name of the photographer. Full photographic credits are given at the end of the book.

Distribution maps

A large number of the distribution maps are adapted from *Owls of the World* (König *et al.* 2009). However, relevant corrections have been made to most of the maps, especially to those of African and European owls. New maps have been provided for the previously unpublished species. Most maps show only one colour: green. This indicates the normal resident distribution as currently known and as accurately as is possible on such small-scale maps. A very few owl species migrate for the winter and in such cases the winter range is indicated in blue and the summer range in yellow, as for Northern Saw-whet Owl; green areas on such maps indicate the year-round occurrence of some subspecies. Brown Hawk Owl's northern populations are migratory, but these overlap so extensively with the sedentary southern races in Malaysia and Indonesia that these movements are not shown on the distribution map and are only indicated in the text. Irregular nomadic movements are shown with arrows, as in the case of Northern Hawk Owl. Arrows are also used to highlight some small islands on which an owl is resident. Question marks are used when the distribution limits are not clear or have changed recently due to population decreases.

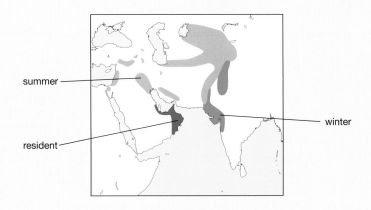

References

To save space, references are not listed at the end of each species account. A number of key sources are listed on page 504. The author has more than 5,000 published papers or books on owls and all of these have been consulted for this book in one way or another. Publishing this list would have added at least 150 pages, making this book uneconomical and more difficult to use. If any reader is seriously concerned about statements in the text or wishes to study a species in greater detail the author is more than willing to share this reference information by email at heimomikkola@yahoo.co.uk

Topography of a typical owl

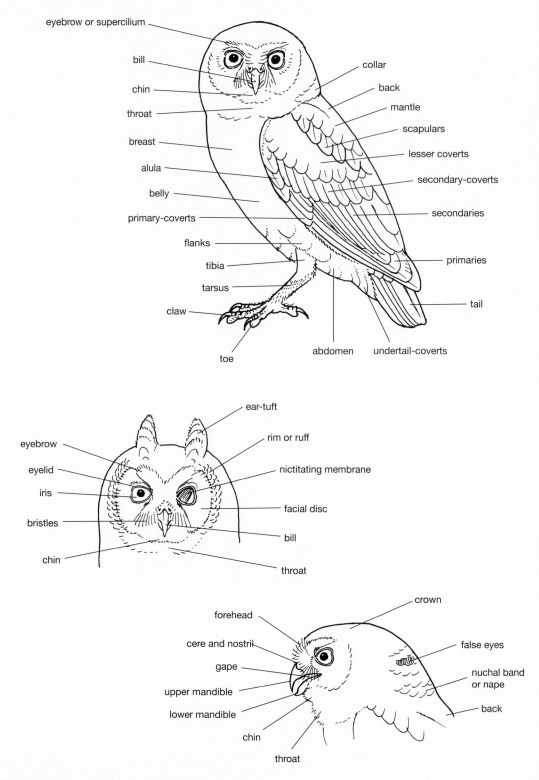

Drawings by Friedhelm Weick, from *Owls of the World* (König *et al.* 2008)

Abbreviations

♀ = female	km = kilometre(s)
♂ = male	mm = millimetre(s)
C. = Central	Mt = Mount
c. = circa (about)	N. = North
cm = centimetre(s)	NC = North-Central
cm² = square centimetre(s)	NE = North-East
DNA = deoxyribonucleic acid	NNW = North-North-West
DR = Democratic Republic	NW = North-West
E. = East	R = Republic
EC = East-Central	S. = South
e.g. = for example	SC = South-Central
etc. = and so on	SE = South-East
g = gram(s)	SW = South-West
ha = hectare(s)	W. = West
kg = kilogram(s)	WC = West-Central

Glossary

Albinism – absence of tyrosinase enzyme (needed to form the pigment melanin) as an inherited condition arising from a genetic defect

Allopatric – mutually exclusive geographically

Allopreening – preening of one bird by another, commonly a mate (often mutual)

Asymmetrical ear – disparity between the ear openings on the two sides

Auriculars – the ear-coverts

Bare parts – bill, legs, eye and (if present) bare facial skin

Concealment display – display posture designed to prevent the owl from being seen or noticed ('tall-thin' and 'stick position' used as synonyms)

Conspecific – being of the same species

Cornea – transparent horny tissue covering eyeball over iris and pupil

Crepuscular – appearing or active in twilight, usually covering dawn and dusk

Desert – a region of extremely low annual rainfall, usually below 25mm

Display – term denoting movements or postures that have become specialised in the course of evolution to serve as 'signals' in social communication (including threat display, courtship display, social displays etc.)

Distraction – general term for those active parental anti-predator strategies which aim to deflect a predator or to divert its attention from eggs or young (terms 'injury-feigning' and 'broken-wing trick' are synonyms)

Diurnal – active in daytime

Endemic – found only in a specific region or country

Erythromelanin – reddish-brown deposition in feathers that normally lack melanin or red-brown pigments

Eumelanin – the most abundant melanin, most likely to be deficient in albinism

Gregarious – living in flocks or communities

Insectivorous – adapted to feeding on insects and similar small creatures

Leucism – complete loss of a particular pigment, or all pigments, affecting melanins, carotenoids and porphyrins differentially, producing anomalous white owls from parents that are normally coloured (may be as slight as a single primary feather on one wing, or as pervasive as an all-white owl with normal eyes, bill and legs)

Lipochrome pigments – pigments that contain a lipid (as carotene or carotenoid) causing yellow, orange or red colours

Mesoptile – second of two nestling down plumages

Melanism – a high concentration of melanin leading to dark coloration of skin, feathers and eyes

Mopane forest – turpentine tree *Colophospermum mopane* forms vast tracts of uninterrupted mopane scrub and woodlands in southern Africa

Morph – an alternative but permanent plumage colour

Nictitating membrane – a transparent fold of skin which is drawn across the eye to form a third eyelid

Nidicolous – of young birds that remain in the nest after hatching

Nomad – a species with no fixed territory when not breeding

Nucleotide sequences – nucleotides are the units of nucleic acids in all cells and are connected by phosphodiester bond, which becomes the backbone of the genetic material (specific nucleotide sequences are discovered as they promote binding of certain enzymes of protein molecule)

Oligocene – epoch from 40 to 25 million years ago

Onomatopoeic – referring to a name created from imitation of sounds made by the animal in question

Paleocene – epoch from 65.5 to 56 million years ago

Parapatry – a situation in which two or more species' ranges do not significantly overlap but are immediately adjacent to each other (the species occur together only in the narrow contact zone, if at all)

Parapatric speciation – parapatric distribution may through time cause speciation into sister-species

Paraspecies – geographically widespread species that have given rise to one or more daughter species as peripheral isolates

Pleistocene – epoch from 2.6 million to 11,700 years ago

Polygamy – having more than one mate at once

Polymorphism – having multiple phenotypes within a population; the switch that determines which phenotype an individual displays can be genetic or environmental

Porphyrins – large group of pigments used by owls and some other birds in plumage and which are easily destroyed by exposure to sunlight (they are most abundant in new feathers, and fluoresce brightly when exposed to ultraviolet light)

Rectrices – the tail feathers

Remiges – the primary and secondary wing feathers

Schemocromatic colour – a physical colour resulting from a colourless submicroscopic body or feather structure which breaks up light rays into component colours

Schizochroic plumage – dilution of only some pigments in the plumage

Sclerophyll forests – multi-aged stands of eucalypts and understorey dominated by hard-leafed shrubs

Sclerotic ring – ring of bone(s) found around the eye

Sexual dimorphism – differences in appearance or size between female and male

Sonogram – standard method of analysing and illustrating vocalisations

Supercilium – area of the head above the eye and corresponding more or less to the eyebrow

Superspecies – a group of at least two more or less distinctive species with approximately parapatric distributions

Sympatric – occurring in the same geographical area (in contrast to allopatric)

Visual acuity – ability to distinguish fine details

Visual sensitivity – ability to distinguish small amounts of light

▼ Leucistic Long-eared Owl *Asio otus*. Completely white like an albino but there is some normal colour in the eyes – see page 30 (*Chris van Rijswijk*).

1. BARN OWL
Tyto alba

L 29–44cm, Wt 254–612g; W 235–323mm, WS 85–98cm

Other names Common Barn Owl, Western Barn Owl
Identification A medium-sized owl without ear-tufts.
The females are sometimes 30–55g heavier than the
males and usually darker and more heavily spotted.
Plumage colour is variably golden-buff on the upperparts and unspotted silvery white on the underparts.
The heart-shaped facial disc is off-white, with rather
small brownish-black eyes and whitish-pink bill.
Notable is the near absence of obvious dark markings
on the underside of the flight and tail feathers. The
upper surface of the flight feathers is covered with a
velvety structure, which absorbs sounds produced by
the moving wings. The white legs are feathered nearly
to the base of the grey-brown toes, which are bare or
lightly bristled; the long needle-sharp claws are dark
grey. *Juvenile* First natal plumage is immaculate white,
and the second one thicker and dull creamy yellowish.
The immature almost identical to the adult. *In flight*
Has a buoyant flight owing to low wing-loading. None
of the primaries is emarginated.
Call Gives a loud hissing scream or screech, sometimes with a tremulous effect, often rendered as
shrrreeee. The screech is most commonly heard in the
very early months of the year while the partners are in
courtship flight. In Britain, the voice gave this owl the

name 'Screech Owl' in 1666, but as early as 1678 it was
changed to Barn Owl.
Food and hunting Preys on small terrestrial rodents
of fields and marshes, mainly common voles and other
colonially living voles. It hunts by searching flights over
open fields and other open and semi-open areas.
Habitat Often lives in very open countryside with scattered trees, mostly near human settlements. It hunts
over open fields interspersed with stands of trees,
bushes, depressions, river valleys and marshes.
Status and distribution This owl is rather common
and widely distributed from C. Europe to Africa and
SE Asia. The total world population of Common and
American Barn Owls has been estimated at 4.9 million
individuals, of which only 50,000 are found in C.
Europe. In England and Wales the population was *c.*
12,000 pairs in 1932, but numbers there have decreased
in the last fifty years by almost 70 per cent. In Finland
the Barn Owl is seen only occasionally, but this may
change if the winter climate becomes much warmer.
It suffers in areas where pesticides are used, and is
increasingly often killed by traffic. Human persecution
is still rampant in Africa and India, where the species
is used for black magic and to remove bad omens. In
Africa, a number of Barn Owls have been rescued from

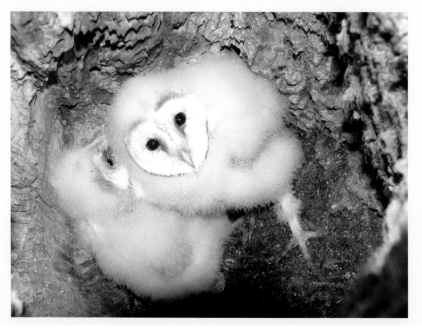

▶ A pair of Barn Owls of
race *erlangeri* in a palm.
The male has only very few
dark spots on the chest but
the sexes seem very similar
in size and colour. Israel,
June (*Eyal Bartov*).

◀ Two pure white very
young Barn Owl chicks in
a nest. Yorkshire, England,
August (*Robin Arundale*).

people who ill-treated them and have been kept for a while in captivity, before being released back to the wild after making a full recovery.

Geographical variation Geographical differences are conspicuous and complicated, relating to body size, strength of bill, legs and talons, colour, and the degree of spotting and other marks on the underparts. Ten subspecies are listed: splendidly white-breasted nominate *alba* from Britain to W. France and Mediterranean countries; heavier *T. a. guttata* from S. Sweden to E. & C. Europe and W. Russia, representing the dark-breasted extreme; *T. a. ernesti*, a very pale Mediterranean race living in Corsica and Sardinia; *T. a. affinis*, an African race found from S. Mauritania east to Sudan and south to C. & S. Africa (and introduced in Seychelles from E. Africa), darker than nominate and with stronger feet and longer tarsi; *T. a. hypermetra* in Madagascar and the Comoros, larger than the African race but with a more distinct ruff; *T. a. stertens* in Pakistan, India and n. Burma to Vietnam and Thailand, very similar to nominate *alba* but with a prominent bluish-grey veil above; *T. a. schmitzi* in Madeira and Porto Santo, small, but with heavy legs and toes, and only four to five wingbars; *T. a. gracilirostris* in the Canary Islands, a small race with a slender bill; *T. a. javanica* of C. & S. Burma east to Laos and south to Indonesia, darker than *stertens* and with the veil greyish-brown (not bluish-grey);

and *T. a. erlangeri* in the Middle East, Iraq and Iran, more golden and less grey above. The taxonomy of *Tyto alba* is not yet fully clarified; e.g. it has been suggested that the very heavy (♂ 555g and ♀ 612g) and darker *javanica* merits treatment as a full species.

Similar species In Europe this is the only owl with a white heart-shaped facial disc. Differences from other *Tyto* species in Africa, America and Australasia are given in the following accounts.

▲ ▶ African subspecies *affinis* of Barn Owl which is similar to *guttata,* but with darker ruff and blackish vertical line down the centre of the face, Namibia.

▶ Barn Owl of race *guttata*. This female with a field vole belongs to the buff-yellow morph or 'adspersa' type, which is pale buff below with small dots. Barneveld, The Netherlands, May (*Chris van Rijswijk*).

▶ ▶ This *guttata* is a dark-morph, or 'obscura' type; grey 'pencil' markings above with white and black spots on the feather tips. Rufous-orange below with many large, dark spots. Barneveld, The Netherlands, June (*Chris van Rijswijk*).

◀ Barn Owl of the nominate race *alba* in flight – note that legs are not hanging behind the tail. Northumberland, England, February (*Ian Fisher*).

2. AMERICAN BARN OWL
Tyto furcata

L 34–38cm, **Wt** 311–700g; **W** 314–370mm, **WS** 90–100cm

Identification A medium-sized owl without ear-tufts. The coloration of the plumage is very variable, the male often paler than the female, which is well over 50g heavier. The upperparts are light yellowish-brown with fine blackish spots, and the underparts are more or less whitish with clear black spots. This species has a whitish facial disc finely rimmed orange-brown, with blackish-brown eyes and a creamy bill, an appearance that has earned it its American name of 'Monkey-faced Owl'. The legs are feathered nearly down to the pale toes, with brownish-black claws. *Juvenile* Similar to young Barn Owl.

Call Emits a drawn-out screech, similar to that of the Barn Owl, from a perch or in flight.

Food and hunting Eats small mammals such as voles, rats and mice. In general it hunts from a perch, but it captures aerial prey, such as bats, on the wing. Is able to locate prey beneath a 25cm-deep snow layer (as can the Great Grey Owl); this behaviour is unknown for any other barn owl.

Habitat Similar to that of the Barn Owl, but S. American subspecies live in tropical and subtropical forests.

Status and distribution This owl is fairly widespread from N. America south to s. Argentina, but in C. & S. America is locally endangered owing to logging of

forest. Traffic kills many owls everywhere. The nominate race is rare or extinct on Grand Cayman.

Geographical variation Taxonomy uncertain; formerly treated as a subspecies of *Tyto aba*. Six races: *T. f. furcata* in Cuba, Caymans and Jamaica is very big,

► Adult nominate of *furcata*, the largest race of American Barn Owl. It has almost uniform white secondaries and tail feathers. Jamaica, November (*Yves-Jacques Rey-Millet*).

◄ Race *tuidara* is smaller than the nominate but has long tarsi and powerful talons. Here the darker female looks slimmer, as it is in the 'tall–thin' position. Both are light morphs. Brazil, September (*José Carlos Motta-Junior*).

the real giant of the species; *T. f. pratincola* of N. & C. America, and also Hispaniola (introduced in Hawaii, where now on all main islands, and also, more strangely, on Lord Howe Island, halfway between Australia and New Zealand) is very similar to nominate race; *T. f. tuidara* of S. America from Brazil south of the Amazon to Tierra del Fuego, in Argentina, also on Falkland Islands, is smaller and paler than the nominate race; *T. f. hellmayri* in Venezuela and the Guianas to the Amazon is very similar to *tuidara*; *T. f. contempta* of Colombia, Ecuador and Peru, the male of which has a much longer wing (312mm) than the female (291mm); and *T. f.* ssp on Bonaire Island, *c.* 50km east of Curaçao and *c.* 90km north of mainland Venezuela, the smallest of the subspecies (wing 285mm). Subspecies *contempta* is similar in plumage to *T. alba guttata*, but because of the size difference from American Barn Owl has been suggested also as possibly a new species, *Tyto contempta*

– 'Colombian Barn Owl', but separation is not yet confirmed by DNA evidence. Even more likely, the unnamed taxon on Bonaire Island deserves specific status, as 'Bonaire Barn Owl', differing clearly from other *Tyto furcata* races and also from the Curaçao Barn Owl.

Similar species This species is similar to the geographically separated Barn Owl (1), but more than 150g heavier and with a larger and stouter head and body, and much more powerful talons. The small, allopatric Curaçao Barn Owl (3) is white below and pale golden-brown above. The geographically isolated Lesser Antilles Barn Owl (4) is also small and has relatively dark plumage, with brownish underparts and a dark greyish veil on the back and wings. The allopatric Galapagos Barn Owl (5) is very dark, with heavily spotted underparts, and the sympatric Ashy-faced Owl (11) is markedly smaller and has an ash-grey face.

▼ A male American Barn Owl of race *pratincola*. Very light compared to the female. The facial disc is pure white but there are some dark dots on the breast and flanks. Arizona, January (*Jim and Deva Burns*).

▼ The darker and heavier female *pratincola*. Its facial disc has slight reddish tinge. Arizona, March (*Jim and Deva Burns*).

▲ Race *contempta* has been proposed as a separate species, Colombian Barn Owl. It is smaller than other subspecies but very variable in coloration. This is a light morph. Facial disc is white with a dark and whitish ruff. Colombia (*Nick Athanas*).

► Race *hellmayri* is one of three races in mainland South America; it is smaller and paler than the nominate race. Venezuela.

3. CURAÇAO BARN OWL
Tyto bargei

L *c.* 29cm, W 246-258mm

Identification A small to medium-sized barn owl without ear-tufts. There are no data on sexual size differences. This species has short wings and a short tail. The upperparts are light yellowish-brown, marked with ochre and golden-brown, the flight and tail feathers with darker bars. The white facial disc has a narrow yellow-ochre rim, and the area around the blackish-brown eyes is white, shading into pale ochre. The bill is ivory-white. The white underparts are coarsely speckled dark on the sides of the breast. The relatively long legs are light greyish and the claws blackish-brown. *Juvenile* Similar to other barn owls.

Call A harsh screeching call is given. Clicking notes in flight suggest possible echolocation skills.

Food and hunting Feeds on small mammals, including bats, also reptiles, small birds and large insects. Catches prey in flight or from perch.

Habitat This owl inhabits rocky areas with thickets and caves, but is found also in semi-open countryside with old buildings or ruins.

Status and distribution Endemic to Curaçao, in the Netherlands Antilles. A visit to the island in 1989 found only 40 pairs, and this owl is, therefore, highly vulnerable to human disturbance, use of pesticides and traffic.

Geographical variation Taxonomy uncertain; formerly treated as a subspecies of *Tyto aba*. Monotypic.

Similar species The only *Tyto* species on the island, the Curaçao Barn Owl is very much smaller than the allopatric American Barn Owl (2); the unnamed race of the latter on nearby Bonaire Island is bigger and has large unfeathered feet and a long tarsus and bill. Similar in size, but geographically separated, the Lesser Antilles Barn Owl (4) has relatively dark plumage with brownish underparts, and a dark greyish veil on the back and wings.

◄ A pair of Curaçao Barn Owls; there is little difference between the sexes, with both female and male having some coarse dark spotting below. Curaçao (*Carmabi Institute/Peter van der Wolf*).

▼ Curaçao Barn Owl has a very dark back (Bonaire Barn Owl has predominantly yellow upperparts, wings and tail). Feet are fully feathered (Bonaire's are unfeathered). Curaçao (*Carmabi Institute/Peter van der Wolf*).

4. LESSER ANTILLES BARN OWL
Tyto insularis

L 27–33cm, **Wt** *c*. 260g; **W** 226–247mm

Identification A small to medium-sized and dark *Tyto* owl without ear-tufts. No sexual weight differences are documented, but females have the wing more than 10mm longer than that of the males. This species has a light brownish face with a black-spotted brown rim and brownish-black eyes; the bill is yellowish. The upperparts and wings have a dark greyish veil, finely white-spotted, and the flight and tail feathers are brownish with darker bars. The underparts are brownish with dark arrow-shaped spots. The long legs are feathered nearly to the base of the toes, the unfeathered parts of the legs and toes being greyish-brown; the claws are blackish-brown. *Juvenile* Similar to other young barn owls.

Call Little known, but a piercing scream and clicking notes have been recorded on Dominica.

Food and hunting Feeds on small mammals, including bats, also reptiles, small birds and larger insects. Hunting habits are little known, but likely to be similar to those of other barn owls.

Habitat Lives in open farmland, and in rocky woodland with bushes, scrub and caves.

Status and distribution Endemic to the Lesser Antilles, where it occurs on Dominica, St Vincent, Bequia, Union, Carriacou and Grenada.

Geographical variation This owl was formerly treated as a subspecies of *Tyto aba*. Its taxonomic status is uncertain and requires DNA evidence, as it is now separated from *T. glaucops* and *T. furcata* on the grounds mainly of its isolated distribution and morphological differences. Two races are recognised: the rare *T. i. insularis* on St Vincent and Grenada and a few islands in the Grenadines, and the rather common *T. i. nigrescens* on Dominica, the latter having hardly any white spots above.

Similar species The allopatric American Barn Owl (2) is much larger, with a white face and underparts. Geographically slightly separated, the Curaçao Barn Owl (3) is similar in size, but also white below and with a white facial disc. The allopatric Ashy-faced Owl (11) differs in having, as its name indicates, an ashy-grey facial disc.

▶ Lesser Antilles Barn Owl of race *nigrescens*. Very similar to Galapagos Barn Owl, it is darker and less spotted above and has smaller dark spots below than nominate *insularis*. Dominica, November (*John Mittermeier*).

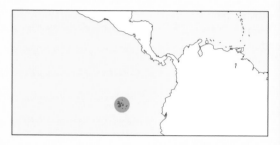

5. GALAPAGOS BARN OWL
Tyto punctatissima

L 33cm; W 229–234mm, WS 68cm

Identification A small to medium-sized *Tyto* without ear-tufts. No data on sexual size differences are available, but the male is much lighter than the female. This species has relatively long legs and powerful talons. Its facial disc is dirty whitish with a partly dark-spotted orange-brownish rim, and with relatively small, blackish eyes. The bill is yellowish-white, but tinged blackish. The cinnamon-brownish upperparts are covered by a dark grey veil with many whitish spots, and the flight and tail feathers are brownish-yellow with darker bars. The dirty whitish or pale cinnamon underparts are densely spotted with dark and white markings, these markings shading paler downwards. The legs are feathered; the toes are greyish-brown and the claws blackish-brown. *Juvenile* Similar to other barn owls. *In flight* Has a moth-like flight, with soft, deep wingbeats.
Call Utters a long and hoarse *kreeeee*, rather high-pitched.
Food and hunting The diet consists of small rats, mice and insects, with an alleged preference for grasshoppers. It occasionally takes small birds. It hunts from a perch or on foot.
Habitat This is an owl of the dry lowlands with little vegetation, at higher altitudes in more open landscapes, but is found also near human settlements.

Status and distribution Endemic in the Galapagos Archipelago, where recorded from the islands of Fernandina, Isabela, San Cristóbal, Santa Cruz and Santiago. On Fernandina it is not so rare, but elsewhere it is scarce. It can be considered vulnerable owing to human persecution and increasing tourism.
Geographical variation Taxonomy uncertain; DNA evidence is still needed in order to confirm that the Galapagos Barn Owl is a full species, and not a race of *T. alba* or *T. furcata*. Monotypic.
Similar species The only *Tyto* owl in the Galapagos Archipelago, *c.* 1000km off the S. American coast, where the mainland American Barn Owl race *T. f. contempta* (2) is larger in size and very variable in coloration.

◀ Galapagos Barn Owl has dark plumage and a dusky brown facial disc. It is dark brown above with numerous white spots, and buffish below with brownish vermiculations and small dark and white spots. Santa Cruz, August (*Seig Kopinitz*).

6. CAPE VERDE BARN OWL
Tyto detorta

L 35cm; W 272–297mm

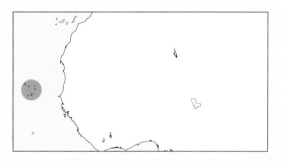

Identification A medium-sized dark *Tyto* owl without ear-tufts. There are no data on sexual size or colour differences. The facial disc is pale yellowish-brown, with blackish-brown eyes, and the bill is ivory-white. It has greyish-brown upperparts with large white dots surrounded by black. The upperwing is dirty golden-brownish without a paler grey veil, the wings being lightly speckled. The underparts are dull yellowish with dark arrow-like spots, the belly and the long legs buffish, with the lower part of the legs and the toes greyish-brown, and the claws brownish-black. *Juvenile* Similar to other barn owls.

Call Said to be similar to that of the Barn Owl.

Food and hunting Feeds on small rats and mice, geckos, and colonial seabirds such as petrels (Procellariidae), and also takes some insects and spiders. It hunts mostly from a perch, but also in flight.

Habitat An owl of open and semi-open landscapes, this species is also found on small islands and in canyons with steep rocky walls and caves. It is often seen near human dwellings and abandoned buildings.

Status and distribution This owl is endemic to the Cape Verde Islands (on St Vincent and Santiago), off the W. African coast, where it is obviously uncommon or rare. Being confined to a small archipelago, it is possibly endangered by destruction of habitats and by direct human persecution.

Geographical variation Often regarded as a sub-species of *T. alba*, this owl has now been separated on account of its geographical isolation and differences in plumage. Nevertheless, further taxonomic study and studies of its ecology and biology are urgently needed. Monotypic.

Similar species This is the only barn owl on the Cape Verde Islands. The African race *T. a. affinis* of the Barn Owl (1) is larger, with a whitish facial disc, and has a greyish veil on the upperparts. The São Tomé Barn Owl (7) is similar in coloration, but smaller and geographically discrete.

▼ Cape Verde Barn Owl is a little darker than the darkest subspecies of Common Barn Owl, *guttata*, and is buffish to cinnamon-buff below, with some large arrow-shaped spots. Facial disc is light rufous-brown and bill is almost whitish Santiago, June (*Eric Didner*).

7. SÃO TOMÉ BARN OWL
Tyto thomensis

L 33cm, **Wt** 380g (1♂); **W** 241–264mm

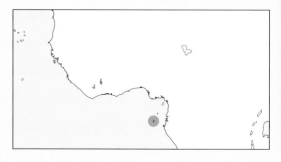

Identification A small to medium-sized and dark *Tyto* owl without ear-tufts. No data on sexual size differences are available. The grey and rufous upperparts have a dark grey veil with conspicuous white and black markings. The flight and tail feathers are reddish-brown with brownish-black bars. The facial disc is yellowish-brown, with a narrow dark zone from the blackish-brown eyes to the dirty-yellow bill. The underparts are golden-brown with dark arrow-shaped spots. The long, feathered legs are yellowish-brown, and the bare parts of the legs and toes brownish-grey, with relatively powerful blackish claws. *Juvenile* Similar to that of other barn owls.

Call The voice is thought to be similar to that of the Barn Owl.

Food and hunting Few data. So far as is known, this species feeds on small mammals and birds, insects, spiders, lizards and frogs.

Habitat Inhabits bushlands and semi-open woodlands, rocky areas and human settlements.

Status and distribution Endemic to São Tomé, in the Gulf of Guinea. Although still fairly common, the fact that it is confined to one small island makes it vulnerable to human persecution and the use of pesticides.

Geographical variation Taxonomy uncertain; further studies are required to confirm the species status, as hitherto it has always been treated as a subspecies of the Barn Owl. Monotypic.

Similar species The allopatric Cape Verde Barn Owl (6) is larger, paler in coloration, and has dull yellowish underparts. The African race *T. a. affinis* of the Barn Owl (1), on the African mainland *c.* 200km away, is larger and has a whitish facial disc and a pale grey (not dark grey) veil.

◀ This São Tomé Barn Owl has captured a large rat with its strong feet and powerful talons. This dark grey to rufous brown owl has a very dark face and yellow-ochre underparts. São Tomé, July (*Fabio Olmos*).

8. AUSTRALIAN BARN OWL
Tyto delicatula

L 30–39cm, **Wt** 230–470g; **W** 247–300mm, **WS** 79–97cm

Other name Eastern Barn Owl

Identification A small to medium-sized white owl without ear-tufts. The female is larger and 25g heavier than the male; females and juveniles are generally more densely spotted than adult males. It has a round face and rather small black eyes. The facial disc is white, sharply edged fawn and black, and the bill is whitish or pale horn-coloured. The upperparts have a mottled pattern of grey, buff, fawn and pale gold, palest on the head and richest on the wings. The flight and tail feathers are faintly barred dark grey over a light yellowish-brown or pale golden background. The pure white underparts have sparse dark spots. The slender white legs are sparsely feathered to beyond the 'ankle', with light greyish-brown toes and dark greyish-brown claws. *Juvenile* Leaves the nest in a plumage barely distinguishable from that of adults, but with heavier spotting on the breast. *In flight* Looks very white and long-winged, with a large head and short tail.

Call A long, hissing screech, *sk-air* or *skee-air,* is delivered at an even pitch and lasts between one and two seconds.

Food and hunting This is a specialist hunter of small mammals, the House Mouse *Mus musculus* often forming the staple diet. Other prey noted include rats, young rabbits, bats, frogs, lizards, small birds and insects. The vast majority of the prey is taken on the ground, but it captures birds and bats also in flight.

Habitat Found in open country, farmland, suburbs, cities, open woodland, heaths and moors, and even on rocky offshore islands.

▼ Female Australian Barn Owl. Female has more dark spots on breast and flanks. Queensland, September (*Rohan Clarke*).

Status and distribution This is a widespread and rather common owl in Australia, but is rare in Tasmania. It also occurs on many Indonesian and New Guinean islands, the Philippines and islands in the W. Pacific Ocean, but its status on these is uncertain. The nominate race *delicatula* has been introduced in New Zealand; an attempted introduction on Lord Howe Island was unsuccessful.

Geographical variation Four subspecies: *T. d. delicatula* in Australia and offshore islands, and in the Lesser Sundas and the Solomon Islands; *T. d. meeki* in Papua New Guinea, and Manam and Karkar (Dampier) Islands, is more buffish-orange above than the nominate; *T. d. sumbaensis* on Sumba Island, Indonesia, is bright cinnamon-orange on the back and has a whitish tail with narrow dark bars; and *T. d. interposita* on Santa Cruz Island, Banks Island and Vanuatu, has the underparts washed orange-ochre.

Similar species The allopatric Golden Masked Owl (13) is golden-rufous and has dark V-shaped markings on the back and wing-coverts. Readily confused

with the partly sympatric Eastern Grass Owl (15) and the pale morph of the Australian Masked Owl (18); for differences, see those two species. The sympatric Lesser Sooty (22) and Greater Sooty Owls (23) are characterised by their very dark, blackish plumage.

◄ Male Australian Barn Owl near the nest – a very white facial disc and underparts, with hardly any dark spots. Victoria, Australia, July (*Rohan Clarke*).

▼ Australian Barn Owl nominate *delicatula* in flight. A very short-tailed, long-winged and pale barn owl. Victoria, Australia, July (*Rohan Clarke*).

► Australian Barn Owl of the race *sumbaensis*. It is a paler grey on the head and nape than in nominate *delicatula*. Eastern Grass Owl *Tyto longimembris* has a dark back and longer legs. Sumba, Indonesia, October (*Ron Hoff*).

9. BOANG BARN OWL
Tyto crassirostris

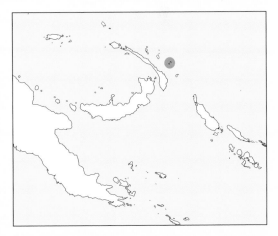

L 33cm; W 285–290mm

Identification A small to medium-sized dark owl without ear-tufts. No data on sexual size differences are available. It has a strong bill and powerful talons. The upperparts are dark and the underparts dirty white, with a brownish wash and dark spots. The white facial disc is slightly tinged dull yellowish, and encloses black-brownish eyes and a black-tinged creamy bill. The bare parts of the legs and toes are greyish-brown, with blackish claws. *Juvenile* No data.

Call Not known.

Food and hunting Not studied.

Habitat Inhabits open and semi-open landscapes with bushes and trees, as well as farmland and grassland.

Status and distribution Endemic to Boang Island, in the Tanga group of the Bismarck Archipelago, in Papua New Guinea. Its isolated island distribution could make it vulnerable, but no data on its status are available.

Geographical variation Taxonomy uncertain. Has hitherto been treated as a race of *Tyto delicatula*; intensive studies of its vocalisations, ecology and biology, and DNA evidence, are required in order to confirm its status as a full species. Monotypic.

Similar species Although similar to the allopatric Australian Barn Owl (8), it is clearly darker in coloration, with a much stronger bill and more powerful talons. The geographically separated Australian Masked Owl (18) has very variable plumage but is much larger, with even stronger talons, and with legs feathered to the base of the toes.

10. ANDAMAN BARN OWL
Tyto deroepstorffi

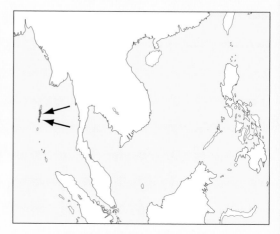

L 33–36cm; W 250–264mm

Other name Andaman Masked Owl

Identification A small to medium-sized *Tyto* owl without ear-tufts. No information is available on sexual size differences. This owl has a facial disc that is wine-reddish with a relatively broad orange-brown rim, and the eyes blackish-brown. The bill is cream-coloured. The upperparts are dark brown with chocolate-coloured patches and orange-buff spots. The breast is golden-rufous with blackish spots, the belly whitish, and the legs feathered whitish-ochre. The toes are dark pinkish-grey with powerful purple-grey claws. *Juvenile* Probably similar to young Barn Owl.

Call Reported as giving a rather high-pitched *sshreeet*, repeated several times.

Food and hunting Bones of rats and mice have been identified in the few pellets collected.

Habitat This owl is found in coastal areas and fields, and also near human settlements.

Status and distribution The status of this species, endemic to the S. Andaman Islands, in the Indian Ocean, is uncertain, as only a few field records and five skins are known.

Geographical variation Taxonomy uncertain; DNA evidence and studies of its ecology and biology are necessary in order to determine if *T. deroepstorffi* is, indeed, the ecological replacement of *T. alba* in the S. Andamans. Monotypic.

Similar species It is similar in shape and size to the Barn Owl (1), but brighter in colour; unlike this species, all races of the Barn Owl have a greyish veil with white-and-black spots. The present species has more powerful feet than Barn Owl.

▼ Andaman Barn Owl is a very dark owl with blackish-brown eyes on a rufous face, an orange-brown ruff, and a chestnut streak before and behind the eyes. South Andaman, December (*Rob Hutchinson*).

▼ A pair of Andaman Barn Owls. South Andaman, December (*Rob Hutchinson*).

11. ASHY-FACED OWL
Tyto glaucops

L 33–35cm, **Wt** 260–535g; **W** 240–280mm

Identification A small to medium-sized owl without ear-tufts. The female is clearly larger and up to 200g heavier than the male. This species is similar in appearance to the Barn Owl, except that the facial disc is ashy grey with an orange-brown rim, and with blackish-brown eyes. The bill is bluish-white. The upperparts are yellowish-brown with fine black streaks, and the pale underparts have dark arrow-shaped spots. The feathered legs, rather long, are yellowish-brown, the toes greyish-brown and the claws blackish-brown. *Juvenile* Similar to Barn Owl.

Call The song is a rasping wheeze of 2–3 seconds' duration, which differs from that of the American Barn Owl.

Food and hunting Feeds on small mammals such as mice and rats, but also takes small birds, frogs, reptiles and insects. No details of hunting habits are available.

Habitat Inhabits open areas and open forests, often near human settlements, from lowlands to above 2000m.

Status and distribution Endemic to Hispaniola and Tortuga Island, in the Caribbean. This species' true status is not known, but it could be endangered by human activities.

Geographical variation On Hispaniola it lives side by side with the American Barn Owl with no evident interbreeding, indirectly supporting the full species status of this taxon. It is not known if any variation

▼ A pair of Ashy-faced Owls. There is little colour or size difference, though the female is said to be clearly larger and maybe a little darker than the male. Dominican Republic, November (*Mario Dávalos P.*).

exists between the two island populations. Treated as monotypic.

Similar species None of the other *Tyto* owls has an ash-grey face. White American Barn Owls (2) living on Hispaniola are much larger. The allopatric Curaçao Barn Owl (3) is a little smaller, and is pale golden-brown above and white below. Also geographically well separated, the Lesser Antilles Barn Owl (4) has darker plumage, with brownish underparts, and a dark greyish veil on the back and wings.

▼ Ashy-faced Owl has yellowish-black upperparts with dense blackish vermiculations and fine black streaks. Edge of wings near wrist orange-brown. Underparts yellowish brown (female) and more yellowish white (male); both have small, dark spots (maybe slightly smaller in the male). Dominican Republic, November (*Mario Dávalos P.*).

▶ The distinctive Ashy-faced Owl is the only *Tyto* with an ash-grey face. It has dark eyes and a strong ruff, dark rufous over the forehead and blackish on sides of face. It has rather long, yellowish-white feathered legs with greyish-brown bristled toes. Dominican Republic, November (*Mario Dávalos P.*).

12. MADAGASCAR RED OWL
Tyto soumagnei

L 28–30cm, Wt 323–435g (2); W 190–222mm

Other name Madagascar Grass Owl

Identification A relatively small barn owl without ear-tufts. There are no data on sexual size differences. The general coloration of both sexes is reddish-ochre to yellow-ochre. The facial disc is yellowish-white, with a brownish wash between the lower edge of the dark sooty-blackish eyes and the base of the light grey bill. The upperparts have fine blackish spots, larger on the wings and the long tail, and the underparts are pale brownish-yellow with scattered fine, blackish dots. The smoky-grey toes are scarcely bristled, with powerful greyish-black claws. *Juvenile* Has white natal down, but at the age of just one month a noticeable facial disc is already developed. *In flight* The wing-bars are especially prominent.

Call Emits a loud hissing screech of *c.* 1.5 seconds in duration, and usually strongly descending in tone (unlike that of *T. alba*).

Food and hunting Feeds on small native rodents, including shrew-like tenrecs (Tenrecidae), but also on other small vertebrates, and insects.

Habitat It inhabits humid rainforest, and also forest edges and plantations, mainly between 900m and 1200m.

Status and distribution Endemic to E. Madagascar, this owl has a very small population and is classified as Endangered. The Peregrine Fund has started intensive studies on its status and distribution, as well as its ecology and behaviour.

Geographical variation Monotypic.

Similar species This species could be confused with the sympatric Barn Owl (1), but the Madagascar subspecies of the latter, *T. a. hypermetra*, is much larger and has a very pale area on the wing-coverts and much paler underparts.

▶ The whitish facial disc has a reddish tinge, with dark soot-black eyes and powerful talons. This female is near the nest cavity. Masoala, Madagascar (*Russell Thorstrom*).

◀ Madagascar Red Owl is a small but relatively long-tailed barn owl with yellowish-ochre or even rufous-ochre coloration. Bealanana, Madagascar (*Russell Thorstrom*).

13. GOLDEN MASKED OWL
Tyto aurantia

L 27–33cm; W 220–230mm

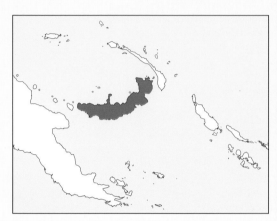

Other name New Britain Masked Owl

Identification One of the smallest owls in the genus *Tyto*. There is no information on possible sexual differences in colour or size. The general plumage coloration is golden-buff, darker above, paler below, mottled with blackish-brown. The facial disc is light dull yellow, and the flight feathers and tail are golden-buff with blackish-brown bars. The iris is dark brown, the pale bill is rather powerful relative to body size, the long toes are yellowish-grey to brownish-grey, and the claws are brown with darker tips. *Juvenile* Undescribed.

Call Utters a long ascending *ka-ka*, repeated about six times per second; the local name 'kakaula' is derived from this call.

Food and hunting Recorded food of this owl includes small rodents, but it probably also takes insects and small birds.

Habitat Frequents primary forests in the lowlands and mountains, up to at least 1830m.

Status and distribution This species is endemic on the island of New Britain, in Papua New Guinea, where it is uncommon or rare. It is affected by logging and deforestation and has been listed as Vulnerable by BirdLife International. Its biology and ecology are poorly known and require investigation.

Geographical variation Monotypic.

Similar species None of the similar *Tyto* species has golden-rufous plumage with V-shaped markings on the back and wing-coverts.

▼ Golden Masked Owl is a relatively small *Tyto* and the only one that has golden-rufous plumage with dark markings, V-shaped on the back and wing-coverts. Legs are rather long but not very strong. No photo of a wild bird is known. Specimen from BMNH (Tring) (*Nigel Redman*).

14. TALIABU MASKED OWL
Tyto nigrobrunnea

L 31cm; W 283mm (1)

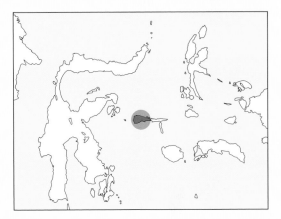

Identification A small to medium-sized owl without ear-tufts. No data on sexual size or colour differences are available. It is similar to other barn owls, but has upperparts dark brown with whitish speckles. The wings are brown, the secondaries with whitish tips, and the brown tail has three dark bars. The facial disc is pale rufous, darker towards the blackish-brown eyes, and the bill is blackish-grey. The underparts are deep golden-brown with dark spots. The reddish-brown legs are feathered to the lower third of the tarsus, with grey toes and powerful blackish claws. *Juvenile* Undocumented.

Call Not known.

Food and hunting Unknown.

Habitat The only specimen was found in a primary lowland forest, but a recent sighting was made in selectively logged lowland forest.

Status and distribution Endemic to Taliabu Island, in the Sula Archipelago, off E. Sulawesi, this owl was for a long time known only from a single female collected more than 60 years ago. In 1991, however, one sight record was reported, and the first photos were taken in 2009, proving that the species still survives on Taliabu. It is listed as Endangered, because its population must be extremely small owing to its very limited range. Studies are needed of this almost totally unknown species.

Geographical variation Monotypic.

Similar species This is the only barn owl with uniformly dark brown wings without any whitish markings on the primaries. The widespread Brown Hawk Owl (221) is slightly smaller, with yellow eyes, streaked underparts and a barred tail.

▼ Taliabu Masked Owl has dark brown upperparts, with small whitish speckles from crown to lower back and on the tips of secondaries, and uniform dark primaries without any pattern. Prior to the taking of this photo in 2009, this endangered owl had been recorded only twice. Taliabu, October (*Bram Demeulemeester*).

15. EASTERN GRASS OWL
Tyto longimembris

L 32–40cm, Wt 250–582g; W 273–360mm, WS 103–116cm

Identification A medium-sized *Tyto* owl without ear-tufts. Males are up to 65g smaller and more slender than the darker females. The male has pale upperparts, a white facial disc and white underparts. The female's upperparts are grey-black or dark brownish-yellow, heavily blotched with lighter orange, each feather with a tiny white spot at the tip, and the underparts light chestnut, finely spotted with black and white. The primaries and tail are broadly barred dark grey-black and orange. The pale reddish-brown or yellowish-brown and white facial disc is more heart-shaped than those of other Australian *Tyto* owls. The eyes are small and black and the bill ivory-coloured. The very long legs are feathered deep reddish-brown or white to the top of the bare tarsus; the tarsus and toes are pale flesh-coloured, with blackish-brown claws. *Juvenile* The first down is white and the second is warm golden-brown; all young appear to fledge in dark plumage, indistinguishable from that of the adult female. *In flight* The barring on the underside of the wings and tail is very obvious; the wings are longer than those of the Barn Owl, and the feet protrude beyond the tail.

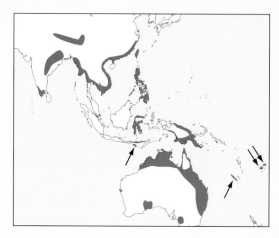

Call This species appears to be more silent than the Barn Owl. A thin, high screeching has been described, but its vocalisations are little studied.

Food and hunting Rodents of numerous species are the favoured prey, but it may also take birds and insects. It hunts entirely in flight, and locates and catches its prey by the combined use of hearing and the very long legs.

Habitat Although this owl prefers open tropical grass-lands with dense and tall grass, it can at times be found also in desert grasslands, especially during plagues of Long-haired Rats *Rattus villosissimus*.

Status and distribution It occurs from India and SE Asia east to Fiji and Australia. It is apparently common

◄ Eastern Grass Owl subspecies *amauronota* is very light and large. Its wings are some 50mm longer than those of Australian Barn Owl. The wings of African Grass Owl seem more rounded, making the wingspan 10cm smaller. Luzon, Philippines, May (*Michael Anton*).

locally but is threatened by the use of agricultural pesticides.

Geographical variation Four or five subspecies: *T. l. longimembris* from India to Australia (New Caledonia and Fiji birds are often separated as race *oustaleti*, but no recent records from Fiji, unless overlooked); *T. l. chinensis*, found in China and Taiwan, has the upperparts much less uniform than other races, and has creamy underparts; *T. l. amauronota* in the Philippines is very large, weighing 582g, and has the upperparts less dark; *T. l. papuensis* in New Guinea differs from the nominate race in having a plain facial disc and is paler above and below.

Similar species This species is distinguished from all other *Tyto* owls by its long wings and legs, with protruding feet in flight. The very similar African Grass Owl (16) is geographically well separated, and has sooty-blackish upperparts without yellowish flecks. Confusion is possible between dark individuals of the present species and largely overlapping, dark Australian Masked Owl (18), but the grass owls are more or less uniformly dark brown above, with fine whitish spots.

▼ Nominate Eastern Grass Owl has a whitish face and creamy underparts. Upperparts are darker than in *amauronota* and there is no brownish veil on the upper breast. The face is clearly heart-shaped with a distinct rim. Cape York, Queensland, August (*Edwin Collaerts*).

▶ Nominate Eastern Grass Owl in flight showing the huge legs hanging far behind the tail, which is fairly short and bears 3–4 dark bars. Victoria, Australia, August (*Rohan Clarke*).

16. AFRICAN GRASS OWL
Tyto capensis

L 34–42cm, **Wt** 355–520g; **W** 283–345mm, **WS** 100–108cm

Identification A medium-sized but long-legged owl without ear-tufts. No data on sexual weight differences are available, but males seem to have longer wings and tail than females. The sexes are alike in plumage, having a whitish-creamy facial disc with a thin yellowish-buff rim densely dark-spotted; the eyes are brown-black and the bill whitish to pale pink. The upperparts are rather uniformly sooty olive-brown, finely flecked or spotted whitish, while the flight feathers are pale brownish-grey with darker bars and yellowish bases above, the underwings being whitish to light golden-brown; the short tail has four dark bars. The underparts are whitish to pale reddish-yellow with small dark brown spots. The feathered legs are whitish, with yellowish-grey toes and blackish claws. *Juvenile* The first coat is white, and the second coat dull yellow or light golden-brown. The heart-shaped facial disc is mainly brown with whitish edges, the eyes being black and the bill creamy-grey. *In flight* Shows long wings and a short tail, with uniformly coloured central tail feathers.
Call The call is very similar to that of the Barn Owl,
but less strident, a high-pitched sibilant tremolo lasting 1–2 seconds. It often gives a clicking *tk-th-tk-tk…*, somewhat like a cricket, during hunting flights.
Food and hunting The food is mainly rodents and shrews weighing 1.5–100g, supplemented with some small birds, bats and insects. It hunts by quartering low over the ground.
Habitat This owl inhabits moist grassland and open savanna at up to 3200m. It is not associated with human habitations or activities.
Status and distribution This is a resident from Cameroon and Ethiopia south to S. Africa, but is absent in large areas of western parts of S. Africa. It is locally rather common, but its numbers are declining owing to habitat destruction.
Geographical variation This species was earlier treated as conspecific with *T. longimembris* of Asia and Australasia. Monotypic.
Similar species Larger and much darker than the sympatric Barn Owl (1), this species is more commonly confused with the similarly grassland-dwelling and

▼ African Grass Owl often lives on the ground, walking easily with its long legs. Rather uniform dusky brown above, flecked and spotted, without buffish markings. South Africa, August (*Eyal Bartov*).

largely overlapping Marsh Owl (249), which is brown overall, with a smaller head lacking a white facial disc. The geographically fully separated Eastern Grass Owl (formerly considered conspecific) has a dark brown back with yellowish markings; the two grass owls are the only *Tyto* species having blackish-brown upperparts and a pale underside.

▲ An African Grass Owl in flight, showing the legs hanging well behind the short tail. South Africa (*Markus Lilje*).

◀ Two young African Grass Owls, still with remnants of their down (*Lucian Coman*).

17. LESSER MASKED OWL
Tyto sororcula

L 29–31cm; W 227–251mm

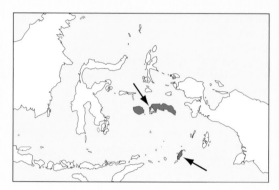

Other name Moluccan Masked Owl

Identification A small to medium-sized *Tyto* owl without ear-tufts. Data on sexual weight differences are lacking, but the female has much longer wings than the male. The sexes are apparently similar in plumage. They have a pale rufous facial disc, and a brownish wash around the blackish-brown eyes extending to the base of the yellowish-creamy bill. The greyish-brown upperparts have orange patches and black-edged white dots. The flight and tail feathers are greyish-brown to reddish-brown with darker bars. Below, the plumage is whitish, but coarsely spotted with brown dots. The feathered feet are also whitish, with yellowish-grey toes and blackish-brown claws. *Juvenile* Undescribed.

Call The call consists of three fast screeching whistles in two seconds, or three or four slower screeches, higher-pitched but more drawn out.

Food and hunting Unknown.

Habitat This species is assumed to live in primary forests and to take shelter in limestone caves, but no habitat was recorded for any collected specimens.

Status and distribution Endemic to the Lesser Sundas, this owl is very poorly known. Only four birds have been collected, two each from the Tanimbar Islands and Buru. It is clearly rare and probably endangered but has recntly been photographed on Tanimbar.

Geographical variation Taxonomy uncertain. Two subspecies have been described, based on just four specimens, but the taxonomy and biology of this owl are totally unstudied. *T. s. sororcula* on Yamdena and Larat, in the Tanimbar Archipelago; *T. s. cayelii*, on Buru and possibly also Seram, is larger than the nominate race and more tawny-buff over its entire plumage, and could even be a separate species.

Similar species The geographically separated Australian Masked Owl (18) is much larger.

▼ Lesser Masked Owl has darker upperparts than the clearly larger Australian Masked Owl, and stronger legs. Wing-coverts contrast with the paler secondary coverts and secondaries. It has a rufous facial disc, pale ochre-buff underparts and a white belly. Tanimbar (*James Eaton*).

18. AUSTRALIAN MASKED OWL
Tyto novaehollandiae

L 33–47cm, Wt 420–673g; W 290–358mm, WS *c.* 90 cm

Identification A medium-sized to fairly large owl without ear-tufts. Females have clearly longer wings than males, but sexed weights are not available. The most variable of Australia's owls, both in size and in colour: Queensland males are hardly bigger than Barn Owls; pale and dark morphs may be distinguished. It is generally larger and darker than Barn Owls, and has feathered tarsi. The facial disc is white with a blackish-brown border, and some chestnut shading around the dark brown eyes. The bill is whitish or pale horn-coloured. The crown and upperparts are dark, finely patterned with black, grey, white and dull yellow, the flight feathers and tail with dark brown and orange-buff bars. The underparts are white, with or without a buffy tinge, and sparsely spotted with black. The whitish or orange-buff feet are fully feathered to the base of the toes, which are yellowish-grey or pale pinkish-grey and slightly bristled; the strong claws are dark greyish-brown with darker tips. *Juvenile* The first down is white and the second down creamy. The immature resembles the adult, and with the colour morph which it will retain throughout its life.

Call It emits a loud version of the Barn Owl's call: a drawn-out rasping *cush-cush-sh-sh* or *quair-sh-sh-sh*. A strange, wild cackling has also been described.

Food and hunting The diet comprises small mammals up to the size of rabbits; it also preys on small birds and lizards. It hunts on the wing or from a perch, and seems to prefer open country.

Habitat Found in open woodlands, isolated stands of large trees, and residual copses and tree-lines in farmland.

Status and distribution This species' status is uncertain owing to its secretive habits, and even the voice is seldom heard. It occurs in Australia, S. New Guinea, and perhaps still on Manus, in the Admiralty Islands, off N. Papua New Guinea, thus having a peculiar distribution; it may possibly be overlooked in some

▼ A totally white-faced Australian Masked Owl – this owl is the most variable in Australia. The light morph lacks the orange-buff ground colour and has large white spots above. Queensland, October (*Adrian Boyle*).

places. Evidence of extinct populations on many New Guinea islands would suggest that masked owls (undescribed races or closely related species) may once have occurred more widely in the Bismarck Archipelago, off Papua New Guinea.

Geographical variation Four subspecies have been described: *T. n. novaehollandiae* from South Australia; *T. n. kimberli* from Northern Territory, which is very variable in coloration, but always white below and pale above; *T. n. calabyi* from New Guinea, with a darker back with yellowish patches; and *T. n. manusi* from Manus Island (Admiralty Islands, in Bismarck Archipelago), which is less dark above and whitish below, becoming pale ochraceous buff towards the belly and feet, and coarsely brown-spotted from the breast to the thighs. The last subspecies is stated to be variously the smallest (33cm) or rather large (*c.* 49cm), these measurements being hardly within the size range of this species. Recently, this race has been proposed as representing a new species, the Manus Masked Owl *Tyto manusi*, although none has been found on the island since 1934; because of obvious size confusion and lack of any new details, this taxon is treated here as a subspecies which is either seriously overlooked or extinct. Further research is needed on the ecology, behaviour and taxonomy of this extremely variable species, which may prove to have a wider distribution, especially in New Guinea.

Similar species The sympatric Australian Barn Owl

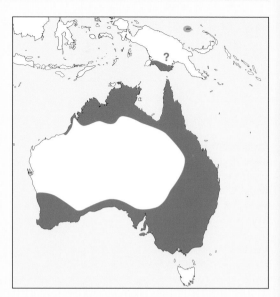

(8) is clearly smaller and has a much shorter tail, and the lower third of the legs is bare and sparsely bristled. The allopatric Tasmanian Masked Owl (20) is similar, but larger and distinctly darker, with generally a darker facial disc, and is more clearly spotted below. Overlapping and allopatric grass owls (15, 16) are more uniformly dark brown above with fine whitish spots. The sympatric sooty owls (22, 23) are wholly sooty-grey with fine white spots.

▼ A very dark-faced Australian Masked Owl; this is a dark morph, which has orange-buff ground colour above and below, with v-shaped spots on breast and belly. Victoria, Australia, December (*Rohan Clarke*).

▼ Race *kimberli* light-morph is somewhat paler than the nominate *novaehollandiae*. Queensland, Australia, October (*Adrian Boyle*).

19. SULAWESI MASKED OWL
Tyto rosenbergii

L 41–46cm; **W** 331–360mm

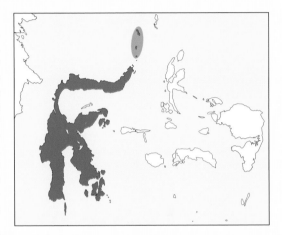

Other names Celebes Masked Owl, Celebes Barn Owl
Identification A medium-sized to fairly large owl without ear-tufts. No information is available on sexual size differences. The female is darker than the male and has heavier spotting on the underparts. Both sexes have a whitish facial disc, becoming darker towards the blackish-brown eyes; the rim is rufous with dark speckles, and the bill is whitish creamy. The greyish-brown upperparts are spotted clearly with black-and-white dots, and the wings and tail have darker bars on a paler brownish-grey background. The underparts are light ochre with some darker spots. The tarsi are pale ochre, feathered nearly to the base of the greyish-brown toes, which are bare and sparsely bristled, with very powerful blackish claws. *Juvenile* Unknown.
Call Utters an eerie, dry, creaky, rasping *chreeeochreoh* or *chreeeeho*, somewhat wavering and with a downward inflection, although lasting only about one second.
Food and hunting Rats and shrews are recorded in pellets from this species.
Habitat It inhabits rainforest and wooded areas from sea level to 1200m, but occurs also in cultivations and grasslands near human dwellings.
Status and distribution This owl is rather widely distributed on Sulawesi, and is obviously not so greatly affected by forest destruction. It needs intensive study, as its numbers and life history are poorly known.

Geographical variation Two races are recognised: *T. r. rosenbergii* from Sulawesi and Sangihe, and *T. r. pelengensis* from Peleng, in the Banggai Islands. The latter is known only from the type specimen, which is much smaller than the nominate race.
Similar species The geographically separated Australian Masked Owl (18) has more numerous bars on the secondaries and a more prominent, reddish-brown rim around the whitish facial disc. The Sulawesi Golden Owl (21) is much smaller with a strikingly small facial disc, even in relation to its much smaller body size.

▼ Sulawesi Masked Owl is similar in size to Australian Masked Owl *Tyto novaehollandiae*, but is generally darker with a more prominent ruff around the greyish-blue facial disc; also more barring on the secondaries. Primary tips have black and white spots Northern Sulawesi, April (*Marcel Holyoak*).

20. TASMANIAN MASKED OWL
Tyto castanops

L 47–55cm, Wt c. 600–1260g; W 310–387mm, WS up to 129cm

Identification A fairly large to large owl without ear-tufts. This is the largest of all *Tyto* owls; the females are darker and larger than the males, sometimes twice the weight and with much longer wings. The female has a pale chestnut to brownish-buff facial disc with an almost black zone around the blackish-brown eyes, and with the bill creamy whitish; above, it is dark greyish-brown with white and black spots, the flight and tail feathers greyish-brown with dark bars; the underparts are fulvous with large dark spots. The male is generally pale and with a brownish-white facial disc, and is whitish below with small dark brown spots. The tarsi are fulvous-brown, feathered to the base of the greyish-brown or yellowish-grey toes, which have powerful blackish-brown claws.

Juvenile The young has whitish downy plumage of the first and second coats; the immature is much darker than the adult.

Call Similar to that of the Australian Masked Owl.

Food and hunting This is a generalist predator, comparable to medium-sized eagle owls of other continents. Its prey ranges in size from 2g invertebrates to mammals of more than 2kg.

Habitat It inhabits dry sclerophyll (hard-leaf) forests and woodlands, and is found also in semi-open bushy areas or 'cleared land'.

Status and distribution This species is apparently not rare in Tasmania and on Maria Island, off the east coast. It has been introduced on Lord Howe Island, where *Tyto furcata pratincola* (2) from California was also

▼ The male is smaller than the female, with the legs being noticeably less powerful. Face and underparts are much paler and less dark-spotted. Here the male droops its wings to look bigger. Southern Tasmania (*Dave Watts*).

introduced (investigation of possible interbreeding desirable).

Geographical variation Formerly regarded as an allopatric subspecies of the Australian Masked Owl, this taxon is here separated on the grounds of the huge size difference and the large difference in sexual dimorphism, which is exhibited far less by the latter species. Monotypic.

Similar species Pale individuals could be confused with smaller and sympatric Australian Barn Owl (8), but have a rounder and more margined facial disc, denser feathering on the legs and massive talons. The geographically separated Australian Masked Owl (18) is very similar, but smaller, distinctly paler and less boldly spotted below.

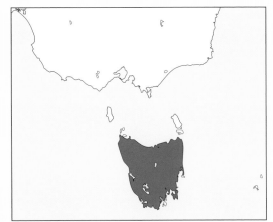

▼ Tasmanian Masked Owl is darker and larger than Australian Masked Owl. It has the most powerful talons of all *Tyto* species. Female has a dark facial disc with brownish-black spots below the dark brown eyes. Southern Tasmania (*Dave Watts*).

▼ Before its first moult, the juvenile is very dark and has a dark rufous-brown facial disc. The final white appearance of the male will take at least three years to develop. Tasmania, May (*David Hollands*).

21. SULAWESI GOLDEN OWL
Tyto inexspectata

L 27–31cm; W 239–272mm

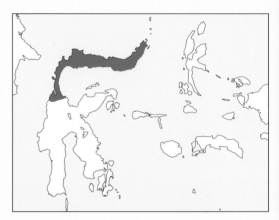

Other name Minahassa Masked Owl

Identification A relatively small barn owl without ear-tufts. There are no data on sexual size or colour differences. It has a small facial disc which is pale creamy, slightly tinged reddish, with brownish shading around the blackish-brown eyes and towards the base of the whitish-creamy bill. The upperparts are greyish-brown with orange-yellow to rusty-red patches, and with relatively large white spots bordered black on the upper edge. The wings and tail are brownish-yellow with several dark bars. The feathered legs are uniformly ochre, with reddish-grey toes and relatively powerful blackish-brown claws. *Juvenile* Unknown. *In flight* Shows rounded wings and numerous tail-bars.

Call The voice has been recorded but details are not yet published. It is said to be similar to that of *Tyto rosenbergii*, but weaker and less deep.

Food and hunting Not known. Small mammals are most likely the main prey.

Habitat It inhabits tropical rainforest, and also degraded forests, from 100m to 1500m.

Status and distribution This owl is known only from eleven collected specimens and a few field sightings on N. Sulawesi, in Indonesia. Its biology and behaviour are totally unknown. It is listed as Vulnerable owing to the continuous habitat destruction in its range.

Geographical variation Monotypic.

Similar species It occurs partly alongside the much larger Sulawesi Masked Owl (19), which also has a much larger facial disc. The inter-relationships of these sympatric species are undocumented.

◄◄ Sulawesi Golden Owl has a cinnamon-brown head and relatively small and pale rufous facial disc, with a dark-speckled ruff. Tangkoko, Sulawesi, October (*Bram Demeulemeester*).

◄ Sulawesi Golden Owl has very strong claws that may facilitate small mammal capture. Tangkoko, Sulawesi, September (*Rob Hutchinson*).

22. LESSER SOOTY OWL
Tyto multipunctata

L 31–38cm, **Wt** 430–540g; **W** 237–266mm, **WS** 86cm

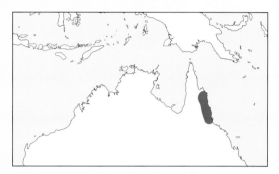

Identification A medium-sized barn owl without ear-tufts. The female is up to 100g heavier than the male. The facial disc is greyish-white, becoming sooty towards and around the black eyes, which are relatively large for a barn owl; the bill is greyish-brown with a much darker tip. The rest of the plumage is sooty-grey to dark silvery grey, a little paler below, and densely spotted and dotted whitish above. The flight feathers and the very short tail are grey with some distinct darker bars. The underparts are silvery white with many fine dark grey chevrons, especially on the breast. The legs are greyish-feathered to the base of the greyish-brown toes, with somewhat darker claws. *Juvenile* The first and second coats are sooty-grey, and the immature is similar to the adult. *In flight* Shows pale underwing-coverts and rounded wings, with the dangling legs extending far behind the short tail.

Call It emits a piercing descending whistle, less powerful than that of the Greater Sooty Owl.

Food and hunting The diet consists of small animals, including insects and some birds. It hunts from low perches, and takes prey on the ground more than does the Greater Sooty Owl.

Habitat This owl lives in rainforest and wet eucalypt forests, from sea level to some 300m.

Status and distribution The small population (*c.* 2,000 pairs) is confined to NE Queensland south of Cape York Peninsula, in NE Australia. It is probably threatened or even endangered by deforestation.

Geographical variation Often regarded only as a race of *T. tenebricosa*, this species has now been separated on the grounds of size and coloration differences. More research is needed, however, to confirm this treatment. Monotypic.

Similar species It exhibits many similarities to the allopatric Greater Sooty Owl (23), but is lighter in colour and clearly smaller. The Greater Sooty Owl can weigh more than twice as much as a Lesser Sooty.

◀ Lesser Sooty Owl is pale sooty grey, and densely spotted and dotted above and below. It has a dark ruff, large blackish eyes and a greyish-brown, very dark-tipped bill. Northern Queensland, Australia, April (*David Hollands*).

23. GREATER SOOTY OWL
Tyto tenebricosa

L 37–43cm, Wt 500–1160g; W 243–343mm, WS 103cm

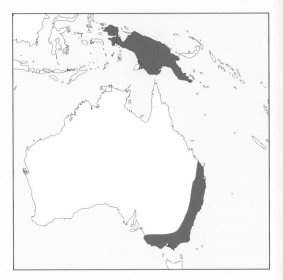

Identification A medium-sized owl without ear-tufts. Females are usually up to 350g heavier than males, but one pair exhibited reversed sexual size dimorphism. Some individuals are reported as being paler and barred, but the sexes are usually similar in plumage. The plumage is in general sooty-black, with very large dark eyes set in a rounded facial disc which is greyish-white to sooty, darkest around the eye, palest towards the margin. The crown, neck, upperparts and breast are brownish-black, spangled with small white spots; the flight feathers and tail are blackish. The lower belly and thighs are dusky with fine irregular darker barring. The iris is blackish-brown and the bill whitish-creamy; the legs are densely feathered to the dark grey toes, which have massive black or dark brown claws. *Juvenile* The downy young is whitish to greyish-white; the immature resembles the adult, but has a darker face. *In flight* Shows rounded and relatively uniform wings.

Call The best-known call is a drawn-out descending whistle, which has been aptly likened to the sound of a falling bomb.

Food and hunting This is a generalist which preys on any available small and medium-sized mammals, including possums, bats and giant rats, and occasionally on small birds and reptiles. It captures prey from the forest canopy to the ground.

Habitat It frequents rainforest, tall moist eucalypt forests and forest edges, from the lowlands up to at least 3660m.

Status and distribution This owl occurs in New Guinea, including Yapen Island, and E. Australia except most of Queensland. It is probably rare and endangered in SE Australia, but is widely distributed in New Guinea.

Geographical variation Two races are known: *T. t. tenebricosa*, from Australia, is generally browner and much smaller (500–750g); *T. t. arfaki* occurs almost throughout New Guinea.

Similar species The geographically separated Lesser Sooty Owl (22) is smaller and paler, but some birds from Papua New Guinea are similar in plumage coloration to the Lesser Sooty. Both species require more detailed study.

◀ Greater Sooty Owl is in general sooty-black – perhaps this is an adaptation to life in a fire-prone forest. Underparts are somewhat paler. Note the distinctive whitish-cream bill. This is the nominate race. Victoria, Australia, July (*Rohan Clarke*).

▲ Greater Sooty Owl nominate *tenebricosa* is very large, and pretends to be even bigger by spreading its tail and drooping its rounded and uniform wings. Queensland, June (*Rohan Clarke*).

▶ Greater Sooty Owl race *arfaki* lives in New Guinea and is much smaller and generally browner than the Australian nominate race. Note the huge legs and distinctly barred tarsi. Tari, PNG, June (*Nik Borrow*).

24. ITOMBWE OWL
Tyto prigoginei

L 23–25cm, **Wt** 195g (1♀); **W** 192mm (1♀), **WS** 63cm

Other name Congo Bay Owl *Phodilus prigoginei*

Identification A small owl without ear-tufts. There are no known sexual size differences. The facial disc is pale, reddish-tinged creamy and very similar to the heart-shaped face of other barn owls. The upperparts are generally deep reddish-brown and the underside russet-creamy, the belly with blackish stripes. The blackish eyes are relatively small, and the yellowish-horn bill is laterally compressed. The very long legs are yellowish-white, the feet feathered to the base of the yellowish-grey toes, and the claws are greyish-brown with darker tips. *Juvenile* Has never been seen.

Call Not well known. A call believed to have been made by this species, tape-recorded in Nyungwe Forest, Rwanda, in 1990, is a long, mournful whistle.

Food and hunting Undocumented, but the long legs would indicate that this owl spends much of its hunting time on the ground.

Habitat This species lives in Itombwe mountain forests and upper slopes with grass and light bush, but is found also in tea estates and slightly degraded forests.

Status and distribution The Itombwe Owl has a very small known range in the Itombwe Mountains, E. DR Congo, and in Nyungwe Forest, SW Rwanda, and thus qualifies as Endangered. It possibly occurs also in Burundi.

Geographical variation First described as 'Congo Bay Owl' and belonging to the genus *Phodilus*, this owl has now been renamed and moved to the genus *Tyto* because of its morphological features; DNA of the Itombwe Owl is required in order to confirm its correct taxonomic status. Monotypic.

Similar species The sympatric sub-Saharan Barn Owl *T. a. affinis* is much larger and has a white facial disc. Allopatric bay owls in Asia have huge eyes and a very peculiarly shaped facial disc.

◄ & ▲ The Itombwe Owl is superficially a little like the Asian bay owls, although it is obviously a *Tyto*, with a heart-shaped face. Also the eyes are much smaller than those of bay owls, the feet are smaller, and the bill is more compressed. Itombwe, May (*Thomas M. Butynski*).

25. ORIENTAL BAY OWL
Phodilus badius

L 22.5–29cm, Wt 220–300g; W 172–237mm

Identification A fairly small owl with short rounded wings and without real ear-tufts, although the ears can look slightly tufted owing to facial-disc movements. There are no data on sexual size or colour differences. The head is very broad, with no narrowing at the neck. It has a vertically elongated, whitish-creamy facial disc, with a very dark brown V-shaped marking running down the centre of the face through the large, blackish-brown eyes; whitish eyelids and the slightly darker forehead form a distinct triangle above the creamy-yellow or pinkish-horn bill. The upperparts are bright reddish-brown, sparsely spotted with black and yellowish-white. The underparts are paler buffish with a rufous tinge, and thinly speckled with black. The feathered feet, long and powerful, have yellowish-brown toes and dark horn-coloured claws. *Juvenile* Has whitish natal downy. *In flight* It has a rapid flight, skillfully avoiding any obstacles such as a maze of lianas or dense stands of young trees.

Call Utters melancholic, loud whistles in series of four to seven, a series lasting 2–8 seconds at times.

Food and hunting Feeds on rats, mice, bats, birds, lizards, frogs and insects. Hunts from a perch, often near water.

▼ Race *badius* from Sumatra has hardly any tail, but very strong talons. It has a greyish shield above the bill and pale tan underparts. Kambas, Sumatra, June (*Ian Merrill*).

Habitat This species inhabits dense lowland forests and montane forests at up to 2300m, and is found also in densely foliaged groves near farmlands and rice fields.

Status and distribution Although widely distributed from Nepal and NE India to Java, this owl is obviously fairly scarce. It occurs also in Bali, where it was rediscovered in 1990, the previous record dating as far back as 1911. In 1927 it was observed on Samar, in the Philippines, but there are no known recent records from there.

Geographical variation Four known subspecies: *P. b. badius* occurs from S. Thailand to Bali; *P. b. saturatus* from Nepal and NE India east to N. Thailand and Vietnam; *P. b. parvus*, known only from eight museum specimens from Belitung Island, Indonesia, is smaller than Javan birds; and *P. b. arixuthus* from North Natunas, Indonesia, is known only from the holotype.

Similar species All Barn Owl races (1) in Asia are much larger, with a more rounded, heart-shaped silky-white facial disc, and have white or slightly dull yellow underparts only sparsely spotted with dark brown. The allopatric Sri Lanka Bay Owl (26) is dark chestnut above, with dense dark and white spots, and heavily barred flight and tail feathers.

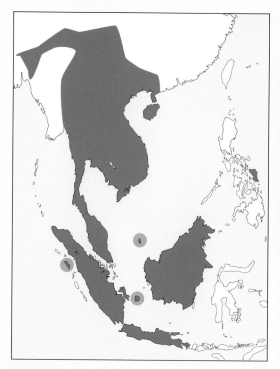

▶ Male and female Oriental Bay Owls on Borneo; nominate *badius* is paler than Sri Lanka Bay Owl, but darker than *parvus* or *saturatus*. Sabah, Malaysia, May (*James Eaton*).

◀ The Burmese race *saturatus*, with its eyes closed and 'false ear-tufts' erect. It has distinctly ochre-coloured scapulars, and a pinkish tinge on the face and underparts. Burma, September (*Martjan Lammertink*).

26. SRI LANKA BAY OWL
Phodilus assimilis

L *c.* 29cm; W 197–208mm

Identification A relatively small owl without real ear-tufts, although some facial expressions give impression that the owl has slightly tufted ears. There are no data on sexual size or colour differences. The slightly heart-shaped and dark facial disc has dark vertical patches through the large, brownish-black eyes, and a V-shaped frontal shield leading down to a yellowish bill. The vermiculated upperparts are heavily spotted black and white, and the dark chestnut-brown flight and tail feathers are distinctly dark-barred and spotted with black and yellow. The underparts are pale buffish with many black-and-white flecks. The relatively short, buffish feathered legs end in pale greyish-brown toes with dirty whitish to pale grey claws. *Juvenile* Undescribed. *In flight* Shows a prominent ochre wing patch in flight.

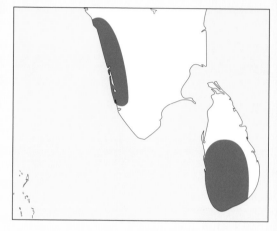

Call Emits a loud but sliding whistle, *whew-ee-yo*, reminiscent of a human being calling his or her dog; the middle syllable is much higher in pitch than the other two, and the whole phrase is usually repeated three or four times before a long pause. The single notes are two to three times longer than those of the Oriental Bay Owl.

Food and hunting Poorly known. Four recently found pellets contained only bones and fur of small rodents.

Habitat Occupies dense evergreen and mixed forests, mainly in the highest hills at up to 2200m.

Geographical variation Two allopatric subspecies are known: *P. a. assimilis* in Sri Lanka and *P. a. ripleyi* in the Western Ghats of India. Compared with *ripleyi* from south India, subspecies *assimilis* has a darker border around the facial mask, a darker colour around the eyes, and stronger, more extensive barring on the primaries, and the markings on the back and breast are elongated, rather than rounded. The voices of the two also differ: *ripleyi* repeats its whistle only twice, with a total duration of *c.* 4 seconds before a long pause. Only after DNA studies will it be known if these allopatric subspecies represent two different bay owls, endemic in Sri Lanka and south India, respectively.

Status and distribution Found only in Kerala, S. India, and in Sri Lanka. Although poorly studied, especially in India, this species is obviously not so rare as had previously been thought. It could, however, become endangered by habitat destruction.

Similar species The allopatric Barn Owl (1) is clearly larger. While similar in terms of size, the Oriental Bay Owl (25) is geographically separated and is much less spotted on its bright rufous back, with tail and wings weakly dark-barred.

▶ The Indian race *ripleyi* resembles the nominate *assimilis* in plumage but has a weaker ruff, and the dark slash through the eyes is lighter. Barring of primaries is less extensive and weaker, and the markings on the back and underparts are rounded, not elongated. This bird has a different voice to those of Sri Lanka, and the two races could be specifically distinct. Kerala, India, January (*András Mazula*).

◀ Sri Lanka Bay Owl with a huge rat. Head, back and bend of wing are darker than in Oriental Bay Owl, and there are darker spots in front of and behind the eyes.

27. COMMON SCOPS OWL
Otus scops

L 16–21cm, **Wt** 60–135g; **W** 145–168mm, **WS** 50–64cm

Other name Eurasian Scops Owl

Identification A very small owl with small ear-tufts, these almost invisible when the plumage is held loose and the head appears rounded and 'earless'. The female is 15–25g heavier than the male. Plumage is greyish-brown with blackish streaks above, the crown with blackish shaft-streaks. The scapulars have white outer webs, a blackish central streak and a black tip, and the flight and tail feathers are dark with pale bars. It has relatively long wings and a short tail. Grizzly-brown underparts have a few thin cross-bars, dark vermicular markings, and blackish shaft-streaks, some much broader than others, with heavier horizontal vermiculations. The greyish-brown facial disc has fine mottling, and is surrounded by a not very prominent ruff. The eyes are yellow, and the bill grey. The tarsi are feathered to the base of the grey toes, which have dark-tipped greyish-brown claws. *Juvenile* Downy chick is whitish. Mesoptile and juvenile resemble adult, but plumage is more fluffy, with very prominent vermicular markings on the crown, upper

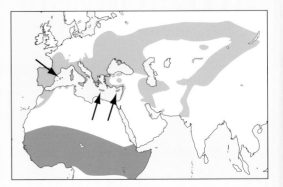

back and breast. *In flight* The normal flight is direct and rapid.

Call Gives a low, short, but disyllabic note, *tyeu*, repeated at intervals of three seconds, sometimes for tens of minutes.

Food and hunting Feeds chiefly on insects, such as beetles and moths, but also on spiders, earthworms, geckos and frogs. Small birds and mammals make up

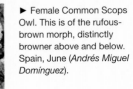

▶ Female Common Scops Owl. This is of the rufous-brown morph, distinctly browner above and below. Spain, June (*Andrés Miguel Domínguez*).

◀ Common Scops Owl is the only *Otus* owl in Europe. This is the nominate race, *scops*, a male of the grey-brown morph, with a cryptic pattern and a less rufous tinge. Spain, June (*Andrés Miguel Domínguez*).

only 1–2 per cent of the diet.

Habitat Prefers fairly open country with scattered trees or small woods, including cultivated areas, gardens, and human settlements. Occurs at up to 2500m above sea level.

Status and distribution Breeds locally in C., E. & W. Europe, Africa north of the Sahara, Asia Minor and eastwards to C. Asia. Winters chiefly in African savannas, but also crosses the sea to Seychelles (e.g. seven records between January and March 2010). Rare in north of range, but fairly common in the Mediterranean. Estimated C. European population is as low as 679 pairs, but this is likely to increase with climate change. One individual was singing in June–July 2011 in Finland, some 500km north of the 'normal' range limit. Also, four territorial individuals were seen/heard in Sweden during summer 2011, indicating a clear northward range expansion (a sign of global warming).

Geographical variation Six subspecies are listed: nominate *scops* from France to N. Africa, N. Turkey and Transcaucasia; smaller *O. s. mallorcae* from the Balearic Islands; *O. s. cycladum* from S. Greece, Crete and the Cyclades to S. Asia Minor; *O. s. cyprius*, the silvery-grey race, from Cyprus and Asia Minor; greyish-brown *O. s. turanicus* from Turkmenistan to W. Pakistan; and *O. s. pulchellus*, greyer than nominate, from the Caucasus east to Yenisei, and wintering in Sind and adjacent parts of India.

Similar species This is the only *Otus* owl in Europe. Sympatric Pallid Scops Owl (28) from the Middle East to Pakistan is paler and has more prominent ventral streaks. In the non-breeding range, sympatric Indian Scops Owl (44) has dark brown eyes.

▼ Race *cycladum,* sleeping in daylight. It is a little paler and more rufous than *cyprius* from Cyprus and SW Asia Minor. Lesbos, Greece, April (*Lesley van Loo*).

Race *cyprius* is the darkest Common Scops Owl subspecies, with blackish streaks on the underparts. This bird holds a captured cicada in its talons. Cyprus, January (*Tasso Leventis*).

▼ Immature Common Scops Owl has prominent vermiculated markings on the crown, upper back and breast. Israel, June (*Eyal Bartov*).

Race *pulchellus* is a similar size to the nominate but it is greyer with more white above. Azerbaijan, September (*Kai Gauger*).

28. PALLID SCOPS OWL
Otus brucei

L 18–22cm, **Wt** 90–130g; **W** 145–170mm, **WS** 54–64cm

Other name Striated Scops Owl

Identification A very small to small owl with small ear-tufts which are almost invisible when head plumage is held loose. The females are at least 15g heavier than the males. There are two colour morphs: light with predominance of pale yellowish colour, and brownish-grey, with small dense dark speckles above. Has fine blackish shaft-streaks on the crown. The upperparts are often yellowish-ochre, with thin but visible blackish shaft-streaks. Has an indistinct row of whitish feathers on the scapulars, edged blackish at rear end, and the flight and tail feathers are barred lighter and darker. The underparts are pale greyish or pale ochre-yellow with dark shaft-streaks, fairly faint in the pale morph. The light facial disc has a thin dark ruff. The eyes are pale yellow, and the bill dark greyish-horn. Feathering of the tarsi reaches the basal part of the greyish-brown toes, with blackish-brown claws. *Juvenile* Downy chick is white. At the age of two weeks mesoptile has adult-like feathers, and ear-tufts are well visible; at three weeks the facial disc is forming, and the coloration becomes

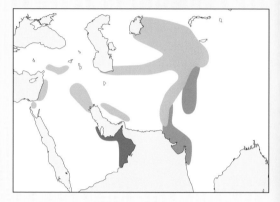

transversally striped, with remnants of down only on back, breast and belly.

Call Utters a fairly quiet, soft, dove-like *kukh-kukh-kukh-kukh*, repeated in long sequences, with up to 67 notes per minute and intervals of half to one second between individual notes. Voice is fairly similar to that of a Stock Dove *Columba oenas*.

Food and hunting Hunts chiefly nocturnal insects, but occasionally also small birds and mammals. Takes reptiles fairly often. Catches its prey in flight or from branches of trees.

Habitat Prefers deciduous forests and riverine habitats, but occurs also in arid, rocky gullies with scrub and comes readily to human settlements and gardens. From 80m to 1800m above sea level.

Status and distribution Distributed from SE Turkey and NE Egypt to SE Arabia and from Aral Sea to NW India. Rather common locally, but suffers from extensive use of pesticides in many countries.

Geographical variation This species is often said to be monotypic, but four subspecies are listed: nominate *brucei* from Kazakhstan, N. Tajikistan and Kyrgyzstan; more sandy-buff *O. b. obsoletus* from NE Egypt and SE Turkey to N Afghanistan and the lowlands of Uzbekistan; deeper ochre-yellow *O. b. semenowi* from

► Pallid Scops Owl from Dubai. This could be *exiguus* based on location but it looks more like the nominate *brucei*, which is migratory. It is paler with more prominent ventral streaks than Common Scops Owl, which has totally bare toes and a bark-like pattern on the upperparts. Dubai, October (*Arto Juvonen*).

◄ This Pallid Scops Owl also looks more like a migrant nominate race. Dubai, December (*Rob Hutchinson*).

N. Pakistan to S. Tajikistan and W. China; and pale sandy *O. b. exiguus* from E. Arabia and C. Iraq to W. Pakistan. The validity of, especially, *obsoletus* has been questioned.

Similar species Partly sympatric Common Scops Owl (27) is generally darker, with totally bare toes, and a bark-like pattern on the wing-coverts and back. Larger Indian Scops Owl (44) overlaps only during the winter migration of northern Pallid Scops Owls, but has hazel-brown to dark brown eyes. Allopatric Oriental Scops Owl (45) is similar in size but less streaked on back and wing-coverts. Allopatric Collared Scops Owl (46) is clearly larger and darker, and has dark brown or orange-brown eyes.

▲ Race *obsoletus* is very similar to nominate *brucei*. This race is more sandy-buff than *exiguus*, and very similar to *semenowi* but with sharper, narrower streaks. Turkey, June (*David Monticelli*).

▲ ◄ Birds in Israel appear to have very pale grey *exiguus*-type coloration while being *obsoletus*-like in size, so racial identity in this locality is questionable. The eyes of Pallid Scops are pale yellow and the small ear-tufts are visible only when fully erected. Eilat, Israel, February (*Marcel Holyoak*).

29. ARABIAN SCOPS OWL
Otus pamelae

L *c.* 18cm, Wt 62–71g; W 134–148mm

Identification A very small owl with small but erectile ear-tufts. No data on sexual size differences. Has pale brownish-grey upperparts indistinctly mottled and streaked dark, with a buffish or rufous tinge on upperwing-coverts, paler bars on light greyish-brown flight and tail feathers, and an indistinct row of paler areas on scapulars, but no prominent scapular stripe. Underparts are sandy-buff to light brownish-grey with a few darker markings. The very pale or dirty white facial disc has a very thin, dark ruff. The eyes are golden-yellow and the bill dark-tipped greyish. Sandy tarsi are feathered to the base of the greyish-brown toes, with dark-tipped horn-coloured claws. *Juvenile* Not described.

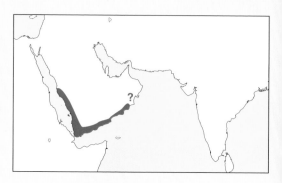

Call Utters a trill with long *kreeeerrh* notes, scratchy and fairly high in pitch, repeated at intervals.

Food and hunting Not well documented, but most likely insects and other invertebrates, and occasionally small vertebrates.

Habitat Found in semi-deserts, but with palms and groups of trees, including rocky areas with little vegetation. Also near human settlements.

Status and distribution Endemic from Saudi Arabia to Oman. Status unknown, as formerly treated as a race of Pallid Scops Owl or African Scops Owl.

Geographical variation Taxonomy uncertain; hitherto regarded as a subspecies of Pallid Scops Owl or of African Scops Owl. DNA studies should confirm the new taxonomic status and interspecific relationships within the genus. Monotypic.

Similar species Partly sympatric Pallid Scops Owl (28) is larger, with a darker facial disc. Geographically separated African Scops Owl (30) is much darker, with very distinct whitish stripe on scapulars, and more prominent streaks on breast. Allopatric Socotra Scops Owl (31) could be a close relative, but is much smaller.

▲▶ Based on voice this greyish owl is thought to be an Arabian Scops Owl although it was in Daallo, Somaliland, south of the normal range. It is possible that it may be an undescribed taxon. Somaliland, September (*Werner Müller*).

▶ Arabian Scops Owl pair mutual preening to strengthen the pair bond. Oman, March (*Hanne and Jens Eriksen*).

◀ Male Arabian Scops Owl calling. It is small and is pale below, with a few indistinct streaks, white blotches and a pale buffish tinge. Oman, March (*Hanne and Jens Eriksen*).

30. AFRICAN SCOPS OWL
Otus senegalensis

L 16–19cm, Wt 45–100g; W 117–144mm, WS 40–45cm

Identification A very small owl with small, but well-developed ear-tufts, blade-shaped when erected. Females are on average some 24g heavier than males. Grey and brown morphs exist, as well as wide individual variation. The crown and forehead have relatively broad shaft-streaks. Upperparts are grey or brownish with fine vermiculations and darker streaks, and with rufous-edged feathers on the mantle and upperwing-coverts. Whitish areas on scapulars form a white band across the shoulder. The flight and tail feathers are barred light and dark, and there are large white spots on the outer webs of the primaries. Paler underparts have dark streaks and fine vermicular markings. The facial disc also has fine vermiculations and a dark ruff. The eyes are yellow and the bill blackish-horn. Tarsi are feathered to the base of the dusky greyish-brown toes, which have blackish-brown claws. *Juvenile* Downy chick is whitish. Mesoptile resembles the adult, but is fluffy and less distinctly marked.

Call A short, frog-like purring trill, *krurrr*, each note lasting half to one second, is given in long sequences at intervals of 5–8 seconds for protracted periods.

Food and hunting Feeds chiefly on insects, but also spiders, scorpions and occasionally small vertebrates. Hunts usually from a perch, also hawking insects in flight.

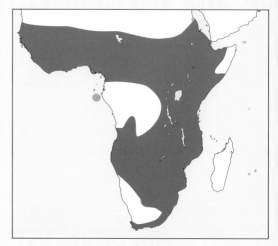

Habitat Semi-open woodlands and savanna with scattered trees and thorny bushes, but often found also in park-like areas and gardens near human habitations. Normally below 2000m above sea level.

Status and distribution Lives south of the Sahara from S. Mauritania across to Ethiopia and Eritrea, and south to the Cape in S. Africa, but partly absent from densely forested areas of the continent. Rather common throughout its range.

Geographical variation Three subspecies are separated: nominate *senegalensis* from entire sub-Saharan Africa; *O. s. nivosus*, a very pale race from SE Kenya; and *O. s. feae* from Annobón (Pagalu) Island, in Equatorial Guinea (but lying off the coast of Gabon), which is much darker than the nominate race. This last island race could warrant species status but has not been studied. Race *nivosus* is known only from three specimens taken along the lower Tana River and in the Lali Hills of SE Kenya.

Similar species During the winter Common Scops Owl (27) overlaps with this species in Africa north of the tropical rainforest, but has relatively long wings (different wing-formula); voice is the best means of separation. Geographically separated São Tomé Scops Owl (35) has nearly invisible ear-tufts and unfeathered lower half of tarsi, and is generally uniform in coloration.

◄ African Scops Owl of the race *senegalensis*, which occurs from Mauritania to South Africa. This individual has a greyer appearance with little rufous tinge. Namibia, August (*Paul Noakes*).

◀ African Scops Owl has brown and grey morphs, but in South Africa grey is the most common. Kruger, South Africa, November (*Rick van der Weijde*).

▼ Brown morph of African Scops Owl showing much grey on the underparts. Kruger, South Africa (*Tom and Pam Gardner*).

31. SOCOTRA SCOPS OWL
Otus socotranus

L 15–16cm, Wt 64–85g; W 124–135mm

Identification A tiny owl with very short, blunt ear-tufts, often hardly visible. Females are little heavier than males. It has pale sandy-grey upperparts, with a slight yellow-ochre tinge, and a fairly conspicuously streaked head, hindneck, mantle and back with dark mottling and vermicular markings. A row of light ochre-grey dots on the scapulars forms an indistinct scapular stripe on the closed wing. The flight and tail feathers are paler grey and barred darker. Underparts are light greyish-sandy with darker streaks and vermiculations and a slight ochre tinge, with dark shaft-streaks more prominent on the upper breast than on the lower underparts, and a nearly plain pale grey belly. The light ochre-grey or greyish-buff facial disc is bordered by an indistinct darker ruff. The eyes are yellow and the bill blackish-horn. Tarsi are feathered to the base of the pale greyish-brown toes, which have greyish-horn claws. *Juvenile* Unknown.

Call Gives a sequence of three or four notes, resembling the voice of Oriental Scops Owl (45).

Food and hunting Very little known, but seen to catch moths at twilight. One stomach contained remains of grasshoppers, one centipede and two lizards.

Habitat Semi-desert and rocky landscapes with scattered bushes and trees.

Status and distribution Endemic to Socotra Island, off easternmost Somalia. Status is not well known.

Geographical variation Taxonomy uncertain; previously regarded as a subspecies of Pallid or African Scops Owl, and vocal similarity to Oriental Scops Owl is interesting. Monotypic.

Similar species No other *Otus* owls occur on Socotra.

▼ Pallid, Arabian and African Scops Owls are all larger than the cream-grey Socotra Scops Owl, which has numerous speckles above and is very light sandy grey below. Socotra, October (*Ulf Ståhle*).

▼ Socotra Scops Owl is the only resident owl on the island. Socotra, February (*Richard Porter*).

32. CINNAMON SCOPS OWL
Otus icterorhynchus

L 18–20cm, **Wt** 61–80g; **W** 117–144mm

Other name Sandy Scops Owl

Identification A very small owl with rufous, white and blackish mottled and speckled ear-tufts. No sexual size differences according to weight data. Exhibits a wide range of individual colour variation. Two morphs, lighter and darker, exist. The pale yellowish-brown upperparts have buffish and white spots, the latter with blackish tips or edges. Dark-edged whitish webs of the scapulars form a band on the folded wing. Outer webs of primaries are spotted white, cinnamon-coloured inner webs and secondaries are barred dark brown. The yellowish-brown tail has dark-edged rufous bars. Pale underparts have whitish spots, larger and more prominent on the belly; underwing-coverts are pale buff. The yellowish-brown facial disc has prominent white eyebrows and some darker concentric lines. The eyes are pale yellow the bill and cere creamy yellow. Pale buffish tarsi are feathered to the base of the pinkish-cream toes, which have grey-tipped dull whitish claws. *Juvenile* Downy chick unknown. Mesoptile resembles the adult, but is barred on the back and plain below. *In flight* Shows sandy-coloured back, wings and tail contrasting with darker primary coverts and primaries.

Call Poorly known, but said to be a drawn-out whistle, *kweeah* and *kewhurew* or *kewhurr*, uttered at intervals of several seconds, and each phrase lasting some 2.5 seconds.

Food and hunting Not documented, but most likely feeds on insects.

Habitat Prefers humid lowland forest at up to 1000m, but occurs also in forest–scrub–grassland mosaic.

Status and distribution Patchily distributed from Liberia to Gabon and E. DR Congo. It is very poorly studied but is known to occur in all forested areas of Liberia.

Geographical variation Two subspecies listed: nominate *icterorhynchus* from Liberia, Ivory Coast and Ghana, and *O. i. holerythrus* from Cameroon, Gabon and E. & C. DR Congo, which is more rufous and has less breast streaking.

Similar species Common Scops Owl (27) is similar in size, but greyer, with a greyish-brown bill, and is only a winter visitor in Africa. The voice is clearly different.

▼ Cinnamon Scops Owl is poorly known and very rarely photographed – these are the only known photos. It has prominent ear-tufts and white 'eyebrows', and is sandy below, with white longitudinal shaft-marks, but more spotted on belly. Mount Tokadeh, Liberia, March (*K.-D. B. Dijkstra*).

33. SOKOKE SCOPS OWL
Otus ireneae

L 16–18cm, Wt 45–55g; W 112–124mm

Identification A very small owl with small ear-tufts mottled and spotted light and dark. No evidence of sexual size difference. Occurs in three different morphs, but intermediates are also known. Grey and dark brown morphs have a blackish-streaked crown, and greyish or dark brown upperparts with nape pale and dark-spotted, and mantle mottled and spotted light and dark. Whitish spots on outer webs of scapulars form a white stripe on the folded wing. Primaries are white and dark brown, with paler inner primaries and secondaries. The tail has dark bars on slightly paler ground. The light greyish-brown or rufous-buff facial disc has faint, slightly darker concentric lines. Greyish-brown underparts have black-tipped white spots and dark vermicular markings. The eyes are pale yellow, the bill light greenish-yellow and the cere pinkish-grey. The feet are dull greyish-yellow, the toes pale greyish-brown with blackish-tipped dark brown claws. Rufous morph is largely reddish-brown, pale or bright, and quite uniform except for short black crown streaks and white scapular spots. *Juvenile* Said to resemble the adult.

Call Gives a series of high whistling notes, *goohk-goohk*, at rate of some 3 notes every two seconds and 5–9 notes in a series, which is repeated at intervals of several seconds.

Food and hunting Wholly insectivorous, subsisting primarily on beetles.

Habitat Recorded almost exclusively in *Cynometra* woodland, and only rarely in *Brachystegia* woodland, in Kenya between 50m and 170m and in Usambara (Tanzania) between 200m and 400m above sea level.

Status and distribution Endemic to the Sokoke-Arabuko Forest in Kenya and in the East Usambara Mountains in NE Tanzania. Status upgraded to globally endangered, since there are only 800 pairs in Kenya.

Geographical variation Monotypic.

Similar species A winter visitor in Africa, Common Scops Owl (27) is larger and heavier, with a darker bill and blackish shaft-streaks below.

◄ Globally endangered, Sokoke Scops Owl is very small, only a few grams heavier than the smallest owl of all, the Elf Owl. The dark brown morph has a dark brown ground colour, with spots more buffish than white . Sabaki, Kenya, May (*Roy de Haas*).

▼ Four Sokoke Scops Owls bunch together in the presence of an intruder. Sokoke, Kenya, January (*Tasso Leventis*).

34. PEMBA SCOPS OWL
Otus pembaensis

L 17–18cm; W 146–152.5mm

Identification A very small owl with short ear-tufts. No information on sexual size or colour differences. Two morphs exist. Light morph has fairly plain reddish-brown upperparts, with several darker streaks on the crown and sometimes on nape, and fine dusky bars and vermiculations on the forehead, eyebrows and mantle. Dark-tipped whitish-buff outer webs of scapulars form a pale dark-barred line across the shoulder. The flight feathers are barred lighter and darker, and tail feathers have dark-barred outer webs. Underparts are reddish-brown with pale grey tinge, finely vermiculated light rufous, grey and white, with a few narrow streaks on breast and flanks. The pale buff facial disc has a darker but not very prominent ruff. The eyes are yellow, finely black-rimmed, and the bill dark greenish-black. Tarsi are profusely feathered to the base of the grey toes, which have large blackish-tipped dark greyish-brown claws. Rufous morph is more uniformly reddish-brown, with vague broad barring on flanks. *Juvenile* Downy chick not described. Mesoptile has whitish down vaguely barred yellowish-brown, with unstreaked white-barred crown, and somewhat banded tail.

Call Utters a long series of hollow, monosyllabic *hu* notes at intervals of half to one second, the sequence repeated at irregular intervals.

Food and hunting Not well documented, but feeds mainly on insects, hawked in flight or from a perch.

Habitat Semi-open areas with densely foliaged trees, but also near clove plantations.

Status and distribution Endemic to Pemba Island, off N. Tanzania. Not rare, but vulnerable owing to its restricted range.

Geographical variation Formerly considered a subspecies of Madagascar Scops Owl (41), but now separated; biology and ecology require study. Monotypic.

Similar species The only known *Otus* owl on Pemba Island. Differs from allopatric Madagascar Scops Owl (41) in smaller size and much more uniform plumage. Widely geographically separated Flores Scops Owl (64) from Indonesia is similarly plain reddish-brown above, but has fully barred tail.

◀ Pemba Scops Owl is smaller than Madagascar Scops Owl *Otus rutilus,* and has fairly uniform plumage. It is plain tawny-rufous above with a fine pattern on the head and yellow eyes. Underparts are creamy-rufous with a few bars. Pemba Island, May (*Adam Riley*).

35. SÃO TOMÉ SCOPS OWL
Otus hartlaubi

L 17–19cm, **Wt** c. 79g; **W** 123–139mm

Identification A very small owl with very small ear-tufts. Females are a little larger and heavier than males. Two morphs exist. Grey-brown morph has almost solid dark brown upperparts, with indistinct rufous and dark markings, blackish shaft-streaks on crown, nape and back, some dark-edged white spots on nape. Very large whitish areas on scapulars form a clear scapular stripe. Essentially unbarred brownish primaries are mottled whitish and buffish-brown, and brown secondaries have fine vermiculations. Nearly unbanded tail has narrow, indistinct buffish bars and pale reddish-brown vermicular markings. Whitish underparts have fine brown and rufous barring, and blackish shaft-streaks along which the bars converge. Light brown facial disc becomes darker near the deep yellow eyes. Reddish-brown ruff, and yellow bill and cere. Distal quarter of tarsus is unfeathered. Yellowish-ochre tarsi and toes, with brown claws. Rufous morph has dark reddish-brown ground colour with prominent black markings. *Juvenile* Downy chick not described. Mesoptile of both morphs is paler than adult and fully finely barred.

Call Gives a high, whistling trilled *hu-hu-hu* or a low, harsh *kowe* repeated at intervals of 12–20 seconds, the sequence somewhat downward-inflected. Heard by day, as well as at dusk and night.

Food and hunting Feeds chiefly on insects, but also

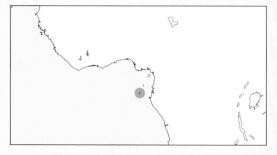

takes small lizards. Hunts mainly in dense foliage, but also on the ground.

Habitat Prefers humid primary cloud forest, but lives also in lowlands in secondary forests and near clove and mango plantations, from sea level to 1300m.

Status and distribution Endemic on São Tomé Island, in the Gulf of Guinea. May be rare and threatened owing to restricted range. Very little studied.

Geographical variation Monotypic.

Similar species The only owl on São Tomé apart from the much larger and very different São Tomé Barn Owl (7). Allopatric African Scops Owl (30) occurs on Annobón (Pagalu) Island, less than 200km to the SW, but is heavily streaked below, with prominently banded wings and tail, and fully feathered tarsi. Voices are totally different.

▲ São Tomé Scops Owl is generally more uniform in coloration than African Scops Owl *Otus senegalensis,* which is not known to cross the sea. São Tomé, July (*Tasso Leventis*).

► The pale underparts have fine brown barring and a rufous tinge. São Tomé, July (*Tasso Leventis*).

◄ São Tomé Scops Owl has deep yellow eyes and large whitish spots on the scapulars. São Tomé, January (*Atle Ivar Olsen*).

36. SEYCHELLES SCOPS OWL
Otus insularis

L 20cm; **W** 162–173mm

Identification A very small owl, with small ear-tufts often practically invisible. No data on sexual size differences. Has dull yellowish-brown crown spotted with white and streaked with black. Buffish-brown upperparts are mottled light and dark, and dark-edged whitish outer webs of scapulars form a white band across shoulder. Upperwing-coverts are yellowish-brown with some vermicular markings and dark shaft-streaks, and flight feathers and tail are barred light and dark. Pale underparts have relatively

▼ Seychelles Scops Owl is practically 'earless' and has greenish-yellow eyes. Underparts are paler than the upperparts, indistinctly whitish barred and with broad blackish streaks. Greyish-yellow bill has darker tip. Relatively long-legged; bare tarsi and toes, with dark horn claws. Mahé, Seychelles, December (*Trevor Hardaker*).

broad blackish streaks and faint whitish and darker barring. Light yellowish-brown facial disc has darker mottling, and prominent ruff. The eyes are yellow and the bill dark-tipped greyish-yellow. Totally bare tarsi and relatively long toes are pale greyish-yellow to greenish-white or pinkish-grey, with dark horn-coloured claws. *Juvenile* Said to resemble adult, but not well known. *In flight* Shows P7 as the longest primary on a rounded wing.

Call Utters a deep, monotonous and rhythmic frog-like grunting at the rate of about one note per second. Sawing or knocking sounds are commonly reported (but these not heard by the author in 1983). Calls very rarely during daylight hours. Males often duet and ruffle all feathers when singing, making them look bigger than females.

Food and hunting Not well documented. Often seen eating geckos and tree-frogs, as well as insects such as beetles and grasshoppers.

Habitat Hilly forests between 250m and 600m above sea level, and occurs also in well-wooded upland gardens.

Status and distribution Endemic to the Seychelles Archipelago. Originally present on most of the islands, but now only on Mahé, and perhaps a few pairs on Praslin. Status listed as Critical by BirdLife International. Biology and ecology practically unknown.

Geographical variation Monotypic.

Similar species Wintering Common Scops Owl (27) has been seen in Seychelles, but it has feathered tarsi to the base of the toes, and short but erectile ear-tufts. Geographically separated Moluccan Scops Owl (68) was earlier linked with this owl because of similar voice; it has tarsi feathered nearly to base of toes, and well-developed ear-tufts.

37. MAYOTTE SCOPS OWL
Otus mayottensis

L 24cm, Wt c. 120g; W 166–178mm

Identification A small owl with short ear-tufts, mottled light and dark greyish-brown. No data on sexual size differences. Has greyish-brown upperparts with dark streaks and vermicular markings, and a fairly prominent nuchal collar spotted lighter and darker. Scapulars have only a few whitish areas, not forming distinct scapular stripe. Darker greyish-brown flight feathers have paler bars, and brownish-grey tail has rather indistinct paler bars. White throat is prominently dark-streaked and dark-barred. Underparts are brownish with blackish shaft-streaks and dark and white vermiculations. The greyish-brown facial disc has a not very distinct dark ruff. The eyes are yellow and the bill greyish-horn. Tarsi are heavily feathered to the base of the pale greyish-brown toes, with dark greyish-brown claws. *Juvenile* Unknown.

Call A clear, hollow *woohp-woohp* is given at long intervals, without any purring sounds. Notes are a bit longer and deeper, and sequence slower, than those of Madagascar Scops Owl.

Food and hunting Not documented.

Habitat Evergreen forest.

Status and distribution Endemic to Mayotte Island, in the Comoros Archipelago. Status not studied, but may be endangered by forest destruction.

Geographical variation Long regarded as only a subspecies of Madagascar Scops Owl; DNA evidence required to confirm its taxonomic status. Monotypic.

Similar species Geographically separated Madagascar (41) and Torotoroka Scops Owls (42) are smaller, have shorter ear-tufts, and lack distinct white throat and nuchal collar. Vocalisations also differ.

▼ Mayotte Scops Owl is very similar to the brown morph of Torotoka Scops Owl *Otus madagascariensis* but it is larger and somewhat darker, with a whitish area from throat to belly. Very dark face, and greenish-yellow eyes. Mayotte, December (*Dubi Shapiro*).

38. GRANDE COMORE SCOPS OWL
Otus pauliani

L 18–20cm, Wt 70g (1♂); W 138–144mm

Other name Karthala Scops Owl

Identification A very small owl, with invisible ear-tufts. No data on sexual size differences. Two colour morphs exist. Light morph has remarkably uniform dark greyish-brown upperparts with some paler spots and fine barring or darker vermiculation. Dull yellow scapulars have indistinct dark bars. Underparts are ochre-buff with a few dark shaft-streaks and a dense pattern of darker vermicular markings. Underwing-coverts and undertail-coverts are whitish. The grey-brown facial disc has some darker concentric lines around the yellow eyes; brown-eyed individuals have also been seen recently, and eye colour is therefore variable, maybe age-dependent. The bill and cere are greyish-brown. Has rather weak talons, and distal part of tarsus unfeathered; tarsi and toes are yellowish-grey to pale greyish-brown, with blackish-brown claws. Dark morph is rather uniformly dark brown or earth-brown with fine buffish speckles. *Juvenile* Not well documented.

Call Gives a long series of short *gluk-gluk-gluk-gluk* notes, about two per second, the phrase beginning with a few drawn-out *choo* notes. Whole series is repeated often

for more than ten minutes.

Food and hunting Weak talons would suggest that it preys mainly on insects and other arthropods.

Habitat Primary forest and also degraded mountain forest, between 460m and 1900m.

Status and distribution Endemic to Mt Karthala, on Grande Comore Island. Population estimated as some 1000 pairs. Status listed as Critical by BirdLife International.

Geographical variation Monotypic.

Similar species No other *Otus* owl in the Comoros Archipelago has such a finely barred and vermiculated appearance.

▼ An almost 'earless' scops owl with remarkably uniform upperparts and a finely marked crown. Scapulars have some minor buff markings; primaries have four broken white spot-lines, visible only when the feathers are fully spread in the hand. Grande Comore (*René-Marie Lafontaine*).

▼ Grande Comore Scops Owl is very similar to Tororotoka Scops Owl *Otus madagascariensis* and light morph of Anjouan Scops Owl *O. capnodes*, but is distinctly smaller with yellow eyes. Grande Comore (*René-Marie Lafontaine*).

39. ANJOUAN SCOPS OWL
Otus capnodes

L 22cm, **Wt** 119g (1♂); **W** 153–167mm

Identification A small owl with hardly visible ear-tufts. No data on sexual size differences. Three colour morphs exist. Light morph has greyish-brown upper-parts, the feathers with dark centres and greyish-buff edges, giving a mottled appearance. Dull yellowish and barred scapulars do not form a distinct stripe. Several lighter and darker bars are visible on the flight and tail feathers. Underparts are greyish-brown with fine barring and dark shaft-streaks, forming a herringbone-like pattern. The whitish-creamy facial disc has narrow dark concentric lines around the greenish-yellow eyes, and is bordered by a dark ruff. The bill and cere are horn-coloured. Tarsi are heavily feathered up to the distal third; the bare parts of tarsi and the toes are greenish, and the claws are blackish-brown. Rufous morph is similar to light morph, but less whitish-speckled and in general more reddish-brown. Dark morph is overall dark chocolate-brown to earth-brown, fairly uniform, with fine buffish speckles. *Juvenile* Not described.

Call Gives a series of prolonged whistles, *peeooee* and *peeoo*, repeated three to five times at varying intervals of between half a second and one second. Each series is separated by about ten seconds. Calls sometimes during the day. Voice resembles that of a Grey Plover.

Food and hunting Not documented, but food most likely insects.

Habitat Found in primary evergreen forests on mountain slopes at above 800m.

Status and distribution Endemic to Anjouan Island, in the Comoros Archipelago. Population estimated at between 100 and 200 pairs. Status is listed as Critical by BirdLife International owing to deforestation.

Geographical variation Monotypic.

Similar species This is the only *Otus* owl on Anjouan. Geographically separated Grande Comore Scops Owl (38) is smaller and lighter, with much less powerful talons. Allopatric Mohéli Scops Owl (40) is similar in size, but generally more reddish-brown.

◄ Dark morph Anjouan Scops Owl Anjouan, July (*Charles Marsh*).

▼ Here an extremely dark morph and a dark rufous-brown morph have paired – this helps explain the intermediates. Anjouan, July (*Charles Marsh*).

40. MOHÉLI SCOPS OWL
Otus moheliensis

L 22cm, Wt 95–119g; W 155–164mm

Identification A small owl with short and rather inconspicuous ear-tufts. Female can be up to 20g heavier than the male. Two morphs exist. A rufous morph has reddish-brown upperparts with irregular blackish barring and mottling, the densely blackish-mottled nape forming a dusky collar. Pale cinnamon outer webs of scapulars have one or two fine dark bars. Outer webs of the flight feathers are regularly barred, with buffish dots near the base and reddish-brown towards the tips; the inner webs are uniformly brown. Brown tail is mottled with rufous or indistinctly barred. The throat is whitish with a dull yellowish suffusion, and the underparts are buffish-cinnamon with blackish shaft-streaks, the finely barred belly brown or dull yellowish-brown. The silky, light brown facial disc has a thinly dusky-edged ruff. Thighs and feathered parts of tarsi have scaly brown spots; unfeathered lower parts of tarsus and toes are grey, with dark horn-coloured claws. A redder morph also exists. *Juvenile* Not described. *In flight* Shows rounded wings, on which sixth and seventh primaries are the longest.
Call Not well studied, but said to emit an aspirated

hissing whistle, repeated in a series of up to five notes.
Food and hunting Little studied, but probably feed chiefly on insects and spiders.
Habitat Humid forests with dense epiphytes, between 450m and 700m.
Status and distribution Endemic to Mohéli Island, in the Comoros Archipelago. Population estimated at 200 pairs, and classified as endangered. The biology of this rare owl requires further research.
Geographical variation Monotypic.
Similar species The only scops owl on Mohéli Island. Allopatric Grand Comoro Scops Owl (38) is paler and smaller, with a weaker bill, and feathered tarsi.

▼ The red morph of Mohéli Scops Owl is very rare, and has been photographed just a handful of times. It has a white area on the scapulars and yellowish lines of spots on the primaries. The throat is whitish. Mohéli (*Nick Gardner*).

▼ Mohéli Scops Owl, brown morph. Grande Comore Scops Owl *Otus pauliani* is smaller and paler, and has a weaker bill. Mohéli (*René-Marie Lafontaine*).

41. MADAGASCAR SCOPS OWL
Otus rutilus

L 19–22cm, Wt 85–120g; W 151–161mm, WS 53cm

Other name Rainforest Scops Owl

Identification A small owl with small ear-tufts. Females on average 18g heavier than males. Three morphs exist. Brown morph has pale brown upperparts and crown, ochre-mottled and with whitish markings, sometimes also showing a herringbone pattern caused by black shaft-streaks. Blackish-edged white areas on scapulars form a stripe across the closed wing. The greyish-brown primaries are barred light and dark, the secondaries less distinctly barred. Tail has lighter and darker bars. Underparts are ochre to pale greyish-brown with rusty-brown vermicular markings and blackish shaft-streaks, producing similar herringbone pattern to that above. The facial disc is whitish-brown with a dark brown ruff. The eyes are yellow and the bill and cere greyish-brown. The pale greyish-brown tarsi are feathered to the base of the greyish-brown toes, which have blackish-brown claws. Grey morph is similar, but has ground colour more greyish. Rufous morph is more rusty-brown, often without dark markings. *Juvenile* Unknown.

Call Gives a series of about five to nine short, hollow

but clear *oot* notes, each note lasting 0.16 seconds, and average interval between notes 0.21 seconds. The interval between these series is of several seconds.

Food and hunting Feeds chiefly on insects, and may also take small vertebrates. Often catches moths on the wing but also hunts on the ground.

Habitat Found in rainforest and other humid brushy

◀ Madagascar Scops Owl has a very dark brown morph; the dark markings on the upperside are more obscure than those of Torotoroka Scops Owl *O. madagascariensis* (Roy de Haas).

areas from lowlands up to 1800m, sometimes even higher.

Status and distribution Confined to N. & E. Madagascar. This species is rather common but is not well studied in terms of its population dynamics. There have been recent records of an unidentified scops owl on Réunion; at first these were thought to be vagrant Madagascar Scops Owls, but vocal differences suggest that they may be an as yet undescribed species ('Réunion Scops Owl').

Geographical variation Further studies are needed in order to determine this species' relationships with the Torotoroka and Mayotte Scops Owls, earlier treated as subspecies of *O. rutilus*. Monotypic.

Similar species Geographically well-separated Mayotte Scops Owl (37) is darker and larger, with a white throat and more conspicuous nuchal collar. Only narrowly allopatric Torotoroka Scops Owl (42) is similar to the grey morph of the Madagascar Scops Owl, but has a longer tail and a very different voice.

▶ Madagascar Scops Owl has three morphs. The rufous morph has less clear dark markings below. (*Keith Valentine*).

▼ Madagascar Scops Owl lives in higher-elevation rainforest; it has a shorter tail than Torotoroka Scops Owl. This is the brown morph. September (*Ian Merrill*).

▼▶ Madagascar Scops Owl, dark morph. Ranomafana, September (*Paul Noakes*).

42. TOROTOROKA SCOPS OWL
Otus madagascariensis

L 20–22cm, **Wt** 85–115g; **W** 152–161mm

Identification A small owl with relatively short ear-tufts. Females more than 15g heavier than males. Three morphs exist. The commonest, the grey morph, has pale brownish-grey upperparts with rather prominent, long, dark streaks, and large white areas on the scapulars forming a clear stripe across the closed wing. The brownish-grey flight feathers have finely dark-bordered whitish bars, and the relatively long tail has similar barring. Underparts are light grey with long, thin, dark shaft-streaks and with horizontal vermicular markings and whitish barring. The facial disc is uniformly grey, paler towards the prominent dark ruff. The yellow eyes, and whitish eyebrows, are not so prominent; has pinkish rim of eyelids. The bill is rather blackish. Tarsi are heavily feathered, and the toes are light greyish with dark greyish-brown claws. Brown morph is similar to grey morph, but in general more brownish. Very rare rufous morph is pale orange-tinged reddish-brown overall, with dark streaks above, and with white bars and obscure streaks and vermiculations below. *Juvenile* Downy chick undescribed. Immature has more strongly barred uppertail-coverts; tips of rectrices are narrower and more pointed than those of adult. *In flight* Tail is visibly longer than that of Madagascar Scops Owl but it has identical rounded wings.

Call Gives a disyllabic or trisyllabic purring *gurrok-gurrok-gurrok-...gurrerok* in phrases of three to five notes, the series being repeated many times at varying intervals.

Food and hunting Not documented.

Habitat Central plateau and drier lowland forests, but occurs also in degraded areas and sometimes in villages near human habitation.

Status and distribution Endemic in W. & SW Madagascar. Status totally unknown.

Geographical variation Needs intensive study and DNA confirmation of its taxonomic status; previously treated as a subspecies of Madagascar Scops Owl, but separated on grounds of vocal differences. Monotypic.

Similar species Madagascar Scops Owl (41), an ecological counterpart in the higher elevations of more humid E. Madagascar, is very similar but has a shorter tail and has rear of tarsi practically bare; its dark brown and saturated rufous morphs are more common, and dark markings on its upperside are more obscure. It also differs clearly vocally.

▼ Torotoroka Scops Owl has relatively small ear-tufts. Ankarafantsika, Madagascar, November (*Dubi Shapiro*).

◄ Torotoroka Scops Owl at the nest hole. The grey morph has cryptic plumage with fine vermiculations, on a pale grey ground colour. Madagascar, September (*Ian Merrill*).

► Torotoroka Scops Owl also has a brown morph; this is not as dark as that of Madagascar Scops Owl *O. rutilus*. It has distinct streaking on the upperparts and white barring on a rufous ground colour below. Longer-tailed than Madagascar Scops Owl. Madagascar, October (*Mike Danzenbaker*).

▼ Grey-morph Torotoka Scops Owl, with fine vermiculations on a pale ground colour. Grey birds are more common in this species than in Madagascar Scops Owl *O. rutilus,* as it lives in drier forests. March (*Mike Danzenbaker*).

43. MOUNTAIN SCOPS OWL
Otus spilocephalus

L 17–21cm, Wt 50–112g; W 129–152mm

Identification A very small owl with short and ill-defined ear-tufts. No data on sexual size differences. Nominate race occurs in rufous and more greyish-buff morphs, with intermediate forms. Has dark tawny-brown or reddish-brown upperparts with blackish vermicular markings; many black-edged pale rufous spots on crown, often broadening into bars on neck and back, and forming not well-defined collar on hindneck. Clear white outer webs and bold black tips on inner scapulars form a prominent row of feathers. Median upperwing-coverts are boldly marked with dull yellow and black, and primaries are barred reddish-brown and dark brown. The tail is reddish-brown with blackish barring broken and mottled by chestnut bands. Whitish underparts are barred reddish-brown, with small, triangular, paired black and white spots. The rufous-brown facial disc has conspicuous blackish-tipped bristly feathers with pale bases, blackish-barred ear-coverts and cheeks, and a whitish or rufous-buff ruff with obscure blackish and dark brown bars. The forehead and crown sides, including short ear-tufts, are sometimes paler and dull yellowish. The eyes are golden-yellow or greenish-yellow, and the bill is light horn-coloured or sometimes whitish or wax-yellow.

The tarsi are densely feathered down to and often right over the base of the light flesh-coloured or fleshy-brown toes. *Juvenile* More dull reddish-brown and fluffy than the adult, and has narrow blackish-brown barring on the head and crown, broader on the back, and is only very faintly barred below.

Call A high-pitched, silvery double whistle, *whew-whew*,

► Race *hambroecki* lives on Taiwan. It is slightly darker and more rufous above and below than nominate *spilocephalus* of northern India. This race has no rufous morph. Some distinct races like this may represent good species, but their taxonomy requires further study. Taiwan, September (*Kevin Lin*).

◄ Mountain Scops Owl is highly polymorphic. This is race *vandewateri* from the mountains of Sumatra; it is dark with distinct white scapulars and collar. Sumatra, July (*Rob Hutchinson*).

half to one second between the two notes, is uttered at intervals of 6–12 seconds. Very vocal.

Food and hunting Only a few stomachs have been studied, and those contained beetles, moths and other insects.

Habitat Prefers humid forests and woodlands, usually between 600m and 2700m, mainly above 1200m, and sometimes up to 3000m.

Status and distribution Found from the Himalayan foothills of N. Pakistan to Taiwan in the east and to the Malay Peninsula, Borneo and Sumatra in the south. Locally common, but threatened in many areas by extensive forest-cutting.

Geographical variation Polymorphic; some races may be good species and others only aberrant morphs. Eight subspecies listed: nominate *spilocephalus* occurs from C. Nepal and Assam hills, India, to Burma; much paler *O. s. huttoni* is found from Pakistan to C. Nepal; more strongly rufous *O. s. latouchi* occurs from SE China to Laos; *O. s. hambroecki* from the mountains

of Taiwan has no rufous morph; dark *O. s. siamensis* is found from Thailand, Laos and S. Annam to the Malay Peninsula; *O. s. vulpes* from the mountains of the S. Malay Peninsula is dark rufous; *O. s. luciae* from Borneo is heavily marked and dark overall; and *O. s. vandewateri* from Sumatra is very dark brown. The taxonomy and biology of this complex species require further study and revision.

Similar species There are a number of sympatric *Otus* owls within this species' range. The same-sized Oriental Scops Owl (45) prefers riversides at lower altitudes, and is very boldly streaked below; the much larger Collared Scops Owl (46) has dark brown or orange eyes; also larger, the Sunda Scops Owl (49) has longer ear-tufts, no barring below, and normally brown eyes; the long-tailed White-fronted Scops Owl (58) is much larger, with a white forehead, no nuchal collar, and brown eyes; and the same-sized Reddish Scops Owl (59) has chestnut or amber-brown eyes, and slightly shorter tail and wings.

44. INDIAN SCOPS OWL
Otus bakkamoena

L 20–22cm, **Wt** 125–152g; **W** 143–185mm, **WS** 61–66cm

Identification A small owl having large conspicuous ear-tufts with dark outer margins. Female is clearly larger and slightly heavier than male. Grey-brown and rufous-buff morphs exist. Has a distinct nuchal collar and also a second collar on nape. Crown is almost uniformly blackish, clearly darker than the mantle; upperparts are fairly uniformly greyish-brown with paler and darker markings and long black streaks. Dirty cream or whitish-buff outer webs of scapulars do not form a distinct scapular stripe. The flight and tail feathers are barred lighter and darker, but not prominently so. Underparts are dull yellow-ochre, paler towards the belly, with a few dark shaft-streaks and wavy cross-bars, particularly on the upper breast and flanks. The light greyish-brown facial disc has a distinct blackish ruff. The forehead and eyebrows are paler than the surrounding plumage. The eyes are hazel-brown to dark brown, rarely yellowish-brown, the bill greenish horn-brown and the cere dusky green. Tarsi are feathered to the base of the brownish-flesh

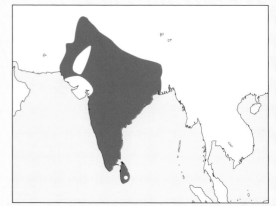

to greenish-yellow toes, which have light horn-brown claws. *Juvenile* Mesoptile plumage is pale grey or warm reddish-yellow, thinly barred dusky brown all over, the dark bars covering also the face and chin.

Call Emits a frog-like series of interrogative *what?* or *wuatt?* notes, regularly spaced with pauses.

Food and hunting Feeds mostly on insects, such as beetles and grasshoppers, but occasionally takes lizards, small rodents and birds.

Habitat Occupies dry forests and secondary wood-lands, also gardens and fruit-tree orchards near human habitations, from lowlands up to 2400m.

Status and distribution Found from S. Pakistan and C. Nepal to Sri Lanka in the south and W. Bengal in the east. This species is locally common, but its ecology and biology are not well known.

Geographical variation Four subspecies listed: nominate *bakkamoena* occurs in S. India and Sri Lanka; the more grey and less rufous *O. b. marathae* is resident from C. India to SW Bengal; *O. b. gangeticus* from NW India is paler than the nominate; and *O. b. deserticolor* from S. Pakistan and SE Arabia is paler sandy, with more golden eyes.

Similar species Partly sympatric Oriental Scops Owl (45) is smaller, with yellow eyes. Geographically separated Collared (46) and Japanese Scops Owls (47) are similar to the Indian Scops Owl in size and coloration, but have less prominent ear-tufts and a very different voice.

◀ The buffish morph (here of race *gangeticus*) has conspicuous ear-tufts and a very distinct rufous-brown nuchal collar. Gujarat, India, March (*Arpit Deomurari*).

▲ A pair of *gangeticus* Indian Scops Owls. This race is larger and paler than the nominate, and smaller and paler than Collared Scops Owl *Otus lettia*. Rajasthan, India, January (*Harri Taavetti*).

► Indian Scops Owl's grey morph has fairly uniform greyish-brown upperparts and a short tail. This is the nominate *bakkamoena*. Karnataka, India, April (*Niranjan Sant*).

45. ORIENTAL SCOPS OWL
Otus sunia

L 17–21cm, **Wt** 75–95g; **W** 119–158mm, **WS** 51–53cm

Identification A very small owl with visible ear-tufts. Female is on average 6g heavier than the male. Three colour morphs exist. Rufous morph has plain reddish-brown upperparts, with dark streaking on forehead and crown, and blackish-edged whitish-buff spots on scapulars. Wings and tail are banded pale and dark. Underparts pale, dull yellowish to white towards belly; herringbone pattern on neck and breast is formed by dark shaft-streaks with fine horizontal vermicular markings. The facial disc is pale rufous, with a narrow dark ruff bordered reddish-brown; whitish around blackish-grey bill. Has white eyebrows and yellow eyes, and a not so distinct reddish-brown hindneck collar spotted black and white. Tarsi are feathered to the base of the greyish-brown toes, which have blackish-brown claws. Grey-brown and reddish-grey morphs are much more spotted and streaked above; under-part markings are similar to those of rufous morph, but with greyish-brown or reddish-grey background. *Juvenile* Downy chick is whitish. Mesoptile has less prominent markings, with fluffy feathers, and faint barring on back and below.

Call Gives a trisyllabic, throaty *kroik ku kjooh. ..* which is audible to several hundred metres.

Food and hunting Prefers insects and spiders, but also takes small vertebrates. Hunts often on the ground, and will also take insects in the canopy of a tree.

Habitat Found in semi-open and open woodlands, parks, savannas and wooded riversides; also in urban areas. Occurs from lowlands to mountains, and to 2300m in Himalayas.

Status and distribution Distributed from W. India, Sri Lanka and N. Pakistan to extreme E. Siberia, Japan, Taiwan and the Malay Peninsula, including Andaman and Nicobar Islands in the Indian Ocean.

Geographical variation Seven subspecies described: nominate *sunia* is resident from Pakistan and Nepal to Assam and Bangladesh; *O. s. stictonotus*, from Manchuria to Korea, is the largest and palest subspecies; *O. s. japonicus*, from Hokkaido and Kyushu, in Japan, occurs very frequently in the rufous morph, and the grey morph is similar to the previous race; *O. s. modestus* from Assam to Burma, NW Thailand and Indochina, and also in the Andaman and Nicobar Islands, differs

◀ Race *modestus* from the Andaman Islands resembles brown-morph *rufipennis*. This may well prove to be a separate species altogether. Chidiya Tapu, Andaman Islands, January (*Niranjan Sant*).

◄ A dark individual of the greyish-brown morph of race *stictonotus*, in a position of alertness; note rather long ear-tufts, with blackish outer and reddish-brown inner webs. China, May (*Roy de Haas*).

▲ Grey-brown morph of race *stictonotus* from China. Has similar pale coloration to *japonicus* but is larger and more coarsely and boldly patterned below. Beijing, China, August (*Christian Artuso*).

▼ Race *malayanus* occurs from southern China south to the Malay Peninsula. This light greyish-brown bird is perhaps intermediate between dark brown and chestnut-rufous morphs. Cambodia, January (*James Eaton*).

in wing shape from the others; *O. s. malayanus*, from S. China to the Malay Peninsula, has the tarsus less feathered and the plumage much more chestnut-rufous; *O. s. rufipennis*, from Bombay to Madras, in India, is a dark race resembling *malayanus*; and *O. s. leggei* from Sri Lanka is the smallest and darkest race, also having a cinnamon-bay morph. The taxonomy is not very clear, as owls in Thailand and Sri Lanka have different voices; in addition, the taxa on the Andaman and Nicobar Islands would require further investigation.

Similar species Needs to be distinguished from several partly sympatric scops owls. Mountain Scops Owl (43) is similar in size, but is not streaked below and is finely vermiculated; larger Indian Scops Owl (44) is darker, with dark brown eyes, and also larger Collared Scops Owls (46) has orange or dark brown eyes and a well-developed collar around hindneck.

◄ Dark-brown morph of Oriental Scops Owl race *malayanus* is similar in plumage to dark brown *rufipennis*. Hong Kong, November (*John and Jemi Holmes*).

◀ Race *nicobaricus* has often been combined with *modestus*. However, this rufous morph from Nancowry Island in the Nicobar Islands shows that this taxon forms a very distinct race (if not species). It has no shaft-streaks below and no dark streaks on the head; also, there is no pale neck collar. Eyes are not pale yellow as in Nicobar Scops Owl *Otus alius*. Very distinct white whiskers occur around the dark horn-coloured bill (*S. P. Vijayakumar*).

◀▼ Race *rufipennis* lives in India, from Mumbai and Chennai southwards. Brown morph is very similar to rufous *modestus* of northern India. Goa, India, March (*Amano Samarpan*).

▼ Oriental Scops Owl of Japanese race *japonicus*. Rufous morph is very frequent in this race (*Toyonari Tanaka*).

46. COLLARED SCOPS OWL
Otus lettia

L 23–25cm, Wt 100–170g; W 158–188mm

Identification A small owl with rather long, dark-spotted ear-tufts. Females are larger and up to 30g heavier than the males. Two morphs exist. Grey-brown morph has light grey-buff upperparts, and rufous morph has more reddish-brown to dull yellow upperparts. Plumage is generally light yellowish-brown above, freckled, mottled and spotted with black and dull yellow, and with pale buffish feathers on scapulars forming an indistinct stripe on wing; two light-coloured collars are visible on the hindneck. Underparts are pale brown with small arrowhead-like shaft-streaks. The facial disc is dull yellowish with some faint darker concentric markings. The eyes are dark brown to orange, and the bill is light-tipped greenish-horn with a pale dusky yellow lower mandible. Tarsi are fully feathered to the base of the dusky to fleshy-grey toes and claws, the toes having yellowish-white pads. *Juvenile* Mesoptile is covered with narrow, irregular dark barring above and below.

Call Emits a single, mellow *buuo*, repeated at intervals of about 12–20 seconds. Singing can continue for 15 minutes or longer.

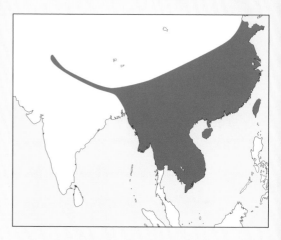

Food and hunting Varied diet includes insects, lizards, mice and small birds.

Habitat Found in many types of forest, including second growth and bamboo stands not far from human habitations; also enters towns. Occurs in lowlands and

▼ This Collared Scops Owl of race *erythrocampe* looks very whitish below. Hong Kong, January (*Michelle and Peter Wong*).

▼ Race *glabripes* is endemic to Taiwan. Grey morph is fairly similar to nominate *lettia*. Also resembles Japanese Scops Owl *O. semitorques* in plumage, but has totally bare toes (*Pei-Wen Chang*).

plains, and at up to 1200m in Pakistan and to 2400m in Himalayas.

Status and distribution Found from Himalayas to Assam, and from E. Bengal and Thailand to S. China and Taiwan. Unconfirmed winter visitor to peninsular Malaysia. Locally not uncommon.

Geographical variation Five subspecies listed: nominate *lettia* occurs from E. Himalayas and E. Assam to Burma and Thailand; generally less grey *O. l. erythrocampe* is found in S China; paler *O. l. glabripes* is resident in Taiwan; darker *O. l. umbratilis* is restricted to Hainan, in S China; and *O. l. plumipes*, from NW Himalayas, has densely feathered toes, as in Japanese Scops Owl. Some DNA evidence indicates the possibility that the Chinese race *erythrocampe* could be specifically distinct.

Similar species Geographically separated Common (27), Indian (44), Japanese (47), Sunda (49) and Singapore Scops Owls (55) are similar in size and coloration, but all differ in voice.

▼ Collared Scops Owl race *lettia*; grey-brown morph with black bill. India, February (*Hugh Harrop*).

47. JAPANESE SCOPS OWL
Otus semitorques

L 21–26cm, Wt *c.* 130g; W 153–196mm, WS 60–66cm

Identification A small owl with long, prominent ear-tufts. No data on sexual size differences. Has greyish-brown upperparts with blackish and dull yellowish markings, and a grey double collar, one band on hind-neck and one on nape. Has small paler areas on outer webs of scapulars, but no distinct scapular stripe. Flight and tail feathers are barred light and dark. Underparts are pale greyish-buff with a dark herringbone-like pattern. The facial disc is light greyish-brown with minute dark flecks, and a narrow prominent ruff. The forehead is pale, the whitish eyebrows extend almost to the tips of the ear-tufts. The eyes are fiery to dark red, and the bill and cere greyish-horn. Tarsi and toes are feathered, with horn-coloured claws. *Juvenile* Downy chick is whitish. Mesoptile is greyish to dull yellow, diffusely barred above and below. Juveniles have more yellowish eyes than the adults. *In flight* Wings are pointed, with 7th primary the longest.

Call Gives a fairly deep, mournful *whook*, uttered at long intervals.

Food and hunting Feeds mainly on larger insects, but also on spiders, frogs, small mammals and birds.

Habitat Occurs in forests and wooded gardens, and often found near villages and suburbs. From lowlands to 900m. Especially during the winter, comes to areas near human settlements at lower altitudes.

Status and distribution Found from SE Siberia to Japan. It is locally not uncommon, but requires more study.

Geographical variation Three subspecies listed: nominate *semitorques* is found in S. Kuril Islands, Russia and Hokkaido and Kyushu, in Japan; paler *O. s. ussuriensis*, from Ussuriland and Sakhalin, winters (and perhaps also breeds) in C. & S. Korea and N. China; and *O. s. pryeri*, from Izu Islands (Hachijo) and Ryukyu Islands (Okinawa), has dark yellow eyes and is less greyish above and below; photos show two very different colour forms of this race.

Similar species Geographically separated Indian (44), Collared (46) and Sunda Scops Owls (49) are similar in size and coloration, but all have brown eyes and a very different voice. Elegant Scops Owl (48), sympatric with present species in Ryukyu Islands, is slightly smaller and darker with yellow eyes.

◄ Japanese Scops Owl nominate *semitorques* at the nest. This race is more greyish than *pryeri*. Honshu, Japan, March (*Kazuyasu Kisaichi*).

◄ Japanese Scops Owl
has fiery red-orange or dark
yellow eyes. Ear-tufts can be
hidden almost completely.
The nape has a distinct pale
collar. This is a very pale and
buffish individual of race *pryeri*.
Okinawa, Japan, September
(*Kenji Takehara*).

► Race *pryeri* differs from the
nominate in having bare toes,
dark yellow eyes, and a strong
ferruginous wash above and
below; generally less greyish in
appearance. The ear-tufts are
large if fully erected. Okinawa,
Japan, May (*Kenji Takehara*).

48. ELEGANT SCOPS OWL
Otus elegans

L 20cm, Wt 100–107g; W 165–178mm

Other name Ryukyu Scops Owl

Identification A very small owl having relatively long ear-tufts with blackish outer and reddish-brown inner webs. No data on sexual size differences. Two morphs exist. Buffish-grey morph has dull yellowish to grey-brown upperparts with fine vermiculations and dark shaft-streaks, and some whitish spots on the mantle. Scapulars are dull-yellowish to white with dark lower edge, and form a whitish line across the shoulder. Flight and tail feathers are barred light and dark. Pale underparts become whitish towards the belly, and have a herringbone-like pattern created by dark shaft-streaks and fine dark horizontal vermicular markings. Crown and forehead streaked dark. Facial disc light greyish-brown with a narrow dark ruff, and whitish eyebrows. The eyes are yellow and the bill dark horn-coloured. Tarsi are heavily feathered nearly to the base of the greyish-brown toes, which have blackish-horn claws; the toes and claws are relatively large. Rufous morph is in general darker reddish-brown, with clearer markings. *Juvenile* Downy chick is whitish. Mesoptile is similar to adult. *In flight* Shows rounded wing, with 7th primary the longest.

Call A fairly hoarse, cough-like *kew-guruk*, is repeated at regular intervals, some 15–30 times per minute.

Food and hunting Hunts mainly insects, such as beetles, crickets and grasshoppers, but also spiders and small vertebrates.

Habitat Inhabits evergreen forests, originally primary forests with old trees, but nowadays adapted to logged forests, sometimes also near villages. On Lanyu from sea level up to the highest elevation (550m).

Status and distribution Found from Japanese Ryukyu Islands south to small islands north of Luzon, in Philippines. It is common on some Japanese Islands, but population on Lanyu, off SE Taiwan, is only 150–230 individuals. Populations on Batan, Calayan and other small islands north of Luzon have been very little studied.

Geographical variation Three subspecies listed: nominate *elegans* occurs in Ryukyu and Daito Islands, south of Japan; *O. e. botelensis*, from Lanyu Island (off Taiwan), is somewhat paler and longer-winged than nominate; and *O. e. calayensis*, from Batan, Calayan and maybe some other small islands north of Luzon (Philippines) is short-winged and ochre-coloured. It is possible that the last subspecies, which is morphologically very distinct, may not belong to this species.

Similar species Geographically separated Oriental Scops Owl (45) is smaller, with short wings and ear-tufts. Much paler Japanese Scops Owl (47) is sympatric in Ryukyu Islands, but has deep orange-red eyes, very long ear-tufts, and a nuchal collar.

◄ The nominate *elegans* is reddish brown below and very similar in plumage to the Lanyu subspecies *botelensis*. Okinawa, Japan, July (*Kazuyasu Kisaichi*).

Nominate Elegant Scops Owl is similar to Oriental Scops Owl *O. sunia* with an identical wing-formula, but it is larger and longer-winged. Okinawa, Japan, August (*Mike Danzenbaker*).

▼ Elegant Scops Owl subspecies *botelensis* resembles both rufous and darker buffish-grey morphs of the nominate *elegans*, but it is generally paler, more finely marked and less streaked. Lanyu Island, April (*James Eaton*).

Race *calayensis* is distinctly ochre-coloured, with fleshy toes, and smaller and paler claws than the nominate. It is short-winged but the wing-formula is the same, with P7 the longest. Calayan, Philippines, April (*Desmond Allen*).

49. SUNDA SCOPS OWL
Otus lempiji

L 19–21cm, **Wt** 90–140g; **W** 136–157mm

Identification A small owl with prominent, dull yellowish ear-tufts broadly black-edged on outer webs. No sexed measurements are available, but the range of weights would indicate clear sexual size difference (female larger?). Three morphs exist, as well as much individual variation; grey-brown, grey-buff and rufous morphs may also intergrade. Sandy-brown upperparts are mottled and freckled black and dull yellow, and black-blotched. Has an indistinct sandy-buff collar on hindneck. Wings are dark brown with sandy vermiculated barring. Underparts are paler grey-buff or reddish-brown to dull yellow (depending on the morph), often densely peppered with minute black spots and with arrowhead-like or chevron-like black shaft markings. The facial disc is rufous-buff, with a blackish ruff, the crown is blackish, and the forehead and eyebrows are whitish to dull yellow, more black-variegated posteriorly. The eyes are usually dark brown, but sometimes orange-yellow, and have light pinkish-brown eyelids. The bill is whitish-yellow. Tarsi are fully feathered to the base of the horn-white to greyish-pink toes, with the claws a little darker or more brownish. *Juvenile* Downy plumage is generally bright

reddish-brown, but pure white in Malay Peninsula and dark grey or rufous-grey in Java. Mesoptiles have clear cross-bars, especially on head, and elsewhere irregular and indistinct markings. *In flight* Shows rounded wings and dark dorsal plumage.

Call Utters a musical, interrogative *wooup wooup...*, with rising inflection, at fairly long intervals of 10–15 seconds, this repeated for long periods.

Food and hunting Feeds chiefly on insects, and occasionally takes small birds. Known to hunt insects attracted by cow dung or poultry droppings near human habitations.

Habitat This species is less forest-dependent than many others of the genus and frequents wooded gardens and plantations, including suburban and urban areas with some trees. Occurs from lowlands to 2000m, sometimes even up to 2400m.

Status and distribution Found from the Malay Peninsula to Bali and Borneo. It is rather common, even in cities of the region.

Geographical variation Four subspecies described: nominate *lempiji* occurs from Malay Peninsula to Bali and Borneo (Kalimantan); *O. l. hypnodes*, from Pulau Padang, off E. Sumatra, is darker, especially the rufous morph; *O. l. kangeana*, from Kangean Island, north of Bali, is smaller than nominate; and *O. l. lemurum* is confined to N. Borneo (Sarawak).

Similar species Sympatric Singapore Scops Owl (55) is very similar, but has a darker facial disc and less prominent eyebrows.

◄ Sunda Scops Owl subspecies *lemurum* is similar in size to the nominate. Sarawak, Malaysia, September (*Ch'ien C. Lee*).

50. NICOBAR SCOPS OWL
Otus alius

L 19–20cm; W 160–167mm

Identification A small owl having medium-sized ear-tufts with rounded tips. No data on sexual size differences. Has fine dark and light bars on crown and nape, and no nuchal collar. Upperparts are warm brown, fairly densely dusky-barred, with wider dark and pale bars from mantle to lower back. Large, blackish-framed, rounded white spots on outer webs of scapulars form a distinct row across shoulder. Flight and tail feathers are barred lighter and darker brown. The throat is pale yellowish-brown, and the upper breast cinnamon-brown with fuscous bars and a few dark shaft-streaks; there are heavy whitish, yellowish-brown and dusky-brown bars and indistinct shaft-streaks on lower breast, flanks and belly. The facial disc is pale with some darker vermicular markings and with a thin, indistinct, darker brown ruff, and the eyebrows are pale and finely dark-mottled. The eyes are pale yellow with a narrow bare pink orbital ring. The bill and cere are yellowish-brown, bill with darker cutting edges and tip. Has sparsely feathered tarsi, with bare rear edge and lower part, and bare toes dark yellowish-brown, with relatively large, dusky horn-coloured claws. *Juvenile* Unknown.

Call Emits a long series of piping *weeyu weeyu* notes, each lasting less than half a second and repeated every four seconds.

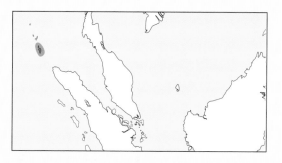

Food and hunting Two individuals were seen to take a spider, a beetle and a gecko.

Habitat Found in wooded areas near sea level.

Status and distribution Endemic on Great Nicobar Island, in the Bay of Bengal; possibly occurs also on Little Nicobar. This species is certainly rare, but extremely little known. It was first described in 1998, from two specimens collected on Great Nicobar. Subsequently, several more individuals have been observed.

Geographical variation Monotypic.

Similar species Geographically separated Simeulue Scops Owl (51) is smaller, much more reddish-brown and uniform, but with dark streaks on the forecrown and below. All other SE Asian scops owls, except Biak Scops Owl (71), have dark shaft-streaks on the upperparts.

▼ Nicobar Scops Owl has clear dark stripes on the warm brown head. Rufous below, with thin tricoloured marks. Eyes are very pale yellow and whiskers are not as well developed as in *nicobaricus* Oriental Scops Owl. Teressa Island, Nicobars, April (*S. P. Vijayakumar*).

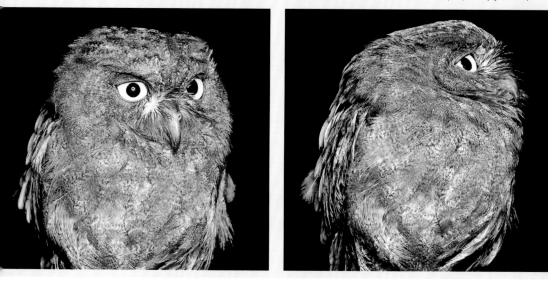

51. SIMEULUE SCOPS OWL
Otus umbra

L 16–18cm, Wt 90–100g; W 142–145mm

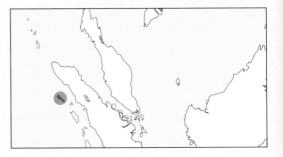

Identification A very small owl with short, but fairly prominent ear-tufts. No data on sexual size differences. Plumage is dark reddish-brown with vague dark vermicular markings above, black-edged dull yellowish and whitish outer webs of scapulars forming a short whitish stripe across the shoulder. Wings are rufous with light, darker-edged barring, and reddish-brown tail has pale barring. Underparts are pale, with some white barring and narrow dark shaft-streaks. The reddish-brown facial disc has an indistinct ruff. The eyes are yellow and the bill blackish-grey. Tarsi are feathered almost to the base of the grey toes, which have dark-tipped horn-coloured claws. *Juvenile* Not known.

Call Gives a clear *pook pook pupook*, repeated at short but varying intervals.

Food and hunting Not documented.

Habitat Inhabits coastal forest edges, broken forests and clove plantations.

Status and distribution Endemic to Simeulue Island, off the NW coast of Sumatra. Said to be rather frequent, but has a very limited range.

Geographical variation Monotypic.

Similar species The only *Otus* owl on Simeulue Island. Geographically well separated Enggano Scops Owl (52) is similar, but lives more than 1000km to the south; it is also bigger, with larger ear-tufts.

◀ Simeulue Scops Owl is dark reddish-brown with black-rimmed yellow eyes. Rufous below, with fine white and brown bars and black feather edges. It lacks the pale neck collar of Enggano Scops Owl *Otus enganensis*. Simeulue, December (*James Eaton*).

52. ENGGANO SCOPS OWL
Otus enganensis

L 18–20cm; W 160–165mm

Identification A very small owl having fairly long, distinct, tawny-brown ear-tufts with some sparse whitish, dark-shafted and black-rimmed feathers. No data on sexual size differences. Plumage is dark reddish-brown to brownish-olive above, with indistinct nuchal collar; the crown is darker, forehead with a few small black-tipped white feathers, and back has irregular dark patches. Has a pale shoulder line on outer webs of scapulars formed by white windows with light reddish-brown borders and dark rufous edges. Flight feathers are plain dark reddish-brown with rather indistinct, fine blackish vermicular markings; secondaries and greater wing-coverts are clearly paler. Underparts are light yellowish-brown to brownish-olive with double spots and cross-bars, upper breast with a few dark-barred and dark-shafted white feathers; more reddish-brown breast has black shaft-streaks and a few black bars and vermiculations, and belly feathers are edged dark chestnut, becoming darker near flanks and lower belly. Undertail-coverts are white with broad rufous borders. The facial disc is pale reddish-brown, with circles of white-shafted feathers. The eyes are yellow and the bill bluish-horn. The mottled reddish-brown and white tarsus is fully feathered to the base of the bluish-grey toes, which have darker horn-coloured claws. *Juvenile* Not known.

Call Said to differ from that of Simeulue Scops Owl.

Food and hunting Not described, but probably feeds mainly on insects.

Habitat Found in wooded areas and forest.

Status and distribution Endemic to Enggano Island, off the SW coast of Sumatra. Not much is known of this owl, but it is possibly rather common.

Geographical variation Monotypic.

Similar species Geographically well-separated Simeulue Scops Owl (51) is a little smaller, but similar in coloration; it lacks the nuchal collar, has black-rimmed yellow eyes, and a less contrasting pattern on the greater wing-coverts and flight feathers. Also

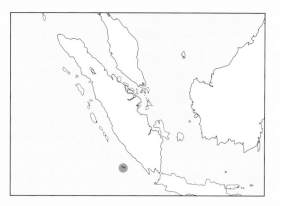

allopatric Mentawai Scops Owl (53) is similar in size, but is blotched and mottled dark brown, with no nuchal collar, and has yellow or brown eyes.

◀ Enggano Scops Owl is generally brown-mottled, varying from cinnamon to brownish-olive below, with cross-barring and double spots. It has a dark back, though the greater coverts and secondaries are paler, contrasting more with the primary coverts and primaries than those of Simeulue Scops Owl *Otus umbra*. Enggano, February (*Filip Verbelen*).

53. MENTAWAI SCOPS OWL
Otus mentawi

L 20cm; W 157–166mm

Identification A small owl with not very prominent dark brown mottled ear-tufts. No data on sexual size differences. Two morphs exist. Rufous morph has dark reddish-brown upperparts, with dark shaft-streaks on back, and with crown and nape darker; has no nuchal collar. Outer scapular feathers are blotched whitish and black, forming a distinct row. The wing-coverts are mottled and freckled dark brown. Underparts are reddish-brown or chestnut, with herringbone markings enclosed by single or paired white ocelli; the belly and undertail-coverts are paler. The reddish-brown facial disc is rimmed with blackish-tipped feathers, and the pale buff eyebrows have dark mottling. The eyes are brown, but sometimes yellow, and the bill yellowish-horn. Tarsus is fully feathered, sometimes

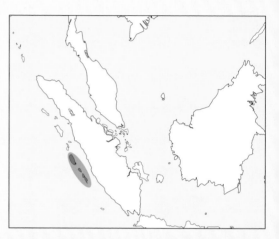

to beyond the toe joint, and the bare toes are flesh-coloured with relatively powerful dark horn-coloured talons. Blackish-brown morph is less reddish, and more blackish-brown. *Juvenile* Not known.

Call Gives a series of rough barking *how-how-how*, three or four notes repeated at intervals of several seconds.

Food and hunting Not studied, but food is most likely insects.

Habitat Inhabits lowland rainforest, but is found also in secondary growth and in villages near human habitations.

Status and distribution Endemic to four islands in the Mentawai group, off W. Sumatra, Indonesia. It is locally common, but very little studied.

Geographical variation This species is sometimes regarded as a subspecies of the Sunda Scops Owl but is separated here on the grounds of its very different vocalisations and the fact that it is isolated on a few small islands. Monotypic.

Similar species No other *Otus* owl inhabits the Mentawai Islands. Geographically separated Sunda Scops Owl (49) is a little smaller, with a very different voice.

▲ Mentawai Scops Owl has a dark brown back, with small white spots on the hindneck. Siberut, Indonesia, January (*Filip Verbelen*).

◄ Mentawai Scops Owl on the ground. This owl is rufous-brown below, with herringbone markings and pale spots. It has brown eyes, a very white throat, a rufous nuchal band, and powerful feet with bare toes. Siberut, Indonesia, January (*Filip Verbelen*).

► Mentawai Scops Owl spreads its wings and tail to look larger when alarmed. This is also typical of *Ninox* owls. Siberut, Indonesia, January (*Filip Verbelen*).

54. RAJAH SCOPS OWL
Otus brookii

L 21–25cm; **W** 162–187mm

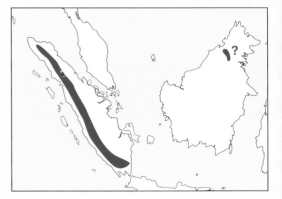

Identification A small owl having long ear-tufts with prominent white inner webs. Females are said to be a little larger than males, although no weights are available. Has a broad white band on the side of the crown extending to ear-tufts, and with a hint of a pale occipital patch. The upperparts are deep dark-coloured or reddish-brown, densely freckled, mottled and speckled with blackish shaft-streaks and dark brown wavy bars mixed with fewer paler spots. The nape and hindneck have a white or whitish double collar with distinct large black feather tips, the lower collar broader and forming a fuller cervical collar. The light reddish-brown underparts are variably mottled and blotched dark rufous and black. The eyes are chrome-yellow or orange-yellow or dark orange, and the bill pale yellow. It has powerful feet, and tarsi heavily feathered to the base of the yellowish-flesh or light grey toes, which have darkish claws. *Juvenile* Mesoptile plumage is reddish-brown with fine dark brown cross-bars, and with the

whitish-rufous underside freckled with darker reddish-brown and dark streaks.

Call Gives an explosively loud, double hoot, *wha-ooo*, both notes some 0.2 seconds long and interval between them 0.7 seconds; this song is repeated at intervals of 7–10 seconds.

Food and hunting Stomachs examined contained insects, and one also held a frog.

Habitat Inhabits rainforests to cloud forests between 900m and 2500m, mainly at 1200–2400m.

Status and distribution Distributed in the mountain regions of Sumatra and Sarawak (Borneo). The very distinctive nominate Borneo race is known only from two specimens collected in 1892–93; it has not been seen since that time and could be extinct. The Sumatra race is also suffering from rampant deforestation in Indonesia.

Geographical variation Two subspecies described: nominate *brookii* is confined to NW Borneo; *O. b. solokensis*, from the highlands of Sumatra, is browner and with a more yellowish tinge above.

Similar species Sympatric Sunda Scops Owl (49) is similar in size, but has a uniformly dark brown back, a less distinct pale nuchal collar, and whitish to dull yellow underparts with black streaks and many very narrow wavy bars. Geographically overlapping Mountain (43) and Reddish Scops Owls (59) are smaller and generally more reddish, with small speckles and spots.

◄ Brown morph of Rajah Scops Owl race *solokensis* is very similar to the nominate *brookii* but browner dorsally, with three distinct collars on the hindneck and nuchal area. Below, has broader blackish shaft-streaks. Rufous-brown face and orange-yellow eyes. Gunung Kerinci, Sumatra, August (*Sander Lagerveld*).

55. SINGAPORE SCOPS OWL
Otus cnephaeus

L *c.* 20cm; **W** 143–157mm

Identification A very small owl with blackish, relatively short and blunted ear-tufts. No data on sexual size differences. Lacks a distinct collar on the hindneck. Has dark earth-brown upperparts and wing-coverts mottled dull yellowish and dark brown, with blackish streaks and sometimes a buffish tinge. Large yellowish-buff spots on outer webs of scapulars form a scapular line. The primaries are earth-brown, with dull yellowish bars on the outer webs, but the inner webs are nearly uniform. The tail is dark brown with narrow paler bars. The underparts are brownish to dull yellowish, densely dark-vermiculated, with some blackish shaft-streaks, the sides of the breast having a few incomplete vertical rows of dull yellowish feathers with rhomboid-shaped black centres. The facial disc is greyish-brown, paler towards the light greyish-brown cere and ivory bill; a distinct dark ruff is well visible only at both sides of the disc. The eyes are dark brown, with dark eyelids rimmed greyish-flesh. The pale brownish-buff tarsi are densely feathered to the base of the pinkish-grey toes, which have greyish-brown claws. *Juvenile* Mesoptile is barred all over.

Call Emits a resonant hoot, *kwookk*, repeated at intervals of *c.* 14 seconds.

Food and hunting Not studied, but probably feeds on insects, spiders, geckos and even mice.

Habitat Found in evergreen forest, but also in

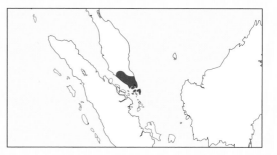

secondary growth, plantations, parks, gardens and trees near cities and other human habitations.

Status and distribution Occurs in S. Malay Peninsula between Kuala Lumpur (Malaysia) and Singapore. This owl is not rare, but is little known.

Geographical variation This species has been treated as a subspecies of Sunda Scops Owl, but is now separated on the grounds of vocal differences; molecular studies should confirm this new taxonomic status. Monotypic.

Similar species Geographically overlapping Mountain Scops Owl (43) is much smaller, with yellow eyes, and normally lives at higher elevations. Sunda Scops Owl (49), also sympatric, has a paler facial disc with a broad dark ruff, more distinctly whitish eyebrows, and much more prominent ear-tufts, as well as paler underparts; its bill is whitish-yellow.

▼ Singapore Scops Owl has large, dark eyes. Singapore, March (*Rob Hutchinson*).

56. PHILIPPINE SCOPS OWL
Otus megalotis

L 23–28cm, Wt 180–310g; W 142–205mm

Identification A small to medium-sized owl with long ear-tufts, this is the largest of the Old World *Otus* owls. Females are larger and some 30g heavier than males. Two morphs occur. Rufous morph is light reddish-brown to yellowish-brown all over, with discrete dull blackish vermicular markings and mottling, these somewhat coarse on ear-tufts and wing-coverts and much less prominent below. The rufous-brown to yellowish-brown greater and primary wing-coverts have very coarse, thick blackish vermiculations, forming five or six indistinct bars. Six or seven light yellowish-brown bars are visible on the blackish flight and tail feathers, these bars being faint on the secondaries and almost obscured on the tail by blackish freckling. The underparts are grey with dark arrowhead-like shaft-streaks with some cross-markings; the ash-brown underwing has reddish-yellow bands. The facial disc has a broad whitish to dull yellow ruff of dark-tipped feathers. The eyes are warm orange-brown, the cere pinkish, and the

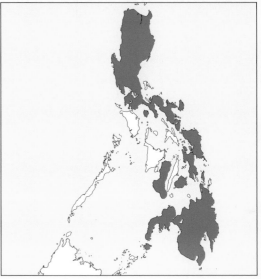

bill pale horn-pink or yellowish-pink, more yellowish on lower mandible. Tarsi are feathered to the base of the pale flesh-coloured, yellowish-brown or whitish-grey toes, with yellowish-pink claws faintly tinged with olive or dark grey. Grey morph is in general greyish-brown, but has a blackish crown, whitish eyebrows extending to the ear-tufts, and a narrow whitish-buff nuchal collar; the upperparts are greyish-tinged dark brown, heavily mottled and barred blackish, with some white and reddish-brown. *Juvenile* Has top of head and neck rufous-buff, finely black-barred, rufous face, faint white shaft-streaks on facial disc; reddish-brown to dull yellow throat is faintly black-barred, and rufous-buff underparts become whitish-buff with fine dusky bars on lower belly and thighs. The tarsi are whitish with obscure dusky bars.

Call Utters an explosive series of three to six descending notes, each with rising inflection. Call is similar to that of Sunda Scops Owl, but with notes longer.

Food and hunting According to stomach contents of some individuals, this species eats insects.

Habitat Lives in dense tropical forests and secondary woodlands from 300m to 1600m, sometimes up to 2000m.

◄ Visayan Lowland Scops Owl *O.* (*m.*) *nigrorum* is the smallest form and has a bright rusty head. Negros, Philippines, June (*Alain Pascua*).

Status and distribution Confined to the Philippines, where it is said to be not so rare.

Geographical variation Three subspecies listed: nominate *megalotis* is found on Luzon, Marinduque and Catanduanes; *O. m. everetti*, from Samar, Leyte, Dinagat, Bohol, Mindanao and Basilan, is smaller than nominate; *O. m. nigrorum*, from Negros, is even smaller, with a bright rusty head and whiter underparts. These three subspecies were earlier treated as races of the Indian Scops Owl (44), but are now separated on the basis of their different vocalisations; recent DNA studies suggest that they should now be treated as three separate species.

Similar species No other similar-sized and similarly coloured owl occurs in the Philippines. Geographically separated Reddish Scops Owl (59), the nearest population of which is in the Sulu Islands, is smaller and much more reddish-brown. Sympatric Giant Scops Owl (108) is much larger, and has a pale reddish-brown breast and belly with well-demarcated dark brown streaks and drop-like markings.

◀ Mindanao Lowland Scops Owl *O. (m.) everetti* is smaller than the nominate. Mindanao, Philippines, February (*Rob Hutchinson*).

◀▼ Mindanao Lowland Scops Owl showing the blackish crown. Mindanao, Philippines, February (*Rob Hutchinson*).

▼ Luzon Lowland Scops Owl: nominate *megalotis* grey morph has whitish eyebrows extending up to the ear tufts. Luzon, Philippines, December (*Alain Pascua*).

57. PALAWAN SCOPS OWL
Otus fuliginosus

L 19–20cm; **W** 139–147mm

Identification A very small owl having short but fairly prominent ear-tufts with a whitish inner edge. No data on sexual size differences. The plumage is rich brown overall, vermiculated and spotted, with a darker brown crown and a prominent pale nape collar. Fairly large whitish areas on the outer webs of the scapulars form an equally prominent whitish scapular band, and contrasting very light bars on the outer webs of the dark primaries are well visible. The dark-vermiculated reddish-brown underparts have a number of arrow-shaped blackish markings. The facial disc is brownish-rufous with several dark flecks; the ruff is not prominent, but the whitish eyebrows and forehead,

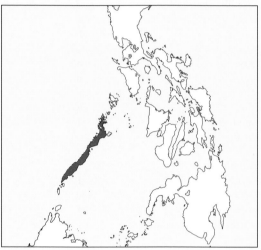

finely dark-flecked, are. The eyes are pale orange-brown, and the bill and cere light brownish-horn. The tarsi are feathered to the base of the yellowish-grey toes, which have dark horn-coloured claws. *Juvenile* Not yet described.

Call A handsaw-like, rasping, disyllabic *krarr-kruarr* is repeated at intervals of several seconds.

Food and hunting Little studied, but feeds mostly on insects.

Habitat Inhabits tropical lowland forests and secondary woodlands, and also mixed cultivations with trees.

Status and distribution Restricted to the island of Palawan, in the Philippines. This species is rare, but has not been studied. It is listed as Vulnerable by BirdLife International.

Geographical variation Monotypic.

Similar species No other *Otus* owl is known to occur on Palawan. This species resembles the allopatric race *everetti* of the Philippine Scops Owl (56) but that species is a little bigger, with a narrow whitish to dull yellow nuchal collar and a whitish-buff throat.

◄ Palawan Scops Owl is similar in size and colour to Philippine Scops Owl *Otus megalotis* of race *everetti* and differs only in vocalisation. It has a broad nuchal collar, distinctly pale scapulars and contrasting pale greater-coverts. Palawan, February (*Ian Merrill*).

58. WHITE-FRONTED SCOPS OWL
Otus sagittatus

L 25–28cm, **Wt** 110–140g; **W** 173–192mm

Identification A small to medium-sized owl with a prominent white forehead, white extending laterally into large ear-tufts. No data on sexual size differences. The upperparts are deep reddish-brown or yellowish-brown, with small, triangular, dull yellowish to white spots with black lower margin on the mantle and back. The inner scapulars have larger white spots, and the outer scapulars have yellowish-white to reddish-brown outer webs, with three or four not so large black spots on the shafts. The rounded wings have some darker and paler brown bars. The relatively long, chestnut-rufous tail is marked with about ten transverse blackish bars, these more distinct towards the base and on the outer feathers. The underparts are light reddish-brown, finely brown-vermiculated on the breast and throat, with a roundish black shaft-spot on the pale or whitish centre of each feather, the belly centre with the largest spots. The facial disc is pale reddish-brown, bordered by black-tipped feathers, and with a broad, deep rufous-chestnut ring around the deep brown to dark honey-brown eyes; it has pink eyelids and whitish loral bristles with black tips. The bill is bluish-white and the cere light bluish-green. The tarsi are feathered reddish-brown, and the flesh-pink toes have bluish-white claws. *Juvenile* Not described.

Call Utters a hollow whistle, *hoooo*, repeated at long intervals, and starting and ending abruptly.

Food and hunting Some stomachs contained only insects, mainly moths.

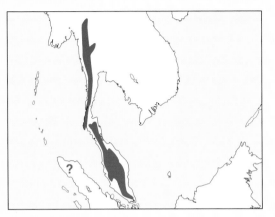

Habitat Inhabits evergreen and lowland rainforests, but also degraded swampy forests, at up to 700m.

Status and distribution Found from Tenasserim, in S. Burma, to Thailand and the Malay Peninsula. An unconfirmed record from Aceh, in Sumatra, could refer to a vagrant bird. Believed to be rare, but poorly known. Listed as Vulnerable by BirdLife International owing to the extensive deforestation of the regional lowlands.

Geographical variation Monotypic.

Similar species Larger than any other *Otus* in SE Asia, with a long tail and prominent white forehead. Geographically separated Javan Scops Owl (62) has white supercilia, but is much smaller with orange-yellow eyes.

◄ White-fronted Scops Owl is a large, dark brown, long-tailed scops owl with brown eyes and pink eyelid-rims. As the name suggests, it has a white forehead and supercilia; even the ear-tufts are partially white. Lacks a neck-collar. Legs are feathered to base of the toes. Thailand, May (*James Eaton*).

59. REDDISH SCOPS OWL
Otus rufescens

L 15–18cm, Wt 70–83g; W 121–137mm

Identification A tiny to very small owl with prominent ear-tufts. No data on sexual size differences. Two colour morphs exist. Light morph has tawny reddish-brown upperparts with triangular or elongated light fulvous spots bordered (on one or both sides) with black; larger spots on mantle and wing-coverts, and more like shaft-streaks on back and rump. The ochre-coloured primaries have very visible blackish bars, the ochre-brown secondaries having darker brown bars. The reddish-brown tail is black-mottled and indistinctly pale-barred. The underparts are yellowish-brown to dull yellow, with sparse spots (as on upperparts). The facial disc is cinnamon-buff, becoming lighter towards the dark brown ruff. The eyes are chestnut-brown to amber-brown, with pink to light reddish-brown eyelids, and the bill is horn-white. The fairly large, pale buff feet are feathered nearly to the base of the yellowish toes, which have horn-coloured claws. Dark morph is much darker yellowish-brown above and below, with patterning similar to that of the light morph. *Juvenile* Downy chick is reddish-brown, darker on crown and mantle and paler on rump. Mesoptile resembles adult, but is far less speckled.

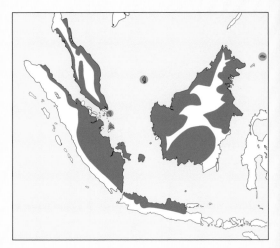

Call Not well documented, but said to be a series of hollow whistles, *wüh-wüh-wüh...*, repeated at regular intervals, and single notes uttered every half-second. Very similar to that of White-fronted Scops Owl (58), but with less abrupt start and finish.

Food and hunting Stomachs studied contained insects, mainly grasshoppers. This owl also takes crabs.

Habitat Found in lowland rainforests and evergreen forests up to 1350m, mainly between 600m and 1000m. Also in logged primary and secondary forests.

Status and distribution Occurs from S. Thailand and Malay Peninsula to Sumatra, Java and Borneo, on some adjacent smaller islands in Indonesia, Brunei and Malaysian Borneo, and in Tapul Islands (in Sulu Archipelago), in the Philippines. It is fairly rare, but is also very little studied.

Geographical variation Two subspecies listed: nominate *rufescens* occurs in Sumatra (including Bangka), Java and Borneo; *O. r. malayensis*, from S. Thailand to Malay Peninsula, is slightly more reddish-brown above and more rufous-ochre below. Sulu Archipelago owls have been named as *O. r. burbidgei*, but this race, and maybe even the species identification, is questionable.

Similar species Partly sympatric Mountain Scops Owl (43) and Sunda Scops Owl (49) are less distinctly spotted.

◄ Nominate *rufescens* has cinnamon-buff underparts with small black spots. Eyes dull amber with pink eyelids. Bill and toes whitish. Way Kambas, Sumatra, July (*Sander Lagerveld*).

Race *malayensis* from the Thai-Malay Peninsula is similar to the nominate, but a little more rufous, and yellowish below. Kedah, Malaysia, November (*James Eaton*).

60. SERENDIB SCOPS OWL
Otus thilohoffmanni

L 16–17cm; **W** 128–140mm

Identification A very small owl without true ear-tufts; pseudo ear-tufts visible only if individual is alarmed. No data on sexual size differences. Has a uniformly reddish-brown head with whitish supercilia. Rufous upperparts are blackish-spotted all over; some paler areas are evident around the blackish spots, but no white flecks. Wing and tail feathers have reddish-brown outer webs and mostly blackish inner webs, the remiges and rectrices with broad, evenly spaced rufous and blackish barring. The breast is pale reddish-brown, sprinkled with triangular blackish spots, and the belly and undertail-coverts are unspotted and paler than breast. The not very prominent facial disc is brownish-rufous, lacking a distinct ruff. The large eyes are orange-yellow surrounded by a striking black ring. The relatively long bill is yellowish and the cere fleshy pink. Light reddish-brown legs are feathered to about the middle of the tarsus, with the bare parts of the thin tarsi and toes pinkish-white, with ivory-yellow claws. *Juvenile* Resembles the adult, but pseudo ear-tufts are more pronounced above the partially formed facial disc.

Call Gives a series of short, tremulous notes, *wuhüwwo*, each one some 0.3 seconds long, and repeated at

intervals of 22–35 seconds.

Food and hunting Not studied, but believed to eat mainly insects, such as beetles and moths.

Habitat Found in lowland rainforests from 30m to 500m, often in secondary forest with rich undergrowth.

Status and distribution Endemic to SW Sri Lanka, where an estimated 200–250 individuals exist. Vulnerable to habitat destruction.

Geographical variation Monotypic.

Similar species Geographically overlaps with Indian (44) and Oriental Scops Owls (45), both of which have prominent ear-tufts.

▼ The male Serendib Scops Owl has more orange-yellow eyes. Tail short, and upperparts rufescent with small black spots. Sri Lanka (*Uditha Hettige*).

▼ Serendib Scops Owl's underparts are pale reddish-brown with some blackish spots. Sri Lanka, November (*Rob Hutchinson*).

61. ANDAMAN SCOPS OWL
Otus balli

L 18–19cm; W 133–143mm

Identification A very small owl with small ear-tufts. No data on sexual size differences. Two morphs exist. Rufous morph is less boldly patterned than brown morph. Has brown upperparts with some rufous tinge, crown spotted whitish to dull yellow and black, nape the same and with fine vermicular markings; mantle feathers have black-tipped dull yellowish to whitish spots. Outer webs of scapulars are dark-edged whitish to pale buff, and the flight and tail feathers are barred dark and light. Brown underparts finely darker-vermiculated, with fairly large whitish spots with black, often arrowhead-shaped lower edges, and unstreaked. The pale brown facial disc has some concentric lines, and a faint dark ruff. The eyes are yellow, and the bill yellowish-horn. Tarsi are partially feathered, with the distal third bare, and the dirty yellow toes have dark-tipped horn-coloured claws. *Juvenile* Has fine close barring on the crown, breast and wing-coverts. *In flight* Shows contrast between rufous coverts and secondaries and browner primaries.

Call Utters a phrase of several *wup* notes, some four or five per phrase, lasting about five seconds.

Food and hunting Practically unknown but eats caterpillars.

Habitat Found in forests and semi-open areas; also in gardens near human habitations, even entering bungalows.

Geographical variation Monotypic.

Status and distribution Endemic to Andaman Islands, in the Bay of Bengal. Little studied but probably rare.

Similar species Geographically overlapping Oriental Scops Owl (45) occurs in the Andamans as a winter visitor, but could also be breeding there. It has a blackish-streaked crown, boldly streaked underparts with no large white spots, and its voice is very different.

◄ The rufous morph of Andaman Scops Owl has a less vermiculated back and smaller pale spots on the underparts than the brown morph; the spots are more buffish and less white. December (*James Eaton*).

62. JAVAN SCOPS OWL
Otus angelinae

L 16–18cm, Wt 75–91g; W 135–149mm

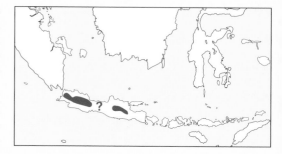

Identification A very small owl with rather long ear-tufts. No data on sexual size differences. Lighter and darker individuals are known, but no real morphs; darker individuals are more common. The prominent white ear-tuft feathers are often black-edged and with some transverse dark reddish-brown bars, and have obvious white inner webs, as well as dark brown to black outer webs with some paler spots. The dark brown or reddish-brown crown has dark feather centres, and there is often a prominent nuchal collar formed by a row of whitish or pale dull yellow feathers with bold black tips. The upperparts are dark reddish-brown with a scattering of pale vermicular markings and dark and pale rufous spots and freckles. The black-tipped and black-edged white outer webs of scapulars form a distinct line. Dark brown wings have about five fairly broad dull-yellowish cross-bars, and the dark reddish-brown tail has indistinct bars and mottling. Underparts are pale rufous or whitish to dull yellow with slightly darker, reddish-brown fine vermiculations; conspicuous black herringbone markings are visible on the breast sides and flanks. The uniformly reddish-brown facial disc, sometimes almost light chestnut-brown, contrasts the prominent white eyebrows, white extending along the forehead to the ear-tufts. The eyes are golden-yellow or orange-yellow, with reddish-brown eyelids. Black-tipped white bristly feathers are visible around the dark straw-yellow or pale greyish-yellow bill. Tarsi are fully

▼ Javan Scops Owl's dark brown morph is more common than the light morph. It is much darker, but still has distinct white supercilia and nuchal collar. Also has scapular spots, but inner webs of the primaries are unspotted. Gunung Gede, Java, June (*James Eaton*).

feathered up to or even well over the toe joint, and the flesh-coloured or pinkish feet (soles a little darker) have relatively long toes, with dark flesh-coloured claws browner near the tip. *Juvenile* Plumage is dark reddish-brown all over, with fine dark cross-bars on the crown, broader on the mantle, back and rump; below darker rufous, with some faint bars and a few broad shaft-stripes.

Call Said to be generally silent, and only when agitated does it produce an explosive *poo-poo* hoot, 0.6 seconds between the two notes, repeated a number of times at intervals of several seconds. Far less vocal than Mountain Scops Owl.

Food and hunting Prefers larger insects, such as beetles, grasshoppers and praying mantises. The prey is taken with the claws from a branch or from the ground. Not seen to hawk insects in flight.

Habitat Occupies humid primary forests with luxu-riant undergrowth from 900m to 2500m, but mainly at 1500–1600m.

Status and distribution Lives in the mountains of Java; unknown from C. Java, although it is likely to occur in entire mountain range of the island. Considered rare and Vulnerable by BirdLife International.

Geographical variation This owl has been regarded as monotypic but individuals in E. Java are perhaps larger (one specimen had wing 163mm) than those in W. Java. These two isolated populations could possibly, therefore, belong to different subspecies.

Similar species Geographically separated Mountain Scops Owl (43) is similar in size, with smaller ear-tufts and characteristic triangular bicoloured twinned spots on the breast and belly. Sympatric Sunda Scops Owl (49) is a little bigger, heavily mottled and vermiculated, with sharply contrasting black and dark brown mark-ings on a pale sandy or grey-brown background.

63. WALLACE'S SCOPS OWL
Otus silvicola

L 23–27cm, Wt 212g (1); W 202–231mm

Identification A small owl with rather long ear-tufts lacking white on inner border, and mottled dark brown and dull yellow. No data on sexual size differences. The upperparts are light grey-brown, with some darker vermiculations and black herringbone shaft-streaks on a reddish or dull yellowish-brown background; has no pale collar. Has ochre-white markings on the scapulars and dark brown and dull yellowish barring on the remiges. The brown tail has fairly indistinct narrow buffish bars. Underparts are whitish to dull yellow with sparse, bold blackish herringbone streaks and well-defined wavy dark brown cross-bars. The face is rather light in colour with whitish eyebrows, white not extending to the ear-tufts. The eyes are dull orange-yellow, and the bill and cere greyish-horn. Tarsi are densely feathered to the proximal phalanges of the toes, usually also covering second phalanges, and the rest of the toes are pale greyish-brown and bare, with horn-coloured claws. *Juvenile* Overall paler and more fluffy, and with ground colour more reddish-brown, and feather patterns fairly inconspicuous above and below, including ear-tufts.

Call Gives a series of deep *hwomph* notes, repeated 9–18 times.

Food and hunting Little studied, but is mainly insectivorous.

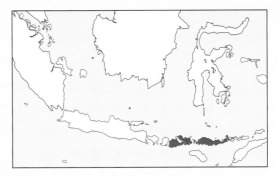

Habitat Lives in tropical forests, also secondary wood-lands and bamboo thickets, often not far from farms and even near cities. From lowlands to 2000m.

Status and distribution Occurs on Sumbawa and Flores, in the Lesser Sunda Islands. It is locally not rare but is in great need of further study.

Geographical variation Monotypic.

Similar species Geographically overlapping Flores Scops Owl (64) is much smaller, yellowish-brown with relatively small ear-tufts. Also sympatric, the Moluccan Scops Owl (68) is smaller, and has bright yellow eyes, small ear-tufts and more whitish lores. Allopatric, but also occurring in Lesser Sundas, Wetar Scops Owl (69) is clearly smaller, with sulphur-yellow eyes.

64. FLORES SCOPS OWL
Otus alfredi

L 19–21cm; W 137–160mm

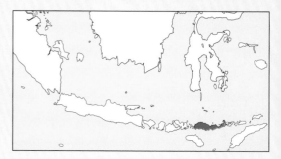

Identification A very small owl with small ear-tufts. No data on sexual size differences. Has a yellowish-brown crown with very fine pale vermicular markings, and almost uniformly foxy yellowish-brown upperparts. The ear-tufts are of the same colour as the forehead and crown. White outer webs of scapulars form a white band across the shoulder. Flight feathers are barred whitish-buff and yellowish-brown, with similar banding obscure on tail. Dirty white underparts are densely marked with indistinct yellowish-brown bars and fine vermiculations, but lack dark shaft-streaks; the neck and upper breast sides are sparsely black-flecked. The facial disc is cinnamon, the ruff mostly obscured by elongated auriculars. The eyes are yellow, with pinkish rims surrounded by a small dark area. The bill and cere are orange-yellow. Distal quarter of tarsus is bare, this and the toes being dull yellowish, and the claws yellowish-horn. *Juvenile* Almost uniformly pale reddish-brown with vague barring, but has more distinct tail-bands than adult.

Call Completely unknown; this species is believed, therefore, to be largely silent.

Food and hunting Not documented.

Habitat Lives in humid mountain forests above 1000m.

Status and distribution Found in Ruteng and Todo Mountains, in W. Flores, in the Lesser Sundas. Formerly known from three old specimens, but in 1995 another individual was collected, confirming the morphological differences compared with Moluccan Scops Owl. This species remains very little known and is likely to be extremely rare and endangered owing to habitat losses.

Geographical variation Monotypic.

Similar species Sympatric Wallace's Scops Owl (63) is clearly larger, and fairly dull, pale greyish-brown, with brownish barring and blackish shaft-streaks below. Also overlapping geographically, Moluccan Scops Owl (68) is more barred and streaked below, and not yellowish-brown above.

◄ Flores Scops Owl is a small rufous-cinnamon owl with small ear-tufts. Darker back with some cross-marks, and white scapulars with black margins. The belly is very white. Flores, July (*James Eaton*).

65. MINDANAO SCOPS OWL
Otus mirus

L 19–20cm, Wt 65g; W 127–132mm

Identification A very small owl with small ear-tufts. No data on sexual size differences. Has greyish-brown upperparts spotted brownish and black. Whitish colour on eyebrows, this continuing to near the tips of the ear-tufts. Rather large whitish areas on outer webs of the scapulars, but not forming a clear whitish row across the shoulder. Flight and tail feathers are barred light and dark; wingts are short and rounded. Dull yellowish to whitish-cream underparts have blackish spots and herringbone pattern of fine bars and streaks. The pale greyish-brown facial disc has concentric rings of blackish spots, with some dark bristles reaching beyond the ruff. The eyes are normally brown, but can sometimes be yellow, and the bill is dark greenish-grey and the cere greenish-yellow to greyish. Bare distal third of tarsi and the toes are light greyish to whitish-yellow, with greyish-brown claws. *Juvenile* Not described.

Call A melancholy and soft, double *pli-piooh* whistle is uttered in long series, at intervals of 10–15 seconds.

Food and hunting Not studied, but probably feeds on insects and other arthropods.

Habitat Inhabits mountain rainforests from 650m upwards, and usually more common above 1500m.

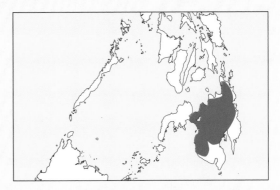

Status and distribution Endemic to Mindanao, in S. Philippines. This owl is apparently rare as a result of habitat destruction. It is listed as Vulnerable by BirdLife International. Further studies are urgently needed.

Geographical variation Monotypic.

Similar species The only *Otus* owl on Mindanao. Geographically separated Luzon Scops Owl (66) resembles it but is more reddish-brown and finely marked, with longer ear-tufts.

◀ Mindanao Scops Owl is a small dark owl with small ear-tufts and strong facial bristles extending over the ruff. Otherwise fairly similar to Mountain *Otus spilocephalus* and Javan *O. angelinae* Scops Owls, but voices are very different. Normally *O. mirus* has brown eyes, but sometimes (as in this photo) they are yellow. Mindanao, February (*Doug Wechsler*).

66. LUZON SCOPS OWL
Otus longicornis

L 19–20cm; W 136–152mm

Identification A very small owl with long ear-tufts. Females are said to be a little larger than males, although no weights are available. Has a complete white collar on nape, the feathers with blackish tips. The bright ochre-buff upperparts have blackish-brown streaks and irregular bars, mainly near the feather tips. The wings are blackish with black and rusty-brown mottling and speckling, and the tertials and tail have thin and indistinct bars. Whitish chin and white throat, with some black-tipped rufous feathers, contrasts with rich reddish-brown breast, mottled boldly with black and less so with white; the largely white belly and flanks are mottled with black and rusty-brown. The forehead is pale and the eyebrows white, but the ear-tufts are coloured like the head. The eyes are orange-yellow,

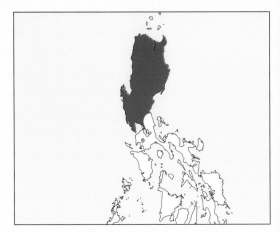

and the bill (slender and laterally compressed) is dingy dull green with dark brown tip and cutting edges; the dirty flesh-coloured cere becomes dull yellowish-green over the nostrils. Thin tarsi are bare for 10–11mm above toe joint, the bare areas and toes whitish-fleshy with slender but long grey claws. *Juvenile* Natal down is pure light grey. Mesoptile has brown-barred soft grey plumage, darker on head and upperparts.

Call Utters a melancholy but drawn-out whistle, *wheehuw wheehuw wheehuw*, at intervals of three to five seconds.

Food and hunting Little known, but feeds mainly on insects.

Habitat Occupies humid closed-canopy forests in foothills and pine forests on mountains at up to 2200m, but observed mainly between 700m and 1500m.

Status and distribution Endemic to Luzon, in N. Philippines. This species is not uncommon, but is becoming rarer as a result of habitat destruction. It is listed as Vulnerable by BirdLife International. Its biology and ecology need study.

Geographical variation Monotypic.

Similar species Geographically overlapping Philippine Scops Owl (56) is much larger and has orange-brown eyes, its nuchal collar contrasts clearly with the back, and its powerful legs and toes are well feathered. Allopatric Mindoro Scops Owl (67) is smaller, buffier below, with shorter ear-tufts.

◀ Luzon Scops Owl is a small, buffish-brown owl with rather long ear-tufts – here lowered almost flat. The breast is rufous with dark spots, the belly paler with rufous and dark pattern. Mount Data, Luzon, April (*Bram Demeulemeester*).

67. MINDORO SCOPS OWL
Otus mindorensis

L 18–19cm; W 133–136mm

Identification A very small owl with medium-length ear-tufts. No data on sexual size differences. Has a uniformly pale buff or whitish forehead and area above eyes. The reddish-brown crown has black shaft markings and spots, and a dull yellowish nuchal collar is narrow and nearly invisible. Brown mantle, back and rump have black shaft-streaks, forming irregular bars (due to short lateral branches). Mainly dull yellowish underparts are marked with thin white bars. The eyes are bright yellow, the bill greenish-yellow with dark brown cutting edges, and the cere dirty flesh. Tarsi are feathered for half their length, and the bare part of legs is whitish-fleshy, with long toes and grey claws. *Juvenile* Not described.

Call Gives a series of short *whoo* whistles at variable intervals of at least five seconds. Single whistle lasts for a little less than half a second.

Food and hunting Not studied.

Habitat Found in closed-canopy mountain forests above 870m.

Status and distribution Endemic to Mindoro, in C. Philippines. This owl is locally common, but suffers from habitat destruction. It is listed as Vulnerable by

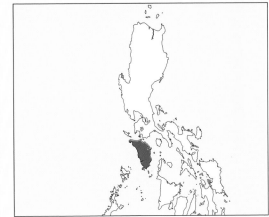

BirdLife International.

Geographical variation Monotypic.

Similar species Geographically separated Mindanao Scops Owl (65) is more or less similar in size, but normally has brown eyes and shorter ear-tufts. Allopatric Luzon Scops Owl (66) is larger and more clearly marked overall, with longer ear-tufts.

▼ Mindoro Scops Owl is similar to Luzon Scops Owl *Otus longicornis* in plumage but is smaller with short ear-tufts; more buffish-orange below with different-shaped markings, and lacking white belly. Specimen from BMNH (Tring), collected in Dulangan, Mindoro in January (*Nigel Redman*).

68. MOLUCCAN SCOPS OWL
Otus magicus

L 23–25cm, **Wt** 114–165g; **W** 153–192mm

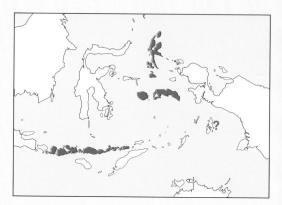

Identification A small owl with short but fairly prominent ear-tufts. No data on sexual size differences. Extremely wide variation in size and in ground colour. At least five morphs are known: more or less brown, yellowish-brown, reddish-brown, greyish-brown, and sepia-brown. Barring can vary from weak to very pronounced. Has bold dark mottling on upperparts, and some larger dull yellowish to white areas on outer webs of scapulars forming indistinct row across shoulder. Flight and tail feathers are barred light and dark; tail-bars are not distinct, and are interspersed with dense vermiculations. Some horizontal vermicular markings are visible on underparts, which have blackish shaft-streaks; the belly is paler than the breast. The facial disc is tinged reddish-brown, and a dark ruff around the disc is not very prominent; has whitish loral feathers. The eyes are yellow, and the bill and cere pale greyish-black. The bill and feet are strong. The tarsi are distally unfeathered to 6–9mm above the base of the greyish-brown toes, which have dark horn-coloured claws. *Juvenile* Like adult, but feathers are more fluffy, and has narrow dark brown cross-bars on head, crown and neck; all spots and freckles are fainter or indistinct.

Call A deep, rough, raven-like croak, *kwaark*, is uttered at intervals of several seconds.

Food and hunting Not well documented, but apparently feeds chiefly on insects and other arthropods. May occasionally take small vertebrates.

Habitat Found in lowland forests, secondary growth and mangrove swamps, also in fruit-tree gardens and coconut-palm plantations near human habitations and villages. From foothills to 1500m.

Status and distribution Occurs in Lesser Sundas, the

▼ Race *albiventris* is small with a dark back, and greyish-white below. Flores, Lesser Sundas, July (*James Eaton*).

▼ Race *albiventris* from Lombok looks more brownish than those on Flores. Lombok, Lesser Sundas, July (*Rob Hutchinson*).

Moluccas and possibly the Aru Islands. It is common on some islands but overall status not well known. At least some forms are known to occupy even heavily degraded habitats.

Geographical variation Moluccan forms are much larger than the Lesser Sunda forms, and six subspecies are listed: nominate *magicus* occurs on Seram and Ambon, in S. Moluccas; *O. m. bouruensis*, from Buru Island, in S. Moluccas, is rather uniformly buffy-brown above and white below; *O. m. morotensis*, from Morotai and Ternate, in N. Moluccas, is rather pale-faced and less spotted above; *O. m. leucospilus*, from Halmahera and Bacan, in N. Moluccas, is very like *morotensis* but generally much paler overall; *O. m. obira*, from Obi, in N. Moluccas, is small; *O. m. albiventris*, from Lombok, Sumbawa, Flores and Lomblen, in Lesser Sundas, is small, but has proportionately very long ear-tufts and a white belly. Taxonomic status of several forms of this owl is not yet clear; some insular populations may represent separate species, but more molecular material is required. Note that Wetar (69) and Sula Scops Owls (70) were previously treated as races of this species.

Similar species Also living in Lesser Sundas, the Wetar Scops Owl (69) is clearly smaller and has the tarsi feathered to the base of the toes. Allopatric Sulawesi Scops Owl (72) is smaller, with weaker bill and feet, and likewise geographically separated Sangihe Scops Owl (73) is smaller, with shorter ear-tufts.

▲ This is the nominate form of Moluccan Scops Owl. Seram, September (*Rob Hutchinson*).

▶ Moluccan Scops Owl shoes a lot of variation in ground colour. This is race *leucospilus*. Halmahera, N. Moluccas, October (*Franz Steinhauser*).

69. WETAR SCOPS OWL
Otus tempestatis

L 19–20cm; **W** 150–171mm

Identification A very small owl with relatively short and rounded ear-tufts. No data on sexual size differences. Two morphs exist. The commoner, the rufous morph, is fox-red, heavily streaked and dark-vermiculated above, especially on crown; the breast is pale cinnamon, with bold orange vermiculations and dark blotchy shaft-streaks, and the belly is white with cinnamon and blackish mottling. The eyes are sulphur-yellow, and the bill blackish to dark horn-coloured with pale lower mandible. Tarsi are totally feathered to the base of the greyish-flesh toes, which have dark claws. Grey morph is generally grey or brownish-grey, finely marked overall with light spots and dark streaks. *Juvenile* Undescribed.
Call Unknown.

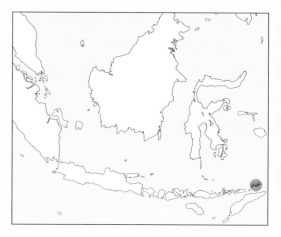

▼ Wetar Scops Owl's reddish-brown morph has bright fox-red plumage, with narrow black shaft-streaks and vermiculations above. The breast is pale cinnamon with blotchy shaft-streaks. Wetar, July (*James Eaton*).

Food and hunting Not documented.

Habitat Inhabits primary and secondary forests, including swampy areas and plantations with trees.

Status and distribution Endemic on Wetar Island, in Lesser Sundas. Nothing is known about this species' biology and conservation needs, and even its taxonomic status is not yet officially clarified.

Geographical variation Monotypic. The holotype specimen of this owl was wrongly described as a subspecies of the Sulawesi Scops Owl (72), and later it was (without research) connected with Moluccan Scops Owl (68). Bioacoustic and genetic evidence, as well as photographs, confirm the status of this owl as a full species.

Similar species Allopatric Moluccan Scops Owl (68) is larger and more distinctly streaked below, with only partly feathered tarsi (7–9mm remaining bare). Also allopatric, the Sulawesi Scops Owl (72) is indistinctly dark-marked above, with more mottled and vermiculated (rather than streaked) crown and forehead.

▼ Wetar Scops Owl is more uniform below than *albiventris* Moluccan Scops Owl *O. magicus*. This is a grey morph with little reddish-brown below. Wetar, July (*Filip Verbelen*).

70. SULA SCOPS OWL
Otus sulaensis

L *c.* 20cm; W 161–175mm

Identification A very small owl with relatively small and rounded ear-tufts. No data on sexual size differences. The crown is very heavily streaked and mottled brownish-black, and the dark brown upperparts are distinctly dark-streaked with whitish cross-bars and light reddish-brown spots. The scapulars have white spots with irregular dark markings. Sepia-brown primaries are uniformly dark on the inner webs, with pale buffish flecks only on the outer webs. Underparts are relatively pale, with blackish and whitish barring and prominent dark shaft-streaks. The pale greyish-brown facial disc has dark, and so not very distinct, concentric lines; a dark ruff is very narrow near the ear-tufts, becoming broader and more visible towards the throat. The eyes are yellow-orange, and emphasised by whitish eyebrows, and the bill is blackish-horn, with ochre

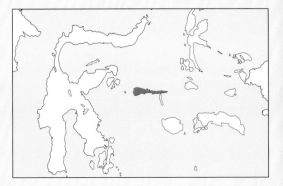

lower mandible and cere. Frontal upper half of tarsus is feathered, with bare lower half; rear of tarsus is totally bare and dirty yellowish, and the claws are dark horn-coloured. *Juvenile* Said to be similar to adult.

Call Emits a rather rapid series of resonant notes, medium-pitched and each a little more than one second in duration. Whole call series can last for up to two minutes.

Food and hunting Not well documented, but feeds mainly on insects and other invertebrates, with some small vertebrates.

Habitat Prefers lowland forests and secondary growth, and occurs also in swampy areas.

Status and distribution Endemic to Taliabu and Mangole Islands, in Sula Archipelago, east of Sulawesi. Its status is unclear but it could be vulnerable. Formerly only two skins of adult owls and one of a juvenile were known but we here include a photo from the wild.

Geographical variation Monotypic. Formerly treated as a subspecies of Sulawesi Scops Owl (72) and later connected with Moluccan Scops Owl (68). Here provisionally accorded full species status but confirmation requires bioacoustic and molecular studies.

Similar species Allopatric Moluccan Scops Owl (68) is larger, with more prominent ear-tufts, barred inner webs of primaries and a less distinct ruff. Also geographically separated, Sulawesi Scops Owl (72) has longer ear-tufts, prominent white eyebrows and clear yellow eyes.

◄ Sula Scops Owl is dark brown above with distinct shaft-streaks and paler rufous-brown spots. Distinctly larger and with shorter ear-tufts than Sulawesi Scops Owl. Taliabu, November (*Bram Demeulemeester*).

71. BIAK SCOPS OWL
Otus beccarii

L 23cm; W 170–172mm

Identification A small owl with moderately long ear-tufts finely dusky-barred. No data on sexual size differences. Two morphs exist. Dark morph has blackish barring all over, alternating with white. Crown is finely blackish-barred, with a few fine streaks on forecrown, and nape has blackish and white or pale buffish barring forming an indistinct collar. Brown upperparts are rather densely barred and mottled blackish and whitish. Outer webs of scapulars are mainly white, with a few black bars and black fringes, forming a row of whitish spots across the shoulder. Flight and tail feathers are barred light and dusky. The white throat is finely vermiculated with blackish. Brownish underparts are practically devoid of shaft-streaks, but densely barred blackish and dull yellow. Light brown facial disc has an indistinct ruff formed by a blackish edge and whitish spots. The eyes are yellow, and the bill blackish with paler cere. Tarsi are feathered nearly to the base of the dirty yellowish toes, which have dusky horn-coloured claws. Rufous-brown morph is in general yellowish-brown and reddish-brown with white barring below, and finely vermiculated and light yellowish-brown above. *Juvenile* Unknown.

Call Utters a sequence of hoarse, corvid-like croaking notes, lower in pitch than those of Moluccan Scops Owl.

Food and hunting Feeds mainly on insects and spiders, but occasionally takes small vertebrates.

Habitat Occupies dense forests and wooded areas but sometimes found near human settlements.

Status and distribution Endemic to Biak Island, in Geelvink Bay, off Irian Jaya, NW New Guinea. This owl is known only from three specimens and a few sight records. Most of the forest on Biak has been logged, and no good-quality forest was left on the island in June 1987. This species is listed as Endangered by BirdLife International. Fortunately, Biak is 'twinned' with another island, Pulau Supiori, which is believed to serve as a refuge for this owl; in 1982, local villagers gave the impression that an owl of this kind had been seen on Supiori from near sea level to 300m. It is apparently not known for certain if the species still exists (or ever has existed) on Supiori.

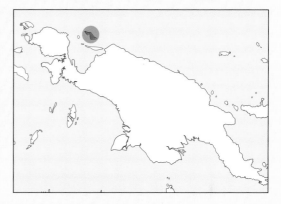

Geographical variation Monotypic.

Similar species Allopatric Moluccan Scops Owl (68) is distinctly dark-streaked above and below. Also geographically separated, the Mantanani Scops Owl (74) is smaller, with a more prominent ruff, and has underparts marked with shaft-streaks.

▶ Biak Scops Owl is extremely poorly known. It has yellow eyes and a blackish bill; the underparts are densely barred. Outer webs of scapulars are mainly white. Flight and tail feathers black-and-whitish or rufous-barred. Biak, May (*Rob Hutchinson*).

72. SULAWESI SCOPS OWL
Otus manadensis

L 19–22cm, Wt 83–93g; W 140–161mm

Identification A small owl with medium-sized but prominent ear-tufts. No data on sexual size differences. There are two morphs, a yellowish-grey one and a very scarce rufous one, apparently with no intergrades. Has rich reddish-brown and dark grey upperparts with dense but irregular dark brown and sepia-brown freckles, mottles and shaft-streaks; banded pattern on interscapular feathers, with dark shaft-streaks and paired buffish ocelli. Flight feathers have broad dull yellow and dark brown barring, the light barring on inner webs of primaries poorly defined. Below, has more reddish-brown breast and boldly but sparsely streaked and barred lower underparts on white background. The brownish or sepia-coloured facial disc is bordered by black-tipped feathers, and there is an often incomplete whitish supercilium ending above the yellow eyes. The lores and chin are whitish. The bill is blackish. Tarsi are feathered down to the yellowish-grey toes, which have horn-coloured claws. Feet and claws are relatively weak. *Juvenile* Has fine dark brown bars on crown and hindneck, otherwise resembles the adult.

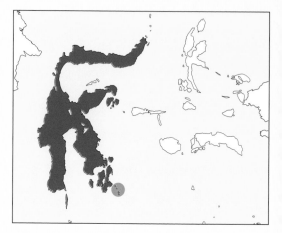

Call Clear, upward-inflected *ooeehk* notes, lasting some 0.4 seconds, are repeated at intervals of about six seconds.

Food and hunting Not studied, but probably takes insects and other arthropods, as well as some small vertebrates.

Habitat Found in humid forests, wooded areas and cultivations from the lowlands up to 2500m.

Status and distribution Occurs on Sulawesi, Peleng, Labobo and Tukangbesi Islands, in Indonesia. This owl is locally common, despite the rampant forest destruction in the region.

Geographical variation Three subspecies are listed: nominate *manadensis* is confined to Sulawesi; *O. m. mendeni*, from Peleng, is very finely speckled above and finely vermiculated below, with black and rufous streaks; slightly larger *O. m. kalidupae* occurs on Tukangbesi Islands. Further studies are required to clarify the taxonomy of these subspecies; it has been proposed that both *mendeni* and *kalidupae* be given full species status, as 'Banggai Scops Owl' and 'Kalidupa Scops Owl', respectively, although not much is known of their ecology, biology, vocalisations and DNA relationships. The entire group of *Otus* species described from the Indonesian Archipelago certainly requires further taxonomic study. There is a good chance that

◀ Nominate Sulawesi Scops Owl has a very dark back with large white scapular spots. Tangkoko, Sulawesi (*Jérôme Micheletta*).

further new species could be found in Indonesia, as indicated by the presence of an unidentified scops owl on Togian Island, off C. Sulawesi.

Similar species Allopatric Moluccan Scops Owl (68) is much larger, with more prominently banded uppertail and tertials. Narrowly separated Sangihe Scops Owl (73) is drabber and more finely marked, with longer and narrower wings and paler cheeks, but darker patches between bill and eyes.

▶ Nominate Sulawesi Scops Owl is much greyer below than *mendeni*. Tangkoko, Sulawesi, July (*Rob Hutchinson*).

▼ Race *mendeni* is distinctly smaller than other races, with smaller feet, tarsi feathered nearly to base of toes, and primaries less distinctly banded on the outer webs; also vocally different. Peleng, March (*Filip Verbelen*).

73. SANGIHE SCOPS OWL
Otus collari

L 19–20cm, Wt 76g (1♂); W 158–166mm

Identification A very small owl having medium-sized ear-tufts with dull yellow spots, black streaks and elliptical tips. No data on sexual size differences. Has drab brownish upperparts with dark shaft-streaks, distinct vermicular markings and prominent yellowish spotting. Pale buff outer webs of scapulars have triangular blackish tips, forming a light and dark row across the shoulder. Flight feathers are dark brown and dull yellow, the tertials less distinctly banded. There are narrow, irregular buff bars and wider dark brown bands on the tail feathers. Throat and underparts are pale, with fine, long shaft-streaks and some cross-bars. The light-coloured facial disc becomes darker between the pale yellow eyes and the bill, the eyelids are dark-rimmed, and short whitish eyebrows meet above the brownish-horn bill. A fairly well-pronounced ruff is formed by light and dark feather tips. Tarsi are feathered to the base of the light brownish-grey toes, which have dark-tipped pale brown claws. The talons and toes are relatively weak. *Juvenile* Unknown.
Call A single note, *peeyuuwit*, lasting 0.7 seconds, is repeated at intervals of 0.3–11 seconds.
Food and hunting Not studied, but diet is most likely to consist mainly of insects.

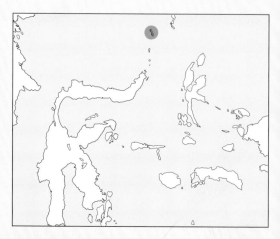

Habitat Found in forests and agricultural land with trees, from sea level to 315m.
Status and distribution Endemic on Sangihe Island, off N. Sulawesi. This species is clearly rather common, but it suffers from habitat destruction.
Geographical variation Monotypic.
Similar species Fairly similar to Sulawesi Scops Owl (72), which has longer eyebrows, but shorter wings.

◀ Sangihe Scops Owl is similar to Sulawesi Scops Owl *Otus manadensis* but with shorter ear-tufts and more contrasting forehead, and a slightly longer tail. Above it is more coarsely vermiculated, with darker lesser wing-coverts, and the primaries have less contrasting bands. Sangihe, Indonesia, November (*Filip Verbelen*).

74. MANTANANI SCOPS OWL
Otus mantananensis

L 18–20cm, Wt 106–110g; W 152–180mm

Identification A very small owl with irregularly marked, prominent ear-tufts. No data on sexual size differences. Two morphs exist, and populations become less rufous from north to south. Greyish-brown morph has greyish-brown upperparts with black pattern of freckles and mottling, and whitish outer webs of scapulars forming row across shoulder; flight and tail feathers are barred light and dark. Pale underparts are peppered black, the breast being darker than the belly. There is a conspicuous dark ruff around the facial disc. The eyes are yellow and the bill greyish-horn. Heavily feathered tarsus with a narrow bare zone above the light greyish-brown toes, which have dark horn-coloured claws. Rufous morph is in general more reddish-brown. *Juvenile* Unknown.

Call A series of deep, nasal, grunting *kwoank, kwoank* notes is uttered normally at intervals of 5–6 seconds.

Food and hunting Feeds mainly on insects and other

▼ A pair of nominate Mantanani Scops Owls. Both adults have very fine pattern above and are mottled dark brown, white and black; there is strong black streaking on the upper breast, with the belly much paler. Rasa, Philippines, February (*Ian Merrill*).

arthropods, occasionally taking small vertebrates. Hunts along forest edges and in clearings.

Habitat Inhabits wooded areas and forests, also coconut groves and casuarina plantations.

Status and distribution Occurs on Mantanani Island, off NW Borneo, and on some islands of WC & SW Philippines. It is locally common, but suffers from forest destruction. Very little studied, including its taxonomy and vocalisations.

Geographical variation Four subspecies are listed: nominate *mantananensis* is found on Mantanani (off Borneo), and on islands of Rasa and Ursula, off S. Palawan; *O. m. romblonis*, from Banton, Romblon, Tablas, Sibuyan, Tres Reyes and Semirara, in WC Philippines, is heavily marked below; *O. m. cuyensis*, from Cuyo and Calamian, in WC Philippines, is large with conspicuous black streaking; *O. m. sibutuensis*, from Sibutu and Tumindao, in Sulu Archipelago of SW Philippines, is usually dull brown with irregular, subdued markings and with reduced pale colour on scapulars. Amount of rufous increases from south to north in these subspecies.

Similar species Philippine (56) and Palawan Scops Owls (57) do not occur on the same small islands inhabited by this owl, and both are darker with long ear-tufts and conspicuous hindneck collar.

► Mantanani Scops Owl of the nominate race; this is a greyish morph with orange-yellow eyes and fairly short ear-tufts. Note whitish throat and some darkish spots on the upper breast. Mantanani, Philippines, January (*James Eaton*).

◄ One adult in the pair is a little more brownish-grey and reddish. Mantanani Scops Owl also has a rufous-brown morph. Rasa, Philippines, February (*Ian Merrill*).

75. FLAMMULATED OWL
Philoscops flammeolus

L 16–18cm, **Wt** 45–63g; **W** 126–148mm

Identification A very small owl with short, erectile ear-tufts. Female is on average only 3g heavier than male. Yellowish-brown toned red and greyish morphs exist, but also intermediates. Greyish morph is more common in the north. Has cryptic greyish-brown upperparts with fine blackish mottles and shaft-streaks. Large rusty-tinged areas on scapulars form an orange-buffish scapular stripe, and large reddish-brown outer webs of scapulars form a distinct row across the shoulder. The contrasting lighter and darker bars on the flight feathers are less prominent on secondaries. The tail has about four similar lighter and darker bars. Underparts are greyish-brown with lighter and darker mottling, and with rusty spots and blackish shaft-streaks. Greyish-brown facial disc is washed with pale chestnut to blackish-brown, and there are rather indistinct whitish eyebrows above the large, dark brown eyes. The relatively weak bill is greyish-brown, as is the cere, and the legs, which are feathered to the base of the toes and have blackish-brown claws. *Juvenile* Downy

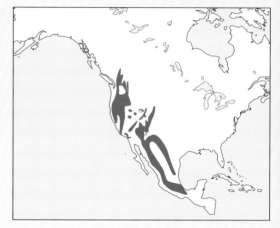

chick is whitish. Mesoptile resembles adult, but has dark barring below, and on crown, nape and back, and rusty-tinged facial disc.

Call A fairly deep, short call, *hoop*, is repeated at intervals of 2–3 seconds.

Food and hunting Takes almost exclusively arthropods, mainly nocturnal insects such as moths. Often hawks moths above the canopy of trees, seizing prey with the bill.

Habitat Prefers mountain forests with ponderosa pine, Douglas fir or mixed oak and aspen, between 400m and 3000m.

Status and distribution Found from Canadian British Columbia in north to Guatemala and El Salvador in south. This owl is still rather common in the south, but northern populations are obviously declining. Studies are required in order to ascertain the reasons for this decline.

Geographical variation Monotypic.

Similar species Sympatric Western Screech Owl (76) and geographically separated Eastern Screech Owl (77) are much larger, with yellow eyes, and lack rusty areas on scapulars.

▶ Flammulated Owl is a small owl with a lot of individual variation in plumage. Rufous and greyish morphs occur, with intermediates. In the north greyish morphs predominate. Colorado, July (*Paul Bannick*).

◀ A greyish-brown Flammulated Owl with a grasshopper as prey. Colorado, July (*Paul Bannick*).

76. WESTERN SCREECH OWL
Megascops kennicottii

L 21–24cm, Wt 87–250g; W 142–190mm

Identification A small owl with short ear-tufts, prominent only when erected. Females are on average over 30g heavier than males. Two colour morphs exist: brown and grey. Brown morph is rare, but more common in the north. Grey morph has brownish-grey upperparts and crown with blackish shaft-streaks and fine vermicular markings. Black-edged whitish outer webs of scapulars, form a white line of spots across the shoulder. Flight feathers are strongly barred light and dark, tail less distinctly barred. The pale underparts have blackish shaft-streaks and irregular cross-barring; some broad spot-like black shaft-streaks on upper breast. Light brownish-grey facial disc is vermiculated darker and finely mottled, with fairly prominent dark brown ruff edged with pale speckles. Not very distinct eyebrows are slightly paler than surrounding area. The eyes are bright yellow, with dark brown eyelids. The bill and cere are blackish, bill with horn-coloured tip and with greyish-brown bristles at the base. Tarsi are feathered to the base of the greyish-brown toes, which have blackish-horn claws. Brown morph is more rufous in general. *Juvenile* Downy chick is whitish. Mesoptile has densely barred head, mantle and underparts.

Call Utters a short trill, *urrr*, followed immediately by a longer one, *uhrrrrrrrrrrr*, which initially rises in pitch, but drops again at the end.

Food and hunting Diet is mostly insects and other arthropods but also takes small mammals, birds, frogs and reptiles; winter food is mainly small mammals and birds. Hunts from a perch but also flies after prey in the air.

Habitat Inhabits riparian woodlands, mesquites, pine and oak forests, including semi-open areas, even cactus

◄ A pale grey-morph *aikeni* Western Screech Owl, with a beetle grub on the ground. Arizona, July (*Jim and Deva Burns*).

deserts, but also gardens and parks in suburban areas. From lowlands to 2500m.

Status and distribution Occurs from northern Canada and Alaska south to central Mexico. This is a locally rather common species which benefits from nestboxes. Occasional interbreeding with Eastern Screech Owl is known, but there is no regular zone of hybridisation; the two interbreed in marginal habitats with low owl densities (as in an area of small mesquites and willows bordering a river crossing the desert in the Rio Grande area).

Geographical variation Eight subspecies have been separated: nominate *kennicottii* occurs from S. Alaska to coastal Oregon, also Vancouver Island; paler and longer-winged *M. k. bendirei* breeds from Washington and Idaho south to S. California and east to Montana and Wyoming; palest grey *M. k. aikeni* is found from SW USA to NC Sonora, in Mexico; *M. k. yumanensis*, from Colorado Desert and Baja California to NW Sonora, is pale pinkish-grey; rather dark *M. k. cardonensis* breeds in S. California and Pacific slope of Lower California; pale *M. k. xanthusi*, from Baja California, is small; somewhat smaller and paler *M. k. vinaceus* breeds from C. Sonora to Sinaloa, in Mexico; *M. k. suttoni*, from Rio Grande, Texas, to Mexican Plateau, is the darkest race. Many of these subspecies are known from only a

few specimens and could be morphs, hybrids or simply examples of individual variation, and taxonomic review is needed.

Similar species Geographically overlapping Flammulated Owl (75) is much smaller, with dark brown eyes and rufous markings in the plumage. Only marginally sympatric Eastern Screech Owl (77) is very similar, but has different call and a greenish bill. Rather common Eastern red morph is easy to separate. More sympatric Whiskered Screech Owl (80) is much smaller and more coarsely streaked below, with a greyish bill and relatively small feet.

▼ Western Screech Owl nominate *kennicottii* from California is coarsely and boldly patterned in blackish- to buffish-fuscous above. Bill is black to grey. California, March (*Mike Danzenbaker*).

▼ Female and a large young Western Screech Owl of race *aikeni*. Pale grey above with broad black streaks; broad streaks and widely spaced cross-bars below. Black bill. Arizona, June (*Jim and Deva Burns*).

▲ A pair of nominate Western Screech Owls at the nest. California, March (*Mike Danzenbaker*).

▼ Western Screech Owl in flight with snake prey. Green Valley, Arizona (*Tom Vezo*).

77. EASTERN SCREECH OWL
Megascops asio

L 18–23cm, **Wt** 125–250g; **W** 145–175mm

Identification A very small to small owl with prominent ear-tufts. Female is on average some 30g heavier than male. Three morphs and intermediates exist; greyish-brown and grey morphs predominate in the north, and red morphs in the southern parts of the range. Greyish-brown morph has these colours on upperparts, with blackish shaft-streaks and thin, dark vermicular markings. Blackish-edged whitish outer webs of scapulars form a white line of spots across the shoulder. Flight feathers are barred light and dark, and the greyish-brown tail has several narrow pale bars. Whitish underparts have a few darker markings, these forming prominent longitudinal lines and very faint bars. Light greyish-brown facial disc is finely mottled or vermiculated darker, and a blackish-brown ruff around disc is much broader on basal half of both sides. Has pale eyebrows and bright yellow eyes, and light greyish-brown whiskers at base of olive-greenish cere and bill, bill with a yellow tip. Tarsi are feathered to the base of the greyish-brown toes, which have yellowish horn-coloured, black-tipped claws. Grey and red morphs are

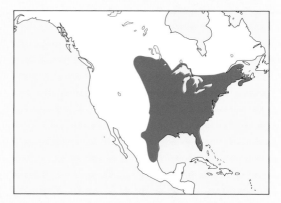

overall grey or reddish, respectively. *Juvenile* Downy chick is whitish. Mesoptile resembles adult, but is thinly barred light and dark on head, mantle and below. *In flight* Shows pointed wings and plain dorsal plumage.
Call A toad-like, quavering trill, *gurrrrrrrrrrrrrt*, lasting some 3–5 seconds, ends abruptly, but is repeated after a few seconds.

▶ Eastern Screech-Owl race *hasbroucki* has dense black streaks and bars below and very white shoulders. Uvalde County, Texas, USA (*Rolf Nussbaumer*).

Food and hunting Feeds mainly on insects and other arthropods, but also small vertebrates, especially in north and during cold periods. Catches insects on the ground or in tree foliage.

Habitat Occupies mixed deciduous or coniferous forests and open woodlands, even suburban gardens and parks. From sea level to 1800m.

Status and distribution Found from E Montana and the Great Lakes of USA south to NE Mexico and from SE Canada to Florida. Overlaps with Western Screech Owl near the US–Mexican border, where hybridisation has been witnessed in Rio Grande area. Some instances of interbreeding also with Whiskered Screech Owl in S. Arizona and adjacent Mexico, but the latter species generally lives at higher elevations (mostly above 1600m).

Geographical variation Six subspecies are listed: nominate *asio* occurs in South Carolina, Georgia, Virginia and E. Oklahoma; pale grey *M. a. maxwelliae* is found west of the Great Lakes in NW USA; *M. a. naevius*, from S. Ontario to NE USA, is very white below, especially on belly; *M. a. floridanus*, from Florida to Louisiana and Arkansas, occurs mainly in red morphs; *M. a. hasbroucki*, from C. Kansas to Oklahoma and Texas, has broad lateral markings on underparts; *M. a. mccalli*, from Lower Rio Grande to NE Mexico, has grey morph much more dark-mottled above and red morph paler.

Similar species Largely geographically separated Flammulated Owl (75) is much smaller and lives normally at higher altitudes; has dark brown eyes and rufous markings on greyish-brown background. Partly sympatric Western Screech Owl (76) is very similar, but has a blackish bill and more coarsely marked underparts. Also partly overlapping in range, Whiskered Screech Owl (80) is smaller and more streaked below, with much smaller talons.

◄ Eastern Screech Owl race *floridanus* has grey and red morphs. Extreme red morphs are similar to the red morph of nominate *asio*. Florida, December (*Paul Bannick*).

Eastern Screech Owl race *aevius* is very white below, especially on the belly, with prominent dark lines and cross-ars. Campbellville, Ontario, Canada (*Scott Linstead*).

▼ Eastern Screech Owl race *maxwelliae* is the palest and east marked subspecies. This male is a grey morph. Manitoba, July (*Christian Artuso*).

78. PACIFIC SCREECH OWL
Megascops cooperi

L 23–26cm, **Wt** 145–170g; **W** 163–183mm

Identification A small owl with rather prominent ear-tufts. Females are on average about 20g heavier than males. No red morph is known. The upperparts are pale greyish-brown with dark mottling and streaks, the crown with fine dark shaft-streaks and coarse dark barring. Blackish-edged whitish outer webs on scapulars form a white band across the shoulder, and whitish-edged wing-coverts produce a second pale band across the closed wing. The primaries are barred light and dark, and the tail and secondaries are less prominently barred. Pale underparts have thin

blackish shaft-streaks and dark vermicular markings, creating herringbone pattern. Light greyish facial disc has fine darker vermiculations and a narrow blackish ruff. The eyes are yellow, with brownish-pink eyelids, and the bill and cere greenish. Tarsi are feathered to the base of the brownish-fleshy toes, which have dark horn-coloured claws. *Juvenile* Undescribed.

Call Utters relatively gruff notes in a rapid, trilled sequence, *u-pu-pu-pu-pu...*, increasing in loudness and becoming slower in the middle; a typical phrase has twelve notes and lasts about 1.5 seconds.

Food and hunting Feeds mostly on insects and other arthropods, but has powerful talons to take small vertebrate prey, too. Hunting not studied.

Habitat Lives in arid to semi-arid woodlands with cacti and palms, and also in mangrove forests. Mainly in lowlands below 330m, but sometimes up to 1000m.

Status and distribution Occurs in Oaxaca and Chiapas, in Mexico, south on Pacific slope to Costa Rica. This is a little-known species and its status is uncertain.

Geographical variation Monotypic. Molecular and biological studies should confirm this owl's taxonomic status, and possible interbreeding with Oaxaca Screech Owl.

Similar species Partly overlapping Oaxaca Screech Owl (79) is smaller, with dark-barred and dark-streaked crown. More sympatric Whiskered Screech Owl (80) is much smaller, with dark-streaked crown and weak talons.

◀ Pacific Screech Owl is a relatively large tawny-grey owl (no red morph is known). It has a fine pattern and narrow stripes above, and ventral cross-bars with very small freckles and broken vermiculations. The eyes are yellow. Note the powerful claws. Costa Rica, March (*Mike Danzenbaker*).

► Pacific Screech Owl eats crickets and other insects, but the powerful talons suggest that vertebrate prey is also taken. Costa Rica (*Fabio Olmos*).

79. OAXACA SCREECH OWL
Megascops lambi

L 20–22cm, **Wt** 115–130g; **W** 148–166mm

Identification A small owl with short but prominent ear-tufts. Female is on average some 20g heavier than the male. Has wine-reddish upperparts and dark crown, contrasting with frosty areas around face and hindneck. Relatively rounded wings (due to reduced outer six primaries). A strong reddish wash is evident on underparts, with herringbone patterns on single feathers mixed with suffused vermicular markings. The greyish facial disc is bordered by a rather strong, dark ruff. The eyes are yellow and the bill and cere olive-green to brownish-olive, bill with yellowish tip. Tarsi are feathered to the base of the pale greyish-brown toes, which have dark horn-coloured claws with blackish tips. *Juvenile* Unknown.

Call Utters a guttural *croarrr* followed by a staccato *gogogogogogok*.

Food and hunting Little known, but diet and behaviour are most likely similar to those of most screech owls.

Habitat Inhabits arid woodland with cacti and palms, but often near coastal mangrove swamps. From sea level to about 1000m.

Status and distribution Endemic in Oaxaca, in SE Mexico. As this species is very little studied, its status

remains unknown. It is not known if this species hybridises with Pacific Screech Owl in areas where the two overlap in range.

Geographical variation Monotypic. The taxonomic status requires clarification.

Similar species Allopatric Western Screech Owl (76) has a blackish bill. Also geographically separated, the Eastern Screech Owl (77) is very similar in size and coloration, but its toes are more feathered than bristled. Partly overlapping Pacific Screech Owl (78) is a little larger, with stronger talons, and said to be vocally different.

80. WHISKERED SCREECH OWL
Megascops trichopsis

L 17–19cm, Wt 70–121g; W 132–160cm

Identification A very small owl with fairly small ear-tufts, prominent only when fully erected. Female is on average about 20g heavier than male in highlands, but only 10g in areas farther south. Two morphs have been separated, but the ground colour also changes gradually as a result of climatic conditions; light grey is more frequent in arid areas and in high-altitude cold areas, and in high humidity plumage becomes increasingly reddish-brown toward the southern limits of the range. Grey morph has greyish to brownish-grey crown and upperparts with blackish shaft-streaks, these having dark horizontal branches and fine vermicular markings. Blackish-edged white outer webs on scapulars form a white line across the shoulder. Wings and tail are barred light and dark. The pale underparts are rather densely vermiculated, with broad blackish shaft-streaks, particularly on upper breast. Light greyish facial disc has fine darker concentric lines around relatively large, yellow eyes, and a prominent blackish ruff.

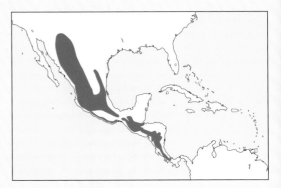

The bill and cere are dark grey. Tarsi are feathered to the base of the greyish-brown toes, with greyish-horn claws; talons are relatively weak. Red morph is less prominently patterned, but reddish-brown all over. *Juvenile* Said to be similar to other screech owls.
Call Voice resembles that of Tengmalm's Owl (Boreal

▼ Whiskered Screech Owl race *aspersus* has only grey morphs; these have a pale grey ground colour, with broad black shaft-streaks and medium ventral cross-bars. Arizona, June (*Jim and Deva Burns*).

▼ Whiskered Screech Owl race *mesamericanus* is more often red or rufous than the two other subspecies to the north. Guatemala, December (*Knut Eisermann*).

Owl): *bubúbububbub*, with emphasis on third note, and falling in pitch at the end. Some five to nine notes are uttered per second, and usually 4–16 notes per series.

Food and hunting Feeds entirely on insects and other arthropods; only once reported to catch a mouse. Hunts rarely from a perch, but more generally in flight among the branches.

Habitat Occupies pine and oak woodlands and mountain forests between 750m and 2500m, mostly above 1600m.

Status and distribution Occurs from SE Arizona, USA, to NC Nicaragua. It is said to be locally frequent, but is not well studied. Has hybridised with Western Screech Owl in Arizona, where it is found locally together with Flammulated Owl and Western Screech Owl in the same habitat.

Geographical variation Three subspecies have been separated: nominate *trichopsis* lives in the highlands of C. Mexico; *M. t. aspersus*, from SE Arizona to N. Mexico, occurs almost entirely as all-grey morphs; *M. t. mesamericanus* is found from S. Mexico to El Salvador and NC Nicaragua, where a red morph is fairly common.

Similar species Partly sympatric Flammulated Owl (75) is similar in size, but has a flammulated appearance and brown eyes. Also partly overlapping in range, Western Screech Owl (76) is larger and has stronger feet and blackish bill. Geographically overlapping Bearded Screech Owl (81) also lives in mountain forests, but has coarsely scalloped underparts, a much darker face, greenish bill and bare pinkish toes.

▼ Female (left) and male *aspersus* Whiskered Screech Owls. This species has yellow eyes (those of Flammulated Owl *Philoscops flammeolus* are brown). Bearded Screech Owl *Megascops barbarus* has a much darker face and coarsely scalloped underparts. Arizona, July (*Jim and Deva Burns*).

81. BEARDED SCREECH OWL
Megascops barbarus

L 17–18cm, Wt 58–79g; W 126–145mm

Identification A very small owl with short ear-tufts, which has also been called Bridled Screech Owl. Female is on average 9g heavier than male. Grey and red morphs are known. Grey morph has greyish-brown upperparts with whitish and dull yellowish spots, also with some darker markings. A coronal rim is formed by rounded whitish and buffish dots on the dark greyish-brown crown. Blackish-edged whitish outer webs of scapulars form a whitish line across the shoulder. Wings are relatively long, extending far beyond the tail, and are barred light and dark. The fairly short tail is similarly barred pale and dark. Light greyish-brown to dirty whitish underparts have dark shaft-streaks, irregular lateral branches, and rounded white spots, especially on upper breast, giving an ocellated appearance. The pale greyish-brown facial disc has darker concentric lines around the yellow eyes, and a dark brown or blackish-brown ruff; the whitish eyebrows are speckled greyish-brown. The bill and cere are greenish. Tarsi are feathered to the base of the bright pinkish, bare toes, which have horn-coloured claws. Red morph

is less clearly marked, with suffused patterns on a reddish background. *Juvenile* Mesoptile has remaining down on side of crown and below, ear-tufts not visible and eyebrows narrow and indistinct.

Call Utters a cricket-like, strident trill, *treerrrrrrrrrrt*, increasing in volume and stopping abruptly. Each calls lasts for about 3–5 seconds, and calls are repeated at intervals of several seconds.

Food and hunting Takes only insects and other arthropods, including spiders and scorpions. Hunts in the understorey, using a sit-and-wait strategy.

Habitat Found mainly in humid habitats of pine–oak and oak forest and cloud forest, locally upwards from 1350m, but mostly between 1800m and 2500m.

Status and distribution Occurs from Chiapas, Mexico, to C. Guatemala. Habitat degradation makes this little-known owl endangered. It is listed as threatened on the Mexican Red List. BirdLife International regards it as globally near-threatened.

Geographical variation Monotypic.

Similar species Partly overlapping Whiskered Screech Owl (80) lacks the scalloped pattern below, its wingtips do not project beyond the tail, the bill is grey, and the bristled toes are greyish-brown.

▲► Bearded Screech Owl from southern Mexico is not generally split subspecifically, though it has large yellow eyes and is maybe slightly browner than Guatemalan birds. Chiapas, Mexico (*Christian Artuso*).

► Grey-morph Bearded Screech Owl. This is a small owl with short ear-tufts, a dark face with dense bristles around the bill, and a spotted appearance. Guatemala, April (*Knut Eisermann*).

◄ Rufous-morph Bearded Screech Owl on the ground. The rufous morph's pattern is identical to the grey morph, with the rufous widely spread. Guatemala, May (*Knut Eisermann*).

82. BALSAS SCREECH OWL
Megascops seductus

L 24–27cm, Wt 150–174g; W 170–185mm

Identification A small to medium-sized owl with short ear-tufts. Female is on average up to 15g heavier than male. No red morph known. Has a greyish-brown crown with blackish-brown shaft-streaks, brown vermicular markings and some white spots. Greyish-brown upperparts are overlaid with wine-reddish to pink, dark vermiculations and streaks. Whitish outer webs of scapulars create a light band across the shoulder, and whitish tips of wing-coverts form a pale second band on the closed wing. Flight and tail feathers are barred light and dark. The pale underparts have narrow dark shaft-streaks, the neck and upper breast having an irregularly spotted appearance. The greyish-brown facial disc, vermiculated and mottled brownish, has a dark brown ruff; brownish-white eyebrows extend to the tips of the ear-tufts, but both are not very prominent. The eyes are tobacco-brown to golden-brown, and the bill and cere are grey-green. Tarsi are feathered to the base of the greyish-brown toes, with horn-coloured claws. *Juvenile* Unknown.

Call Emits a fairly loud accelerating trill, *book-book-bokbokbokbobobobrrrrr.*

Food and hunting Very little studied.

Habitat Inhabits deciduous woods, mesquite, and dense second growth, and is found also in arid areas with scattered trees, cacti and thorny scrubs. From about 600m up to 1500m.

Status and distribution Occurs in lowlands in Colima and Río Balsas drainage of Michoacán and W. Guerrero, in Mexico. Status is poorly known.

Geographical variation Taxonomic status is based on vocalisations, and requires further molecular studies. Monotypic.

Similar species Slightly overlapping Western Screech Owl (76) is a little smaller and has bright yellow eyes. Geographically separated Oaxaca Screech Owl (79) and Bearded Screech Owl (81) are clearly smaller, with yellow eyes.

◄ Balsas Screech Owl is a little larger than Pacific Screech Owl *Megascops cooperi*, with a warm brown ground colour above bearing distinct and broad shaft-streaks (there is no red morph). Below, it is densely and narrowly barred, with bold black dots on the breast. Colima, Mexico, September (*Christian Artuso*).

83. BARE-SHANKED SCREECH OWL
Megascops clarkii

L 23–25cm, Wt 123–190g; W 173–190mm

Identification A small owl with short ear-tufts. No data on sexual size differences. Has rich to dull reddish-brown upperparts heavily spotted, mottled and vermiculated with black, and with dull yellowish hind-neck. Blackish-edged white outer webs of scapulars form a white band across the shoulder. Flight feathers are barred cinnamon-buff and the tail is barred light and dark. The underparts are pale brown, tinged with dull yellowish and a tan-coloured tint, mixed with white on the upper breast, and the lower breast and belly have dusky and reddish-brown horizontal bars or vermicular markings with blackish shaft-streaks. The yellowish-brown facial disc has an indistinct dark ruff. The eyes are pale yellow, the bill greenish-grey or bluish-grey, and the cere horn-coloured. The thighs are mostly dull yellow. Lower third of tarsus is bare, as are the yellowish-pink toes; the dark claws have darker tips. *Juvenile* Downy chick is whitish. Fledglings are cinnamon-buff above, with white speckling and dusky barring, and dull yellowish below, with yellowish-brown barring; no visible ear-tufts.
Call A rather deep *woogh-woogh-woogh* is uttered at intervals of several seconds.
Food and hunting Feeds on large insects, spiders, and also some small vertebrates. Hunts often from the

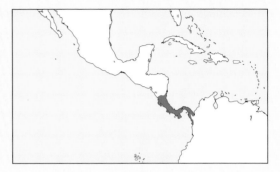

ground or from branches, seizing prey with the talons.
Habitat Inhabiats dense mountain and cloud forests between 900m and 2350m, or at times even higher elevations, to 3200 m.
Status and distribution Found from Costa Rica to extreme NW Colombia. Said to be locally common, but is certain to be at some risk through logging of dense mountain forests. This species is very little studied, and its true real and conservation needs are not known.
Geographical variation Monotypic.
Similar species Tropical Screech Owl (84) and other overlapping *Megascops* have fully feathered tarsi.

▼ Bare-shanked Screech Owl has a more or less ocellated appearance below. Tarsi are bare for more than half their length. San Gerardo de Dota, Costa Rica, August (*Jacques Erard*).

▼ A more reddish-brown individual of Bare-shanked Screech Owl. Costa Rica, April (*Mike Danzenbaker*).

84. TROPICAL SCREECH OWL
Megascops choliba

L 20–25cm, **Wt** 100–160g; **W** 154–182mm

Identification A small owl with short ear-tufts. No data on sexual size differences, but some females are larger than males. Three morphs are known, as well as intermediates. Most common morph has greyish-brown upperparts with dark mottling and streaks; blackish shaft-streaks are present on the crown, mantle and back, but no whitish fringe on the hindneck. Dark-edged whitish or pale ochre outer webs of scapulars form a row of whitish or yellowish spots across the shoulder. Flight feathers are barred light and dark and the mottled tail is indistinctly barred. Whitish-grey underparts have a dark herringbone pattern, narrow shaft-streaks having four or five lateral branches. The facial disc is light greyish-brown with some dark mottling, and a prominent blackish ruff; whitish eyebrows continue towards ear-tufts. The eyes are light yellow to golden-yellow, with blackish eyelids, and the bill and cere are pale greenish-grey. Tarsi are feathered

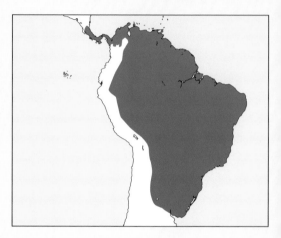

to the base of the grey toes, which have dark horn-coloured claws. The red morph is rusty or cinnamon-buffish with reduced markings; and the brown morph is in general more brownish. *Juvenile* Downy chick is whitish. Mesoptile is distinctly barred dark and light. *In flight* Showing rounded wings and reduced pattern above.

Call Gives a short trill, *gurrrku-kúkúk*, accentuated at end.

Food and hunting Diet consists mainly of insects and spiders, occasionally with some small vertebrates. An individual purchased in 1975 from the local market in Colombia consumed 15–20g of minced meat per day, and managed very well on this. Hunts often from road-side poles and telephone wires, but also hawks moths on the wing.

Habitat Occurs in savanna-type open forests, clearings in rainforest, bamboo stands, farmlands and urban parks. Tends to avoid dense and heavy primary forest. Normally found below 1500m, but locally and in warm climate ascends higher, in Colombia to 3000m and in Argentina locally to 2700m.

Status and distribution Found from Costa Rica south to Argentina and Uruguay; also in Trinidad. This species is quite common, but, being an insect-eater, it suffers from extensive use of pesticides. Its habit of hunting along roadsides makes it a frequent victim of traffic.

Geographical variation Up to nine subspecies are

◄ Tropical Screech Owl of the race *crucigerus*. Here the short ear-tufts are completely invisible. Napo River in Loreto province, Peru, October (*Michael and Patricia Fogden*).

listed: nominate *choliba* is found from S. Mato Grosso and São Paolo, in Brazil, to E. Paraguay; *M. c. luctisomus*, from Costa Rica to NW Colombia, has long wings; *M. c. margaritae*, from N. Colombia to N. Venezuela, including Margarita Island, is paler than nominate; *M. c. crucigerus*, from E. Colombia and Venezuela to E. Peru, NE Brazil, Suriname, Guyana and Trinidad, has fluffy yellowish spots on the body feathers; *M. c. duidae*, from Mt Duida and Mt Neblina, in S. Venezuela, is very dark and has a pale collar on the hindneck; smaller and paler *M. c. decussatus* occurs in CE and S. Brazil; *M. c. uruguaiensis*, from SE Brazil and Uruguay to NE Argentina, has rather prominent shaft-streaks below and the body feathers have buffish-rufous down; *M. c. surutus*, from Bolivia, is brighter rufous with reduced streaks and bars; *M. c. wetmorei*, from the Chaco of Paraguay to Buenos Aires, Argentina, is darkish with more dirty-buff underparts. The taxonomic status of many of these forms listed as subspecies needs further study; e.g. the very dark *duidae* could be a distinct full species, an endemic of the Duida and Neblina cerros, in Venezuela.

Similar species Locally overlapping Montane Forest Screech Owl (88) has pale yellow eyes, a whitish fringe around the hindneck, and a coarser pattern on the underparts with several spade-like shaft-streaks. More sympatric Black-capped Screech Owl (96) is a little larger, with longer ear-tufts, a nearly uniform dark crown, a whitish fringe around the hindneck, and less of a herringbone pattern below. Also geographically overlapping, the Santa Catarina Screech Owl (97) is much larger, with very long ear-tufts.

▼ Red morph nominate Tropical Screech Owl. São Paolo, Brazil, August (*Markus Lagerqvist*).

◄ Nominate grey morph Tropical Screech Owl. São Paolo, Brazil, August (*Markus Lagerqvist*).

▼ Race *wetmorei* of Tropical Screech Owl, Gran Chaco, Paraguay (*Thomas Vinke*).

▼◄ Race *surutus* is slightly rufous with reduced dark bars. Mato Grosso Pantanal, Brazil, July (*Markus Lagerqvist*).

▼ Tropical Screech Owl of race *luctisomus* has pale and brownish morphs. This pair of brownish birds have erected their short ear-tufts. The eyes are not well shown but are pale or yellow, as in other races. Costa Rica, June (*Daniel Martínez-A*).

▼ Race *decussatus* from Brazil has both grey and red morphs. This is a grey morph with ear-tufts erected. Bahia, Brazil, October (*Arthur Grosset*).

▼ The rufous morph of race *decussatus* of Tropical Screech Owl. Another sleeps nearby. Rio de Janeiro, Brazil, September (*Scott Olmstead*).

▼ Tropical Screech Owl race *crucigerus* is more rufous above and below than nominate *choliba*; it has fluffy yellow-ochre spots ventrally, and dense cross-bars on the breast. Tingana, Peru, August (*Christian Artuso*).

85. MARIA KOEPCKE'S SCREECH OWL
Megascops koepckeae

L 24cm, **Wt** 110–148g; **W** 172–183mm

Other name Koepcke's Screech Owl

Identification A small owl having distinct but short ear-tufts with dark shaft-streaks. No data on sexual size differences. The blackish-brown crown has fine pale brown speckling, more ochre-whitish towards the forehead. Dark greyish-brown upperparts have broad but laterally ill-defined shaft-streaks and dark transverse branches; there are hardly any vermicular markings, but some ochre to whitish spots above. Blackish-edged whitish outer webs of scapulars form a broken whitish line across the shoulder. The primaries have uniformly dark tips, and the light and dark barring on the flight feathers is not very clear; the tail is dark brown with thin light yellow bars and small speckles. The greyish-white underparts have broad blackish-brown shaft-streaks and a few irregular lateral branches; the upper breast and sides of neck are washed with pale brownish-ochre. The whitish-grey facial disc, mottled or speckled darker, and lighter towards outer edge, has a prominent blackish ruff. Whitish eyebrows with dark mottling are not very distinct. The eyes are yellow, the dark eyelids with a blackish rim, and the bill and cere are bluish to greenish-horn. Feathered tarsi are light yellow with brown speckles, the toes greyish-brown

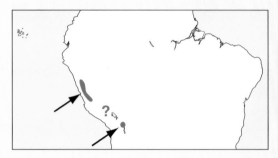

with dark horn-coloured claws. *Juvenile* Not known.

Call Gives a sequence of low *uk* notes in series of eight to ten, rising slightly in pitch, but the last note falling. These calls are repeated at irregular intervals.

Food and hunting Not described.

Habitat Inhabits dry woodland and *Polylepis* woodland on Andean slopes between 2500m and 4500m.

Status and distribution Found from NW to SW Peru, possibly extending also to the La Paz region of WC Bolivia. This owl is locally not uncommon, but is very little studied.

Geographical variation No subspecies have been described, but skins from Apurimac (housed in Lima and Copenhagen collections) are paler than specimens from more northern localities. Recent field observations in Peru reveal that calls of the Apurimac population are similar in pattern to those from birds elsewhere in range, but faster and at higher frequency; moreover, the plumage is colder grey overall, lacking brown tones, and the birds also lack the ochre and greenish tones in the tarsi and have greyer bare-part coloration. It is not yet known how much individual variation could be involved, but the Apurimac population will surely be accorded at least subspecies status.

Similar species Narrowly separated Peruvian Screech Owl (86) is a little smaller, less coarsely patterned below, has a very prominent whitish fringe around the blackish crown, and has distinctive whitish eyebrows around the yellow eyes. Allopatric Tumbes Screech Owl (87) is much smaller and lives mainly in arid lowlands. White-throated Screech Owl (103) is darker and larger, with no ear-tufts, and has orange eyes and a white throat.

◄ Maria Koepcke's Screech Owl has very similar light and dark morphs. Note the line of whitish spots on the scapulars. Amazonas, Peru, October (*Roger Ahlman*).

▲ A pair of 'Apurimac' Screech Owls, from Peru. These birds were suggested to represent a new species but it now seems more likely that they will be named as a race of Maria Koepcke's Screech Owl, based on vocal similarities. Abancay, Peru, September (*Christian Artuso*).

▲ 'Apurimac' Screech Owl is paler than Maria Koepcke's Screech Owls from further north, and has a greyish ground colour lacking brown tones. Abancay, Peru, September (*Christian Artuso*).

▼ Maria Koepcke's Screech Owl is similar to the slightly smaller Peruvian Screech Owl *Megascops roboratus* but with a grey ground colour above and indistinct broad bars. It lacks a nuchal collar. Chachapoyas, Peru, August (*Christian Artuso*).

86. PERUVIAN SCREECH OWL
Megascops roboratus

L 20–22cm, **Wt** 144–162g; **W** 165–175mm

Identification A small owl with short ear-tufts. Females are said to be slightly larger and heavier than males. Two morphs separated. Greyish-brown morph has relatively dark greyish-brown upperparts with blackish shaft-streaks and some dark transverse vermicular markings; a whitish fringe runs around the blackish-brown crown from the whitish eyebrows to the hind-neck. Blackish-edged whitish outer webs of scapulars form a whitish line of spots across the shoulder. Flight feathers are distinctly barred light and dark, and the tail is mottled rather than barred. The underparts are pale greyish-brown with relatively fine shaft-streaks and vermiculations, and irregular transverse bars; some feathers of the upper breast side have blackish shaft-streaks, broader and almost spade-shaped towards the feather tips. The light greyish-brown facial disc is darker-speckled and finely vermiculated, with a dark blackish-brown ruff. The eyes are golden-yellow to pale yellow, with pinkish-olive eyelids, and the bill and cere are greyish-olive. Tarsi are feathered to the base of the brownish toes, which have dark horn-coloured claws. Red morph is in general light rufous, with dark brown rather than blackish markings. *Juvenile* Downy chick is whitish. Mesoptile is densely barred below, the crown is not prominently marked, and it has pale olive-yellow eyes.

Call Utters a long, somewhat undulating trill with equally spaced notes, beginning softly but increasing steadily to higher volume.

Food and hunting Food almost exclusively insects, but not well documented.

Habitat Occupies dry deciduous forests, mesquite and arid woodlands with bushes and cacti between 500m and 1200m, sometimes up to 1800m.

Status and distribution Occurs in S. Ecuador and N. Peru. This owl is locally uncommon or rare as a result of habitat destruction, as it is dependent on trees with holes for breeding. Further investigation of its biology and vocalisations is needed.

Geographical variation Monotypic.

Similar species Narrowly allopatric Maria Koepcke's Screech Owl (85) occurs at much higher altitudes, is much larger, has no whitish fringe around dark crown, and is more coarsely patterned below, with relatively broad shaft-streaks. Also slightly separated geographically, Tumbes Screech Owl (87) lives at lower altitudes and is much smaller, with a short tail.

◄ Peruvian Screech Owl has a dark mantle with only a little brown on the wings. The scapulars have blackish-edged whiter outer webs, and the flight feathers are light- and dark-barred. Tumbes, Peru, June (*Paul Noakes*).

▲ Peruvian Screech Owl grey morph at the nest hole. The grey morph's chest has black rather than brown streaking. Quebrada, Peru, July (*Jonathan Newman*).

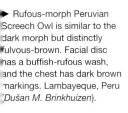

▶ Rufous-morph Peruvian Screech Owl is similar to the dark morph but distinctly fulvous-brown. Facial disc has a buffish-rufous wash, and the chest has dark brown markings. Lambayeque, Peru (*Dušan M. Brinkhuizen*).

87. TUMBES SCREECH OWL
Megascops pacificus

L 20cm, **Wt** 70–90g; **W** 139–150mm

Identification A very small owl with short but pointed ear-tufts. No data on sexual size differences. Two colour morphs exist. Grey morph has greyish-brown upperparts with blackish or dark brown streaks and arrow-shaped markings. A prominent whitish fringe on the dark crown runs from the whitish eyebrows around the head to the nape. Outer webs of scapulars have large whitish, dark rimmed spots, forming a distinct stripe across the wing. Wings and tail are greyish-brown with narrow pale bars, rectrices with darker shafts. Light greyish-brown underparts have darker vermicular markings and distinct dark shaft-streaks with irregular horizontal bars; the breast is darker than the belly. The facial disc is greyish-brown, and bordered by a dark ruff. The eyes are yellow and the bill greenish olive-grey. Tarsi are feathered to the base of the olive-grey toes, with dark horn-coloured claws; talons are relatively weak. The other morph, red, is more common than the grey one, and has generally rufous coloration with less prominent dark markings. *Juvenile* Downy

chick is whitish. Mesoptile is densely barred, but less clearly marked.

Call Emits a rapid purring trill lasting some 1.5–2 seconds, with slight downward inflection.

Food and hunting Little known, but probably feeds mainly on insects and other arthropods.

Habitat Occurs in arid tropical bushlands and dry deciduous woodlands, often near or in human settlements, and mostly in lowlands below 500m.

Status and distribution Found in SW Ecuador and NW Peru. This species is possibly not too rare but is poorly known.

Geographical variation Monotypic. This owl was for a long time considered a subspecies of Peruvian Screech Owl, but has now been separated on the basis of morphology and vocalisations. Molecular and biological studies are needed in order to confirm this separation, which is thought by some to be premature.

Similar species Larger Maria Koepcke's Screech Owl (85) lives above 2000m, and has more coarsely patterned underparts, and no clear white border to the dark crown. Larger Peruvian Screech Owl (86) also lives at higher altitudes (above 500m), and has a relatively long tail, giving it a slimmer appearance. Similarly, the larger White-throated Screech Owl (103) prefers higher altitudes, and has a prominent white throat patch and no ear-tufts. Sympatric Peruvian Pygmy Owl (187) is similar in size, but has a longer, barred tail and no ear-tufts.

◄ Tumbes Screech Owl has fairly uniform dark primary coverts; streaking on breast is similar to Peruvian Screech Owl, but with wider-spaced cross-bars and more uniform ground colour below. Lambayeque, Peru, August (*Christian Artuso*).

▲ Tumbes Screech Owl grey-brown morph is similar to the grey-brown morph of Peruvian Screech Owl *Megascops roboratus*, but it is clearly smaller and paler. Large eyes are yellow to golden-yellow, with small ear-tufts. Jorupe, Ecuador, April (*Ian Merrill*).

► Tumbes Screech Owl is restricted to south-west Ecuador and north-west Peru. Lambayeque, Peru, August (*Christian Artuso*).

88. MONTANE FOREST SCREECH OWL
Megascops hoyi

L 23–24cm, **Wt** 110–145g; **W** 165–177mm

Other names Hoy's Screech Owl, Yungas Screech Owl
Identification A small owl with short ear-tufts. Females are on average over 20g heavier than males. Grey, brown and red morphs are known, brown being the most common. It has greyish-brown crown and upperparts with dark shaft-streaks and vermicular markings, and a narrow whitish line on hindneck extending to back of ear-tufts. Dark-edged whitish outer webs of scapulars form a whitish band across the shoulder. Flight feathers are barred dark and light, and tail is densely barred and vermiculated. Slightly paler underparts are covered with dark shaft-streaks, each streak with two or three horizontal branches; dark shaft-streaks with spade-like tips form two distinct rows on each side of upper breast. The greyish-brown facial disc is finely dark-vermiculated, and has a very prominent blackish ruff on both sides, creating a dark patch on each side of the neck; paler eyebrows are usually notched above

eyes, so that pale line appears broken. The eyes are bright yellow and the bill and cere greenish-yellow. Tarsi are feathered to the base of the yellowish-grey toes, which have dark horn-coloured claws. Grey and red morphs are similar in pattern, but ground colour of latter is rusty-brown. *Juvenile* Fledged young resembles adult, but has fluffy body plumage and ear-tufts are not visible. *In flight* Shows the typical pale neck-band.
Call Gives a long trill with staccato notes, almost eleven notes in a second. Call sequence lasts for 5–20 seconds and is repeated at intervals of several seconds.
Food and hunting Feeds mainly on insects and spiders, and occasionally takes small rodents. Prey is captured with talons from the ground or in the upper storey of large trees.
Habitat Inhabits deciduous mountain forests where epiphytes, such as *Tillandsia*, are common. Occurs at 1000–2600m, locally to 2800m.
Status and distribution Found from SC Bolivia to N. Argentina. This owl is rather common locally, but threatened by deforestation and by overgrazing of cattle.
Geographical variation Monotypic.
Similar species Partly overlapping, smaller Tropical Screech Owl (84) is normally found at lower altitudes, has shorter ear-tufts, weaker talons, unbroken pale band between bill and ear-tufts, and is less coarsely streaked below. Also locally sympatric, the Rio Napo Screech Owl (102) is about the same in size, with very thin shaft-streaks on densely barred and vermiculated underparts, and has brown or yellow eyes.

◄ Montane Forest Screech Owl occurs higher than Rio Napo Screech Owl. Salta, Argentina, January (*Neil Bowman*).

89. RUFESCENT SCREECH OWL
Megascops ingens

L 25–28cm, **Wt** 134–223g; **W** 184–212mm

Identification A small to medium-sized owl with small ear-tufts. Females are on average 28g heavier than males. General colour of upperparts is dark olive-tawny, reddish-brown or grey-brown, with some fine whitish spots and narrow dark vermicular markings; the crown is dark-scalloped. Dull yellowish to white edges of scapulars form a not very prominent pale band across the shoulder. Flight feathers are barred yellowish-brown and dusky, and the tail is barred cinnamon and darker brown. Light underparts have a few narrow shaft-streaks and thin dark and dull yellowish to white vermicular markings. The buffish-brown facial disc is finely marked with darker concentric lines around the honey-brown eyes, and there is an indistinct dull yellowish ruff. The bill and cere are olive-yellow. Densely feathered tarsi to base of yellowish-grey toes have pale horn-coloured claws, which are relatively powerful. *Juvenile* Downy chick is whitish-buff. Mesoptile is dusky-mottled.

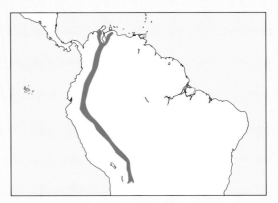

Call Emits a long series of *ul* notes, up to 50 in succession in ten seconds; series begins softly but increases gradually in volume.

Food and hunting Feeds mainly on large insects and spiders but also takes some small vertebrates. Apparently forages mainly in middle storey and canopy of trees.

Habitat Inhabits humid forests on mountain slopes between 1200m and 2250m.

Status and distribution Occurs from Venezuela to N. Bolivia, but only on eastern slopes of Andes. This owl is said to be rare but it may be overlooked. Like all humid-forest owls, it is threatened by extensive logging.

Geographical variation Two subspecies are known: nominate *ingens* is found from E. Ecuador to N. Bolivia; *M. i. venezuelanus*, from Venezuela to E. Colombia, is slightly smaller and paler than nominate. Coastal mountains north of Maracay, in Venezuela, may harbour a third subspecies, but this is as yet undescribed; it remains to be seen how this differs from the newly discovered Santa Marta Screech Owl.

Similar species Geographically separated Colombian Screech Owl (91) is similar in size, with rather prominent pinkish eyelids and blackish-brown eyes, but lives only on western slopes of the Andes. Partly sympatric Cinnamon Screech Owl (92) is a little smaller, and less vermiculated on the breast and belly.

◄ Rufescent Screech Owl has a dark brown ground colour above, more rufous and yellowish below, with some vermiculations. It has an indistinct rufous facial disc and ear-tufts. Eyes are brown, and the eyelid rim is pinkish. Peru, September (*Glenn Bartley*).

90. SANTA MARTA SCREECH OWL
Megascops sp.

L *c.* 30cm

Identification A small owl with short but visible ear-tufts. At least two morphs exist, a rufous one and a grey one. The head is dark brown, but the forehead more whitish, with contrasting dark eyebrows. The wings and tail have reddish-yellow bars, some six bars being visible on the tail. Below, the abdomen is very pale but the lower chest has brownish stripes with an indistinct herringbone pattern. A light area around the large yellow eyes makes the facial disc fairly prominent. The bill is light horn-coloured, and the legs yellowish with dark claws. *Juvenile* Undescribed.

Call Song consists of relatively short phrases (*c.* 2.6–3 seconds in duration) of about 26 staccato notes, *úúúúú… úú*; the series begins softly, reaching its highest volume after one second and becoming fainter towards the end. Song is uttered at intervals of four or five seconds. It resembles that of Montane Forest Screech Owl (88), but has faster sequence with many more notes.

Food and hunting Not yet studied but includes large beetles.

Habitat Found in mountain forests between 1800m and 2150m.

Status and distribution This owl (not yet officially described) was discovered in February 2007 in the El Dorado Bird Reserve, in the Sierra Nevada of Santa Marta mountain range, in N. Colombia. Its status is not known, but at least four or five pairs have been seen within the reserve. Fundacion ProAves is undertaking an urgent comprehensive study of this new owl in the reserve, and has erected 40 nestboxes to encourage breeding.

Geographical variation Presumed monotypic.

Similar species Sympatric Tropical Screech Owl (84) is perhaps a little smaller, has a totally different voice and lives normally at lower elevations. Northern Tawny-bellied Screech Owl (94) could nearly overlap in range if, indeed, it occurs in the Perijá Mountains of NE Colombia/NW Venezuela (to 2100m); it has been suggested that this mountain population could belong instead to this new species.

◀◀ Greyish morph with large yellow eyes. Note that the wings extend beyond the short tail. San Lorenzo, Colombia, October (*Christian Artuso*).

◀ Rufous-morph Santa Marta Screech Owl. Bare toes are shown when the owl hangs from a twig. Santa Marta, Colombia, February (*Jon Hornbuckle*).

91. COLOMBIAN SCREECH OWL
Megascops colombianus

L 26–28cm, Wt 150–210g; W 175–189mm

Identification A small to medium-sized owl with medium-long ear-tufts. Female is on average as much as 50g heavier than male, but no sexual difference in wing length has been noted. A cinnamon-reddish morph and a greyish-brown morph are known. The latter has greyish-brown upperparts finely mottled and vermiculated dusky. Dusky shaft-streaks and dark mottling are evident on the crown, and the hindcrown has a pale yellowish border with darker spots. Outer webs of scapulars are dull yellowish to brownish, thus not forming a pale stripe across the shoulder. Wings and tail are dark brown, the flight feathers barred paler on both webs; primary wing-coverts are almost plain dark brown. Fairly uniform greyish-brown underparts have thin dusky mottling and vermicular markings, also narrow blackish shaft-streaks and more brownish cross-bars. The facial disc is greyish-brown, almost without a darker ruff. There are promi-

nent downward-curved brownish bristles between the dark brown eyes and the base of the bluish-green bill. Only the upper part of the tarsus is sparsely feathered, the remaining parts and the toes being dirty whitish-pink, with whitish-horn claws; the legs are long and relatively powerful. Reddish morph is less distinctly patterned and generally yellowish-brown to reddish-brown. *Juvenile* Downy chick is dull yellowish to whitish. Fluffy mesoptile resembles the adult, but is finely barred on the head and mantle and below.

Call Utters a series of flute-like staccato notes lasting some 18 or more seconds.

Food and hunting Eats larger insects and other arthropods, but also small vertebrates.

Habitat Inhabits cloud forests with rich epiphytic plants and dense undergrowth, between 1300m and 2300m.

Status and distribution Occurs from Colombia to N. Ecuador, but only on western slopes of Andes between 1300m and 2100m. This species may be relatively rare and threatened by forest destruction.

Geographical variation Monotypic. The taxonomic status of this owl may require further research, although it is generally regarded as a full species.

Similar species Geographically separated Rufescent Screech Owl (89) is similar in size, but with shorter legs, and has tarsi densely feathered to base of toes. Locally sympatric Cinnamon Screech Owl (92) is a little smaller but very similar in coloration, with weaker talons and feathered tarsi.

◀ Colombian Screech Owl has grey-brown and reddish morphs. The latter has the same pattern and shaft-streaks as the grey-brown but is rufous-washed. It is smaller than Rufescent Screech Owl *Megascops ingens*, and has no white below. Pichincha, Ecuador, January (*Roger Ahlman*).

92. CINNAMON SCREECH OWL
Megascops petersoni

L 21cm, **Wt** 88–119g; **W** 153–162mm

Identification A small owl with moderately long ear-tufts mottled blackish towards tips. According to weight data, no clear sexual size differences exist. The yellowish-brown upperparts are thinly vermiculated with double wavy bars, darker and paler alternately; darker vermicular markings break up distally into mottling on most feathers of the mantle, and dull yellowish down underlies the back feathers. Pale cinnamon scapulars lack whitish outer webs and are in general indistinctly marked with dusky. The crown is slightly darker than the back, with pale eyebrows, and the wing-coverts are the same as the mantle and back. Flight feathers are barred pale buff and dark yellowish-brown, and the tail is banded brown and blackish (some eight bars of each colour). Below, the plumage is warm yellowish-brown to dull yellow, the throat and breast finely mottled with dark wavy bars or vermicular markings, but the lower breast and belly

almost uniformly cinnamon. The warm yellowish-brown facial disc is finely vermiculated or barred (four or five bars on each feather), and shades gradually to darker brown towards the rather indistinct blackish ruff. The eyes are dark brown and the bill and cere light greyish-blue. Tarsi are feathered to within 5mm of the yellowish-pink toes, which have pale horn-coloured claws. *Juvenile* Not known.

Call Utters a rapid succession of equally spaced *u* notes, some five to seven per second, the series first rising, and then dropping in pitch, before breaking off abruptly.

Food and hunting Not described.

Habitat Inhabits moist cloud forests rich in epiphytic plants and mosses, from 1700m to 2500m.

Status and distribution Occurs in the forested eastern foothills of Andes from S. Ecuador to N. Peru on the E. Andean slopes; possibly present also in the E. Andes of Colombia. Status is not known but it could be threatened by habitat destruction.

Geographical variation Monotypic. Taxonomic status requires further study, as it may be related to the Cloud-forest Screech Owl (93), rather than Rufescent or Colombian Screech Owls as previously suggested. At one time, the present species was treated as a subspecies of Cloud-forest Screech Owl, and all these taxa therefore require more research and molecular data.

Similar species Partly sympatric Rufescent Screech Owl (89) is larger and greyer, with whitish outer edges and spots on the scapulars. Geographically separated Colombian Screech Owl (91) is also larger, with distally unfeathered tarsi, and is clearly more vermiculated on the breast and belly.

◀ Cinnamon Screech Owl is smaller than Colombian *Megascops colombianus*, with warm buffish-brown plumage and a less vermiculated breast and belly. The bill is bluish-grey. Amazonas, Peru, October (*Dušan M. Brinkhuizen*).

 The male Cinnamon Screech Owl is on average just one gram lighter than the female. Note the dark brown eyes. Cajamarca, Peru, August (*Christian Artuso*).

► Cinnamon Screech Owl's head, with short to medium ear-tufts, is a little darker than the back. There are no white scapular spots. Cajamarca, Peru, August (*Christian Artuso*).

93. CLOUD-FOREST SCREECH OWL
Megascops marshalli

L 20–23cm, **Wt** 107–115g; **W** 152–164mm

Identification A small owl with very short ear-tufts. Female is on average almost 10g heavier than male. Has rich chestnut upperparts, irregularly barred and mottled blackish, the deep reddish-brown central back feathers with four wavy blackish cross-bars. Scapulars have whitish or pale yellowish outer webs with a blackish area near tip of each web, forming a clear broken white band. Flight feathers are barred dusky and tawny, and tail has eight reddish-brown and blackish bars. The rufous throat and upper breast have some dark barring and fairly broad dusky shaft-streaks, and the reddish-brown to whitish lower breast has a few cross-bars, dusky shaft-streaks and a scattering of blotchy black spots, giving an irregularly ocellated appearance; the belly is whitish. The reddish-brown facial disc has a broad black ruff, and a dark area around the bill. The crown has reddish-brown and black barring, the feathers marked centrally more with black, and the hindcrown has a prominent whitish border with rufous and dusky tips; whitish superciliary and loral feathers are tipped with rufous. The eyes are dark brown, with pale greyish-ochre eyelids, and the cere and bill greyish-yellow with a greenish tint. Tarsi are feathered to the base of the greyish-fleshy toes, which have pale horn-coloured claws. *Juvenile* Downy chick undescribed. Mesoptile is similar to the adult, but has ten tail-bars (instead of eight).

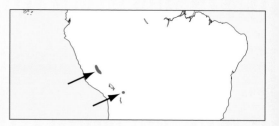

Call Poorly known, but is said to be a monotonous series of single *ii* notes in a rapid sequence and repeated after a short interval.

Food and hunting Poorly studied, but insects and other arthropods probably form the main diet.

Habitat Occupies cloud forests rich in mosses and epiphytes, from 1900m to 2600m.

Status and distribution Occurs from Andes of CE Peru to N. Bolivia. This owl is rather common locally but is virtually unknown.

Geographical variation Monotypic.

Similar species Slightly geographically separated Rufescent Screech Owl (89) is much larger and lives at lower elevations. Allopatric but same-sized Cinnamon Screech Owl (92) may be closely related, but lacks whitish shoulder-band, and is rather uniformly yellowish-brown below, without bold patterning. Sympatric Yungas Pygmy Owl (186) is much smaller, with no ear-tufts.

◄ Cloud-forest Screech Owl has dark brown eyes and an irregularly ocellated appearance below. Yungas forest, Bolivia (*Ross McLeod*).

▼ Cloud-forest Screech Owl has a rufous facial disc with a broad black rim. This bird is from the recently discovered Bolivian population. La Paz, Bolivia, February (*Sebastian K. Herzog*).

94. NORTHERN TAWNY-BELLIED SCREECH OWL
Megascops watsonii

L 19–23cm, **Wt** 114–155g; **W** 164–184mm

Identification A small owl with relatively long ear-tufts. Sexual size differences not documented. The upperparts are dark greyish-brown with vermicular markings, spots and small pale and dark freckles, giving dusty appearance; no whitish outer webs on scapulars. Wings and tail are barred light and dark. The dark earth-brown upper breast is flecked and vermiculated paler; the throat is pale, and the tawny or light rusty-brown lower underparts have paler and darker vermicular markings and fairly thin shaft-streaks and cross-bars. The greyish-brown facial disc has an indistinct darker ruff. The eyes are amber-yellow to pale brownish-orange and the cere and bill are greenish-grey. Tarsi are densely feathered to the base of the light ochre-brown toes, which have horn-coloured claws; talons are relatively weak. *Juvenile* Not known.

Call Emits a long series of regular *u* notes in rapid succession, the series lasting for up to 20 seconds.

Food and hunting Not described.

Habitat Found in lowland rainforests, preferring interior parts of primary forests, from near sea level to about 900m. Some authors mention the occurrence of this owl at elevations of up to 2100m in the Perijá Mountains, on the NE Colombia/NW Venezuela border, and also north-east of the Orinoco River; it has been suggested, however, that these high-altitude owls could be Santa Marta Screech Owls (90), rather than normally lowland-dwelling Northern Tawny-bellied Screech Owls.

Status and distribution Occurs from lowlands of E. Colombia, W. & S. Venezuela and Suriname to NE Ecuador and NE Peru, as well as in N. Amazonian part of Brazil. It is too little studied for its true conservation status to be known. This owl overlaps in range with Southern Tawny-bellied Screech Owl (95), and intermediate songs have been taken as a possible sign of hybridisation between these closely related species. No zone of progressive intergradation is known, however, and the species status of these two owls is not in question.

Geographical variation Monotypic.

Similar species Sympatric Tropical Screech Owl (84) has yellow eyes and a typical herringbone pattern on

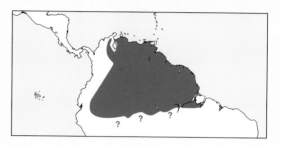

the breast and belly. Narrowly overlapping Santa Marta Screech Owl (90) has bright yellow eyes and occurs at higher elevations. Scarcely overlapping Cinnamon Screech Owl (92) also inhabits higher cloud forests and is overall warm cinnamon-brown. Very similar but largely allopatric Southern Tawny-bellied Screech Owl (95) is a little larger, with darker crown and more reddish-brown upperparts; it also has broader shaft-streaks below and a densely vermiculated dark rufous belly.

▶ The dull greyish Northern Tawny-bellied Screech Owl from north of the Amazon is very similar to the southern species *Megascops usta* but it has finer shaft-treaks on the breast. Belly has a herringbone pattern on a sandy ground colour. The eyes are dull orange to amber. Toes are bare but the tarsi are densely feathered. Yasuní, Ecuador, August (*Nick Athanas*).

95. SOUTHERN TAWNY-BELLIED SCREECH OWL
Megascops usta

L 23–24cm, Wt 115–180g; W 164–187mm

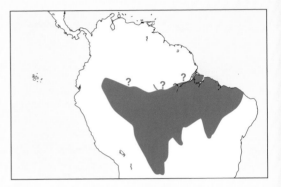

Identification A small owl with blackish, medium-sized, but rather prominent ear-tufts. Sexual size differences not documented. Dark, rufous and light morphs exist, with much individual variation. The upperparts, washed dark earth-brown and reddish-brown, are finely freckled and light-spotted, giving a dusty appearance. Brownish scapulars have indistinctly spotted pale outer webs, and the flight and tail feathers are barred light and dark. Ochre-tawny underparts, becoming paler towards the belly, have lighter and darker vermicular markings and blackish shaft-streaks and cross-bars. The brown facial disc has a distinct blackish ruff. The eyes are warm brown, although older individuals occasionally have dark yellow eyes. The bill and cere are yellowish-brown with a greenish suffusion. Tarsi are densely feathered to the base of the greyish-yellow to brownish toes, which have dark horn-coloured claws. Dark morph is mainly dark greyish-brown, rufous morph dark reddish-brown, and light morph pale brownish overall with a light facial disc. *Juvenile* Mesoptile has barred head, mantle and

underparts. *In flight* Shows rounded wings with 6th primary the longest.

Call Utters a long series of hollow hoots, some two or three *bu* notes per second; series can last for up to 20 seconds, beginning softly, gradually increasing, and fading away at the end.

Food and hunting Not well documented, but feeds mainly on insects and other arthropods, and occasionally on small vertebrates.

Habitat Occupies primary tropical lowland rainforests, often interior parts, but sometimes also near forest edges and clearings.

Status and distribution Occurs from Amazonian Colombia, Ecuador, Peru and Brazil south to lowland forests in N. Bolivia and Brazilian Mato Grosso. This owl is not particularly rare but is very little studied. It may overlap in range with the previous species in SE Ecuador or in Amazonian Peru and Brazil; intermediate songs may indicate occasional hybridisation, but no mixed populations are known to exist, and there is molecular evidence that this owl is not a southern subspecies of the Northern Tawny-bellied Screech Owl (94).

Geographical variation Monotypic.

Similar species Sympatric Tropical Screech Owl (84) has yellow eyes and short ear-tufts, and normally lives in more open parts of the forest. Geographically separated Cinnamon Screech Owl (92) prefers cloud forests and is generally warm cinnamon-brown. Very similar and locally overlapping Northern Tawny-bellied Screech Owl (94) is a little smaller, with relatively narrow dark shaft-streaks and cross-bars, as well as fine vermiculations below. Slightly allopatric Rio Napo Screech Owl (102) has yellow eyes, very short ear-tufts, and densely vermiculated underparts.

▶ Southern Tawny-bellied Screech Owl usually has brown eyes. This dark morph typically has broader shaft-streaks below than the very similar Northern Tawny-bellied Screech Owl. Rio Napo, Amazonian Ecuador (*Glenn Bartley*).

◀ Southern Tawny-bellied Screech Owl has wide individual variation, and is extremely similar to its northern congener *Megascops watsonii*. This one has yellow eyes. Madre de Dios, Peru, August (*Matthias Dehling*).

▼ This Southern Tawny-bellied Screech Owl has a very dark facial disc and also has yellow eyes. Cusco, Peru, September (*Christian Artuso*).

96. BLACK-CAPPED SCREECH OWL
Megascops atricapilla

L 22-23cm, Wt 115-160g; W 170-184mm

Other name Variable Screech Owl

Identification A small owl with rather prominent ear-tufts. Females are up to 25g heavier than males. Three morphs exist. Dark morph, the commonest, has dark earth-brown upperparts mottled and vermiculated, light brown or buff. Flight feathers are barred dark and light, and tail feathers are mottled light and dark but only indistinctly barred. Paler, warm brown underparts are finely dark-vermiculated, and have dark shaft-streaks often widening at feather tip, especially at sides of upper breast, in addition to which two or three dark, often slightly curved branches extend from central part of streak to the edge of the feather; ground colour of lower breast and belly is whitish. Dirty brown facial disc is finely dark-vermiculated and has dark brown or blackish ruff at each side. The crown is almost uniformly dark brown or blackish, with a narrow whitish to light yellowish-ochre margin on hindneck. Light eyebrows are brown-vermiculated. The eyes are normally dark or chestnut, but occasionally bright or amber-yellow; the bill and cere are greenish-yellow. Tarsi are feathered to the base of the brownish-grey to pale flesh-brown toes, with dark horn-coloured claws. Red morph is mainly rufous, with a dark cap and darker upperparts. Grey morph has very prominent dark patterning and lacks brown or reddish-brown in its greyish plumage. *Juvenile* Not known.

Call Gives a long, purring trill lasting usually 8–14 seconds, but which can continue for up to 20 seconds, beginning faintly, gradually increasing, and breaking off abruptly.

Food and hunting Feeds mainly on insects, but also spiders; rarely, may take some small vertebrates. Hunts from a perch, often frequenting low undergrowth.

Habitat Inhabits primary and secondary rainforests with dense undergrowth; sometimes approaches vicinity of roads, forest edges and even human settlements. In warm climates found up to 600m, but in cooler climates in Argentina only to 250m above sea level.

Status and distribution Occurs from E Brazil almost to Uruguay border, and from E. Paraguay to N. Argentina. Being dependent on extensive rainforests to maintain a stable population, this species suffers greatly from increasing forest destruction. Its true status and distribution require study.

Geographical variation Monotypic.

Similar species Sympatric Tropical Screech Owl (84) is smaller with yellow eyes and short ear-tufts. Also geographically overlapping Santa Catarina Screech Owl (97) is larger with more powerful talons and more coarsely patterned underparts.

◄ Red-morph Black-capped Screech Owl has small but prominent ear-tufts and bare flesh-brown toes. NE Brazil, November (*Lee Dingain*).

97. SANTA CATARINA SCREECH OWL
Megascops sanctaecatarinae

L 25–27cm, Wt 155–211g; W 182–210mm

Other name Long-tufted Screech Owl

Identification A small to medium-sized owl having rather bushy ear-tufts (but without long tips). Females are on average 20g heavier than males. Three colour morphs exist. Most common morph has brown upperparts with a slight ochre tinge, and fairly broad dark shaft-streaks on mantle and back, each of these widening into two or three triangular marks on each side. Dark-margined whitish outer webs of scapulars form a row of pale spots across the shoulder. Flight and tail feathers are barred light and dark. Underparts appear to lack clear vermiculations, but otherwise pattern of markings is very coarse. Almost unmarked ochre-brown facial disc has laterally quite prominent dark ruff. The eyes are pale yellow to orange-yellow, but individuals with brown eyes also occur (especially in brown morph). The bill and cere are greenish-grey. Tarsi are feathered to the base of the light greyish-brown toes, which have dark horn-coloured claws. Grey morph has general grey coloration, and red morph is dusky reddish-brown with less prominent dark markings. *Juvenile* Not known.

Call Utters a rapid guttural trill lasting usually 6–8 seconds, initially as a faint grunting sound, gradually increasing in volume and a little in pitch, and ending abruptly; the call is repeated at intervals of several seconds. Similar to voice of Guatemalan Screech Owl (100), but louder, lower and slower.

Food and hunting Feeds mainly on large insects, such as grasshoppers, also taking spiders and small vertebrates. Hunts by swooping down from a perch.

Habitat Prefers humid forests and open woodlands mixed with *Araucaria*, but found also in pastureland with trees; often seen near farmland and even human habitations. Normally occurs between 300m and 1000m above sea level.

Status and distribution Found from SE Brazil, Uruguay and NE Argentina. This species is rather common locally, but it suffers from overgrazing by cattle, logging and forest-burning.

Geographical variation Monotypic.

Similar species Sympatric Tropical Screech Owl (84)

is much smaller, with relatively weak talons and shorter ear-tufts; it also lacks whitish fringe on hindneck, but has a typical herringbone pattern below. Overlapping Black-capped Screech Owl (96) is smaller and paler, with weaker talons, and has dense, fine vermicular markings below, and an almost plain blackish crown. Also geographically overlapping, Buff-fronted Owl (212) has a large, rounded head with no ear-tufts, and plain light cinnamon or ochre-buff underparts.

▶ Santa Catarina Screech Owl is like a large Black-capped Screech Owl *Megascops atricapilla*, with more powerful talons. Eyes vary from yellow to brown. The grey morph is much paler than that of *M. atricapilla*. SE Brazil, August (*Dario Lins*).

98. VERMICULATED SCREECH OWL
Megascops vermiculatus

L 20–23cm, Wt 102–149g; W 150–170mm

Identification A small owl with short ear-tufts mottled and vermiculated dark and light. No data on sexual size differences. Rufous and greyish-brown morphs are known. The latter has greyish-brown upperparts with a warmer brown tinge; the mantle and back have indistinct vermicular markings and whitish or pale yellowish spots, dark shaft-streaks and cross-bars. Outer webs of scapulars are whitish with dark spotting or mottling, forming a whitish stripe across the shoulder. Flight and tail feathers are barred light and dark; wingtips reach beyond tail. Pale greyish-brown underparts have narrow shaft-streaks, cross-bars and thin, dense vermiculations. Reddish to brownish facial disc is finely dark-vermiculated, with very indistinct ruff and inconspicuous light eyebrows. The crown has blackish shaft-streaks and dark mottling or indistinct barring,

and an indistinct pale collar is present on hindcrown. The eyes are bright yellow, with some orange wash towards pupil. Cere and bill are greyish-olive with a greenish tinge. Three-quarters of the tarsus is feathered, the distal quarter of tarsus and the toes are pinkish-brown, with light horn-coloured claws. Rufous morph is overall reddish-brown, with less prominent dark markings. *Juvenile* Undescribed.

Call Gives a very fast purring trill lasting 5–8 seconds, individual *u* notes uttered up to 17 times per second. Song is shorter than that of Guatemalan Screech Owl (100).

Food and hunting Not documented.

Habitat Inhabits tropical forests with high humidity and many epiphytes, mainly in lowlands but can ascend to 1200m.

Status and distribution Occurs from Costa Rica to NW Colombia. Its status is uncertain, but it may be threatened by destruction of humid tropical forest.

Geographical variation Monotypic. Guatemalan Screech Owl (100) is often regarded as conspecific with this species, but the two clearly differ vocally and no interbreeding or hybrids between the two are known.

Similar species Non-overlapping in range, Tumbes Screech Owl (87) lives more in dry habitats; it has a dark crown, and ocellated lower breast and belly, and the wingtips project far beyond the tail. Geographically separated Guatemalan Screech Owl (100) is very similar in plumage, but has prominent whitish eyebrows and distinct dark ruff, is less densely vermiculated but more coarsely streaked and has a greenish bill.

◀ Greyish-brown morph Vermiculated Screech Owl has more rufous on the head and is paler below with light claws. La Selva Biolgocial Station, Costa Rica (*Michael and Patricia Fogden*).

▲ Greyish-brown morph of Vermiculated Screech Owl has a dark rufous face and distinct ruff; more densely vermiculated below than Guatemalan *Megascops guatemalae*. Eyes yellow. Costa Rica, March (*Dan Lockshaw*).

► Vermiculated Screech Owl has a very bright rufous morph, with bright yellow eyes. 'Eyebrows' are not prominent but the scapulars have large white areas on the outer webs. In this morph all dark markings are less prominent. Costa Rica, March (*David W. Nelson*).

99. RORAIMA SCREECH OWL
Megascops roraimae

L 20–23cm, Wt 105g; W 150–168mm

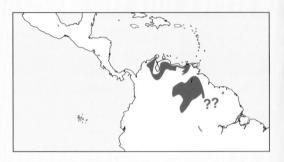

Other name Foothill Screech Owl

Identification A small owl with short, pointed, and mostly dark ear-tufts. No data on sexual size differences. Has fairly dark reddish-brown upperparts with darker shaft-streaks, pale yellowish to almost whitish oval spot on each side of central streak (these spots usually have a dark dot near centre). No distinct pale border is visible on hindcrown. Dark-edged, distinct dull yellowish to white outer webs of scapulars form a pale row across the shoulder. Flight and tail feathers are barred light and dark. Plumage is pale ochre-buff below, with broad, dark shaft-streaks on sides of upper breast and thinner ones on rest of underparts, dark brown and reddish-brown cross-bars on each side of central streak giving underside a coarsely marked appearance. Brownish facial disc, with rufous suffusion and dark mottling, has a prominent ruff. Eyebrows are pale, the eyes bright yellow, and the cere and bill dull olive. Tarsi are feathered nearly to the base of the

pinkish-tinged light brownish toes, which have horn-coloured, dark-tipped claws. *Juvenile* Undescribed.

Call Gives a fairly high-pitched trill with about 50 *u* notes per phrase, lasting 5–8 seconds.

Food and hunting Not documented.

Habitat Inhabits rainforests at lower elevations (300–900m), but also in mountain forests up to 1800m.

Status and distribution Occurs in mountains in S. Venezuela and Guyana, and adjacent N Brazil. It also occurs in the coastal cordilleras of N. Venezuela. This species' status is not known as it is very little studied.

Geographical variation Monotypic. The northern populations have been treated conspecifically with Vermiculated Screech Owl (98), and their taxonomic status requires investigation.

Similar species Sympatric Tropical Screech Owl (84) has a typical herringbone pattern below, and a well-pronounced ruff. Geographically overlapping Northern Tawny-bellied Screech Owl (94) lives in lowland primary forests, and has amber-yellow to brownish-orange eyes.

◀ Roraima Screech Owl is darker than Guatemalan Screech Owl *Megascops guatemalae*, with a coarser plumage pattern and ventral barring. Henri Pittier National Park, Venezuela (*David Ascanio*).

100. GUATEMALAN SCREECH OWL
Megascops guatemalae

L 20–23cm, **Wt** 91–150g; **W** 152–177mm

Identification A small owl with dark, short and pointed ear-tufts. No data on sexual size differences. Two morphs are known. Grey morph has a pale hind-neck without any prominent border. The upper edges of the crown feathers have rather prominent broad and blackish markings, giving a slightly barred appearance. The upperparts are fairly dark greyish-brown with blackish shaft-streaks, cross-bars, flecks and vermicular markings. Blackish-edged whitish outer webs of scapulars form a whitish row across the shoulder. Flight and tail feathers are dark-barred. Pale underparts have blackish shaft-streaks and cross-bars, the upper breast also with brownish vermiculations, the lower areas with fewer or no such bars. The greyish-brown facial disc has fine vermicular markings, and a fine row of blackish spots on the ruff. The eyes are yellow and the bill and cere greenish. Tarsi are feathered to the base of the dusky-flesh, long toes, with horn-coloured claws. Red morph is in general rufous, with less prominent dark markings. *Juvenile* Downy chick is whitish. Mesoptile resembles the adult, but has indistinct barring on the head, nape and mantle and below.

Call Utters a long quavering trill of *u* notes, about 14

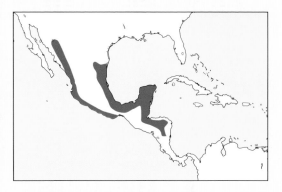

per second; the phrase lasts for up to 20 seconds, and is repeated after a pause of several seconds.

Food and hunting Feeds mostly on insects and larger arthropods; eats small vertebrates occasionally. It takes insects sometimes in flight, but normally by swooping down from a perch.

Habitat Occupies humid as well as semi-arid evergreen or semi-deciduous forests from sea level to 1500m.

Status and distribution Found from Mexico to N. Costa Rica. This species suffers from forest destruction, but at present it is locally not rare.

Geographical variation Four subspecies are described: nominate *guatemalae* from S. Vera Cruz, Mexico, to Guatemala and Honduras; *M. g. hastatus*, from Sonora to Sinaloa, in W. Mexico, has spear-shaped pattern on back, with pagoda-shaped markings (of blacks streaks transversed by rows of dark dots) over a tawny-brown background; *M. g. cassini*, from Tamaulipas to Vera Cruz, in E. Mexico, is the smallest and darkest race; *M. g. dacrysistactus*, from N. Nicaragua to N. Costa Rica, is paler above and more vermiculated below.

Similar species Sympatric Whiskered Screech Owl (80) is smaller, and more coarsely streaked, with bristled toes. Partly overlapping Bearded Screech Owl (81) is smaller, with a much shorter tail and ocellated underside. Geographically separated Vermiculated Screech Owl (98) has a shorter tail, with darker spots or shaft-streaks on the crown (instead of broad barring), and is more finely vermiculated below.

◄ Guatemalan Screech Owl of nominate race *guatemalae*. This greyish morph is less vermiculated than Vermiculated Screech Owl *Megascops vermiculatus*. It has very pale yellow eyes, and the ear-tufts are hardly visible if not erected. Ruff is brownish and not very distinct. La Bajada, Mexico (*Stan Tekiela*).

101. PUERTO RICAN SCREECH OWL
Megascops nudipes

L 20–23cm, **Wt** 103–154g; **W** 154–171mm

Identification A small owl without ear-tufts. No data on sexual size differences. Two morphs are known. Brownish-grey morph is greyish-brown above, slightly barred and spotted, with darker streaks and paler flecks; dark markings are in longitudinal rows on the crown and mantle. Whitish parts of scapulars form a comparatively indistinct pale row across the shoulder. Flight and tail feathers are barred light and dark, the dark bars on the tail being much broader than the whitish bars. Pale underparts become gradually whitish towards the belly; individual feathers on the underside have dark shaft-streaks and cross-bars, and there are many brown vermicular markings on the upper breast, only a few on the lower areas, and hardly any on the whitish belly. Reddish-brown facial disc has darker vermiculations forming almost concentric rows; the ruff is not distinct, but a clear white collar is present around the foreneck. Strong white eyebrows and yellow eyes. The bill and cere are greenish-grey. Only the uppermost part of the tarsus is feathered; the bare parts of tarsus and the toes are greyish-yellow, with blackish-tipped, light or dark horn-coloured claws. Rufous morph is in general

pale reddish-brown or rather foxy ochre-buff. *Juvenile* Downy chick is whitish. Mesoptile resembles the adult, but is less clearly patterned; it is more barred light and dark on the crown and mantle and below.

Call Utters a deep, guttural, toad-like , quavering trill, *rrurrrrrr*, lasting 3–5 seconds.

Food and hunting Feeds chiefly on insects and other arthropods, and occasionally takes small vertebrates.

Habitat Occupies thickets and caves in dense woodlands, and also near human settlements. From sea level to 900m.

Status and distribution Endemic to Puerto Rico and adjacent Culebra and Vieques, as well as Virgin Islands, in the Caribbean. Still common in Puerto Rico but extinct on Isla de Vieques. Since 90 per cent of forest in the Virgin Islands has been lost since the 1600s it is hardly surprising that recent surveys have failed to discover any signs of it there as it is most likely extinct.

Geographical variation Two subspecies are listed: nominate *nudipes* is found in Puerto Rico and adjacent islands; *M. n. newtoni*, from Virgin Islands (St Thomas, St John and St Croix), is less densely vermiculated and streaked below. These differences are sometimes thought to represent only polymorphic variation; this will be difficult to verify if the *newtoni* subspecies is already extinct (see above).

Similar species All other Caribbean screech owls have ear-tufts. Geographically separated Cuban Bare-legged Owl (105) has prominent whitish eyebrows, a relatively long tail, and very long and totally bare tarsi. There are only two true owl species in Puerto Rico, this one and the Short-eared Owl (247), which is much larger, with short ear-tufts and yellow eyes.

◀ Rufous morph of nominate Puerto Rican Screech Owl has pale rufous-brown coloration and is plainer red dorsally. Puerto Rico, April (*Lucas Limonta*).

102. RIO NAPO SCREECH OWL
Megascops napensis

L 20–23cm, Wt 109–129g; W 156–175mm

Identification A small owl with dark, short ear-tufts. No data on sexual size differences. Two morphs are known. Dark morph has a dark yellowish-brown crown densely spotted, streaked and vermiculated with blackish, and no pale border on the hindcrown. Dark cinnamon-brown upperparts have blackish speckles, streaks and vermicular markings. Blackish-edged white outer webs of scapulars form a distinct white row of spots across the shoulder. Flight and tail feathers are barred light and dark; the wingtips do not reach the tip of the tail. Underparts are pale yellowish-brown with narrow dark shaft-streaks and some cross-bars, and with fairly dense vermiculation on the darker upper breast. Brownish to yellowish-brown facial disc is finely darker-vermiculated, with a not very distinct ruff, and the eyebrows are almost whitish. The eyes are normally yellow, but rarely brown, and the bill and cere are light leaden-bluish. Tarsi are feathered to the base of the pale greyish-brown to fleshy toes, with dark-tipped, horn-coloured claws. Pale morph is in general more light cinnamon-buff. *Juvenile* Downy chick is whitish. Mesoptile is similar to the adult, but with indistinct barring on the head, nape, mantle and underparts.

Call Voice is very similar to that of allopatric Guatemalan Screech Owl (100), but a single phrase lasts for only 7–10 seconds.

Food and hunting Not well documented, but hunts mainly insects and other arthropods; locusts said to be favourite prey when available.

Habitat Prefers dense rainforests between 250m and 1500m.

Status and distribution Found from E. Colombia to Peru and N. Bolivia. This owl is locally not rare, but is not well studied.

Geographical variation Three subspecies are listed: nominate *napensis* occurs from E. Colombia to E. Ecuador; *M. n. helleri*, from Peru, is paler; *M. n. bolivianus*, from N. Bolivia, has the upper breast almost without vermiculations.

Similar species Northern Tawny-bellied (94) and Southern Tawny-bellied Screech Owls (95) live at lower elevations in dense primary forests; both are much

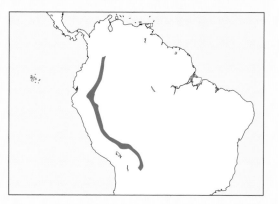

darker, with longer ear-tufts. Geographically separated Vermiculated Screech Owl (98) has relatively longer toes, as well as longer wings (reaching slightly beyond the tip of the tail), and lacks prominent whitish eyebrows. Larger White-throated Screech Owl (103) lives at much higher altitudes, lacks visible ear-tufts, but has, as its name suggests, a conspicuously white throat.

▶ Pale cinnamon-buff morph of nominate race *napensis*. Eyes are yellow, but sometimes brown. Large white spots on the scapulars Pichincha, Ecuador, March (*Scott Olmstead*).

▲ Rio Napo Screech Owl dark morph has dense vermiculations below, fading towards the belly. Napo, Ecuador, May (*Roger Ahlman*).

▶ Rio Napo Screech Owl pale morph has short but clear ear-tufts and a very dark cinnamon-brown back. Eastern Andes, Ecuador, January (*Jon Hornbuckle*).

103. WHITE-THROATED SCREECH OWL
Megascops albogularis

L 20–27cm, **Wt** 130–185g; **W** 190–213mm

Identification A small to medium-sized owl with no ear-tufts, but with a fluffy-feathered head. Females are said to be clearly heavier than males. Above, the plumage is dark fuscous-brown with blackish mottling and some tiny reddish-brown or dull yellowish and whitish spots, but without a row of whitish spots across the shoulder. Wings and relatively long tail are barred light and dark, the pale bars being very narrow on the tail. Tawny-buff underparts have dusky shaft-streaks and cross-bars; the ground colour is darker on the upper breast, becoming paler and more dull yellowish towards the belly. Dark facial disc lacks a distinct ruff. The whitish eyebrows are not particularly prominent. The eyes are orange-yellow to dull orange; the throat has a large oval area of white on each side of the greenish-yellow to greenish-grey bill. Tarsi are feathered to the base of the yellowish-grey to brownish-fleshy toes, with dark-tipped, pale horn-coloured claws. *Juvenile* Downy chick is dirty whitish. Pale yellowish-grey mesoptile has obvious orbital disc and long bristles above the bill; it is evenly dusky-barred on crown, mantle and below. Fledgling is similar to the adult, but less distinctly marked.

Call Gives fairly mellow, gradually descending hoots in rapid succession, four or five notes per second, repeated at intervals of 5–10 seconds.

Food and hunting Not well documented, but diet consists mainly of insects and other arthropods; probably also some small vertebrates.

Habitat Inhabits rainforests and cloud forests rich in epiphytes, but occurs also in stunted alpine forests and bamboo thickets; at 1300–3600m, but mainly between 2000m and 3000m.

Status and distribution Found from N. Colombia and NW Venezuela south to Cochabamba, in Bolivia. It is fairly common in the Andes of Ecuador (both slopes) and Peru, and possibly overlooked elsewhere.

Geographical variation Four subspecies are listed: nominate *albogularis* occurs from Colombian Andes to Ecuador; *M. a. meridensis*, from the Andes of Mérida, in Venezuela, has a whitish forehead and eyebrows; dark

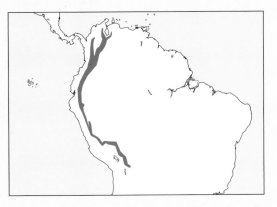

M. a. remotus is found from Peruvian Andes to Bolivia; *M. a. macabrum*, from Colombian W. Andes to W. Peru, has more finely patterned underparts.

Similar species This is the only screech owl in the higher elevations of the Andes with such a prominent white throat.

▶ White-throated Screech Owl is a large, round-headed screech owl without conspicuous ear-tufts; this is *remotus*, the darkest of all subspecies, almost black above and creamy-buff below the dark breast band. Eyes are yellow or orange-yellow. Amazonas, Peru, June (*Paul Noakes*).

▲ White-throated Screech
Owl of the race *macabrum*.
Reserva Yanacocha,
Pichincha, Ecuador
(*Lev Frid*).

◀ Nominate race White-
throated Screech Owl. Rio
Blanco Reserve (Manizales
area), Colombia, August
(*Leif Gabrielsen*).

104. PALAU OWL
Pyrroglaux podarginus

L 22cm; W 155–163mm

Other name Palau Scops Owl

Identification A small owl with no visible ear-tufts. Females are said to be a little heavier than males, although no data on weights are available. Sexes differ in colour: female is darker brown above, with fine blackish vermicular markings. Much darker and lighter rufous morphs exist. The crown and upperparts are reddish-brown, some neck feathers thinly barred brown and white. Sandy-rufous wings have pale reddish-brown to dull yellow bars, and the rufous tail is indistinctly barred dark brown. The throat is whitish to reddish-brown, the pale rufous breast is barred white and black, and the belly is paler reddish-brown. Forehead and supercilia are whitish with rufous-buff tinge, and thinly barred blackish-brown. Light yellowish facial disc has narrow darker, reddish-brown concentric rings; feathers at base of upper mandible have long blackish shafts. The eyes are brown or orange-yellow, the bill dirty whitish, and the tarsi and toes are dull creamy, with blackish claws. *Juvenile* Downy chick is light reddish-brown, with darker breast and back, and paler belly. Immature resembles adult male, but is darker brown above and more heavily barred below, with forehead, crown and back barred ochre and black, and white shaft-streaks on scapulars.

Call Utters a series of clear *kwuk kwuk* notes at one-second intervals. The call resembles that of Palau Fruit Dove *Ptilinopus pelewensis*, which sometimes calls during night.

Food and hunting Feeds chiefly on insects and other arthropods, including centipedes and earthworms.

Habitat Inhabits rainforest and mangrove forests, but is found also near villages in lowlands.

Status and distribution Endemic to the Palau Islands of Micronesia, in the Pacific Ocean. This owl is rather common locally but is obviously endangered on many islands.

Geographical variation Monotypic.

Similar species The only other owl in the Palau Islands is the much larger Short-eared Owl (247), which is very different in appearance.

▶ A small, rufous-coloured owl with barely visible ear-tufts. Palau, February (*Mandy Etpison*).

105. CUBAN BARE-LEGGED OWL
Gymnoglaux lawrencii

L 20–23cm, Wt 80g (1); W 137–154mm

Other names Bare-legged Owl, Cuban Screech Owl

Identification A small, round-headed owl without ear-tufts. No data on sexual size differences. Plumage is brown above, with blackish spots on the crown and hindneck, and whitish tips and darker areas in centre on feathers of mantle and wing-coverts; small whitish areas on outer webs of scapulars form an only very indistinct white row across the shoulder. Primaries have whitish spots on outer webs, but inner webs are uniformly brownish, and secondaries have narrow whitish bars. The tail has ten feathers, with thin whitish bars on outer ones. Creamy-whitish underparts, with a dull yellowish suffusion, are finely black-spotted, with neck and throat washed buffish-brown; distinct, drop-shaped dark shaft-streaks are apparent on entire underparts. The facial disc is whitish to dull yellow, with prominent whitish eyebrows. The eyes are brown and the bill and cere greyish-yellow. Long tarsi and toes are totally bare, yellowish-brown in colour, with dark-tipped horn-coloured claws. *Juvenile* Downy chick is whitish. Mesoptile resembles adult, but is less spotted above. *In flight* Shows short, rounded wings and long tail and legs.

Call Gives a soft, accelerating *cu-cu-cu-cuencuk*, a little higher-pitched towards the end.

Food and hunting Diet consists mainly of insects and other arthropods, but also frogs and snakes, rarely

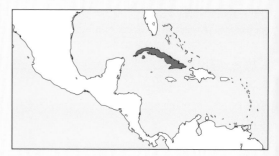

even small birds. Hunts normally on the ground or by swooping down from a perch.

Habitat Inhabits thickets and dense forests, as well as semi-open limestone areas with caves and crevices; found also in large plantations.

Status and distribution Occurs in Cuba and on Isle of Pines (Isla de la Juventud). This species is said to be fairly common.

Geographical variation Monotypic. DNA analysis is needed in order to clarify the true relationships of this genus within Strigidae; it may be more closely related to *Athene*.

Similar species Partly sympatric Burrowing Owl (202), which breeds only in a few parts of Cuba, is larger and is barred below, with yellow eyes and feathered or bristled legs.

◀ Cuban Bare-legged Owl peeking from the nest hole. Note the dark eyes, and white 'eyebrows' and face. Zapata Swamp, Cuba, March (*Oliver Smart*).

▶ Cuban Bare-legged Owl at the nest stump. Bermejas, Cuba, November (*Adam Riley*).

◀ Cuban Bare-legged Owl lacks erectile ear-tufts and has brown eyes. It has a brown throat and drop-shaped shaft-streaks below. Belen, Cuba, December (*Christian Artuso*).

▼ Cuban Bare-legged Owl is spotted above and with a banded tail, which is long but contains only ten rectrices. This owl has long, bare tarsi and toes. Zapata Swamp, Cuba, February (*Arthur Grosset*).

106. NORTHERN WHITE-FACED OWL
Ptilopsis leucotis

L 24–25cm, Wt average 204g (16); W 170–209mm, WS 50cm

Identification A small owl with long ear-tufts, often with blackish tips. Females are on average some 30g heavier than males. Two colour morphs exist. Light morph has whitish facial disc with broad blackish ruff. Fairly pale greyish-brown upperparts have faint vermicular markings and many dark shaft-streaks. Dark-edged white outer webs of scapulars are not conspicuous. Flight and tail feathers are barred light and dark greyish-brown. Underparts are pale with dark shaft-streaks and fine vermiculations. The eyes are deep amber-yellow or orange and the bill yellowish-horn. Basal half of dusky-brown toes is feathered, with blackish claws. Dark morph is much darker, with an ochre tinge, and has brownish-white facial disc, blackish crown, and blackish centres of ear-tufts. *Juvenile* Downy chick is whitish. Mesoptile is greyish-white with greyish-brown feather tips, particularly on crown, nape and back, and has dark grey-brown ruff and yellow eyes. *In flight* Shows rounded wings, with 8th primary the longest.

Call Utters a mellow, disyllabic *po-proo*, first note very short, longer second note following after 0.6 seconds, the phrase repeated several times at intervals of 4–8 seconds.

Food and hunting Hunts invertebrates and small vertebrates. Eats vertebrates more than do any *Otus* owls. Hunts from a perch, swooping down on to its prey on the ground.

Habitat Inhabits dry open forests or savannas with scattered trees, avoiding deserts and dense rainforest. Often seen near settlements, even in suburban gardens and towns. Occurs from near sea level to 1700m.

Status and distribution Occurs in Africa south of the Sahara, from Senegal and The Gambia across the continent to Ethiopia and Somalia, and south to N. Uganda and N. & C. Kenya. This owl is rather common locally, but rare in Somalia.

Geographical variation Monotypic.

Similar species Very similar Southern White-faced Owl (107) overlaps only in Kenya and Uganda; it is greyer in general and darker above, has more powerful talons, and its eye colour varies from orange-red to red. Difference in song is the most reliable method of identification.

▶ Northern White-faced Owl has a broadly dark-rimmed white face, heavily barred plumage, and yellow-orange eyes. Western Nigeria, July (*Tasso Leventis*).

◀ Northern White-faced Owl adopting a 'tall-thin' position. Ear-tufts are up, so the feathers are pointing in many directions. Baringo, Kenya, October (*Eyal Bartov*).

107. SOUTHERN WHITE-FACED OWL
Ptilopsis granti

L 22–24cm, Wt 185–275g; W 191–206mm

Identification A small owl with long ear-tufts vermiculated and streaked with blackish. Females are on average well over 30g heavier than males. Nearly pure white facial disc contrasts clearly with the broad black ruff. Upperparts are fairly dark grey, with well-pronounced black shaft-streaks and many fine vermicular markings on the crown, nape and mantle. Black-edged white outer webs of scapulars form a white line across the shoulder, and black streaks and fine mottling are visible on the upperwing-coverts. Flight and tail feathers are barred light and dark. Paler grey underparts have fine blackish shaft-streaks and fine dark vermiculations. The eyes are orange-red to ruby-red, and the bill creamy-horn. Tarsi are feathered to the basal half of the toes; tarsi are pale grey, and bare parts of toes dusky greyish-brown, with blackish-horn claws. **Juvenile** Downy chick is white. Mesoptile resembles the adult, but is less distinctly marked and has yellowish-grey eyes. Fledgling has yellow eyes, shorter ear-tufts and less pronounced plumage pattern than adult.

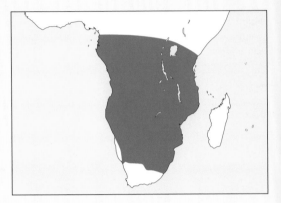

Call The song consists of a rapid, stuttering staccato trill followed by a drawn-out *whhhhhhhu- hóoh*, the last part rising in pitch; the phrase is repeated at intervals of several seconds. This species is very vocal; one rescued in Nampula, Mozambique, stayed with the author for several months, and was found to be the most vocal of all owl species which he had cared for.

Food and hunting Diet consists of large insects, spiders, scorpions, small birds, reptiles and small mammals.

Habitat Lives in dry open woodlands and savanna with scattered trees and thorny scrub, often near human settlements.

Status and distribution Occurs in the southern part of Africa, from Cameroon to Namibia in the west and south from S. Uganda and S. Kenya in the east, but only north of Lesotho in South Africa. This species is rather common locally.

Geographical variation Monotypic.

Similar species Almost geographically separated Northern White-faced Owl (106) is pale greyish-brown with an ochre tinge, with less powerful talons, more prominent vermicular markings below, and less marked contrast between black ruff and ochre-tinged whitish facial disc.

▶ Southern White-faced Owl fully alert, with ear tufts erect and eyes wide open. Namibia (*Adam Riley*).

◀◀ Southern White-faced Owl is darker and greyer than Northern *Ptilopsis leucotis*, but there is large individual variation, and light and dark birds occur in both species. Vocalisations are often the only way to separate them. Namibia (*Adam Riley*).

◀ Southern White-faced Owl immature is similar to the adult but with orange-yellow eyes, shorter, slightly downy ear-tufts, and slightly less pronounced plumage patterns. Namibia, November (*Wil Leurs*).

108. GIANT SCOPS OWL
Mimizuku gurneyi

L 30–35cm; **W** 217–274mm

Other name Lesser Eagle Owl

Identification A medium-sized owl with long, black-spotted ear-tufts. Females have much longer wings and are said to be heavier than males. Pale reddish-brown facial disc has a narrow ruff with black spots, and frosty white eyebrows fade into dull yellowish. Upperparts are dark reddish-brown with blackish shaft-streaks. Blackish-edged whitish to dull yellow outer webs of scapulars often form a line of pale spots, and dark brown wing-coverts have obvious black shaft-streaks. Flight and tail feathers are banded light and dark. Whitish-buff underparts become creamy white on the belly, and breast has large drop-shaped or oval black spots. The eyes are brown and the bill greenish-yellow to greyish-white. Tarsi are feathered to base of the light greyish-brown toes, which have dark-tipped pale horn-coloured claws; talons powerful. *Juvenile* Undescribed.

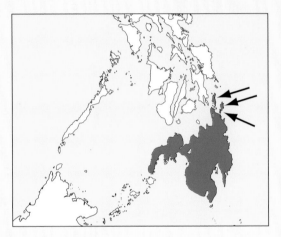

In flight Shows rounded wings with 6th or 7th primary the longest.

Call Utters a mournful, growling *wuoohk wuoohk*, often in series of five to ten notes, and repeated at intervals of 10–20 seconds.

Food and hunting Not well documented but probably eats small vertebrates and larger insects.

Habitat Inhabits rainforest and secondary growth from sea level to 1200m, but recently reported from much higher elevations (to 3000m). It is sometimes found far from forest, in small groves in grasslands.

Status and distribution Found on Mindanao, Siargo and Dinagat, in S. Philippines. This species is normally rare, occurring only at low densities, but is poorly studied. It is listed as Endangered by BirdLife International owing to forest destruction in the Philippines.

Geographical variation Monotypic. A recent study has shown that this species is closely allied to *Otus*.

Similar species All *Otus* owls are much smaller, and *Bubo* owls larger and much heavier.

▶ Giant Scops Owl is a large and attractive owl with long, slightly curved ear-tufts. Unmarked face and undersides rufous with black shaft-streaks on the breast and flanks. Mindanao, February (*James Eaton*).

◀ Giant Scops Owl has an unmarked dark-rufous face, and very light rufous to white underparts with dark streaks on the flanks. The curved ear-tufts are not erected here. Mindanao, February (*Rob Hutchinson*).

109. SNOWY OWL
Bubo scandiacus

L 53–70cm, **Wt** 710–2950g; **W** 384–462mm, **WS** 142–166cm

Identification A fairly large to very large owl with normally invisible ear-tufts. Female is some 300–400g heavier than male; sexes differ also in the degree of dusky patterning on the white plumage. Adult male has plain white upperparts, while female is spotted and slightly barred, and underparts of male are all white, whereas those of female have brown spotting and barring. The female has faintly brown-barred flight and tail feathers, while the male has only a few dusky spots on the alula and at the tips of some primaries and secondaries. The ill-defined facial disc is white, the miniature ear-tufts with a few dusky spots. The eyes are bright yellow rimmed by blackish edges of eyelids, and the base of the blackish bill is densely feathered. Both the tarsi and the toes are thickly white-feathered, with blackish claws. *Juvenile* Downy chicks is greyish-white. Mesoptile is dark greyish-brown, gradually becoming darker with barring on white plumage. Fledgling looks irregularly mottled and blotched, with a mostly dark

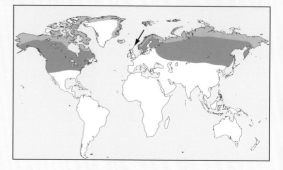

head including contrasting white face and eyebrows. *In flight* Flies with rowing wingbeats interrupted by gliding on stretched wings.

Call The deep and loud, booming song contains two to six *goo* notes; when disturbed, this owl emits a loud, harsh, grating bark, *kre-kre-kre-kre*.

Food and hunting Feeds mainly on small mammals

◀ Male Snowy Owl becomes whiter with age. This male is still fairly young: perhaps 5 years old. Michigan, USA (*Steve Gettle*).

▶ Female Snowy Owl is spotted and barred but has a white face. Michigan, USA (*Steve Gettle*).

(not only lemmings) and birds, and occasionally takes frogs, fish and large insects. Hunts from a slight elevation, such as a hummock or rock, gliding down on to prey.

Habitat Inhabits arctic tundra with mosses, lichens and rocks; prefers areas with slight elevations, often near the sea coast to 300m, but in Norway sometimes up to 1000m.

Status and distribution Has a circumpolar distribution in the very north, and in years with good vole numbers it can be locally common, but is totally absent when food is not available. In the UK no Snowy Owls have bred since 1975, when, famously, a pair bred in Shetland. In Finland there were 30–35 breeding pairs in 1974, but the next year when there were more than ten nests in that country was 2011, when Norway still held more than 30 breeding pairs. Further studies of the circumpolar movements of this species would be useful, as Norwegian ornithologists believe that, because its populations are highly nomadic, the world population is much smaller than was previously thought. This owl winters irregularly as far south as S. & C. California,

C. Texas and Georgia, in the USA. In Eurasia it stays farther north, even during irruption years.

Geographical variation Monotypic. This species was formerly named *Nyctea scandiaca*, but DNA–DNA hybridisation has shown that *Nyctea* and *Ketupa* owls are genetically closely related and are osteologically identical to *Bubo*, and these three genera have, therefore, been united.

Similar species This is the only all-white owl in the Holarctic. Great Horned Owl (110) is fairly similar in size, but has a dark back and very prominent ear-tufts; very pale and extensively white subspecies *B. v. subarcticus* in Canada could be confused in poor light with the female Snowy Owl, but it has visible dark-spotted ear-tufts. Slightly larger Eurasian Eagle Owl (112) has large, orange eyes and prominent ear-tufts; the palest race, *B. b. sibiricus*, is much larger and somewhat darker, especially above. During the breeding season all these owls are geographically separated, but they overlap in range during the southward migrations of the Snowy Owl.

◄ Snowy Owl male brings a prey item to the nest, where the female and young await. Alaska North Slope, July (*Paul Bannick*).

◄ Second calendar-year male Snowy Owl taking off; its huge, fully feathered legs are hanging below. Finland, March (*Esko Rajala*)

► Female Snowy Owl in flight showing the dense spotting and barring. Quebec, Canada (*Rob McKay*).

110. GREAT HORNED OWL
Bubo virginianus

L 45–64cm, **Wt** 900–2503g; **W** 297–390mm, **WS** 91–152cm

Identification A fairly large to large owl with prominent ear-tufts. Female can be on average 700g heavier than male, but sexes are identical in plumage. Individual variation in coloration occurs. Above, the plumage is warm brownish to dull yellow, mottled and vermiculated with greyish-brown, black and white, the crown finely barred dark and light. Outer webs of scapulars have a fairly large whitish zone marked irregularly with a few dark transverse bars, and with an inconspicuous row of whitish spots across the shoulder. Flight and tail feathers are prominently barred dark and light. The throat is whitish, this being prominent when the bird calls. The brownish to dull yellow underparts become paler towards the abdomen; there are blackish blotches and some cross-bars on upper breast, and the rest of the underparts are coarsely barred light and dark. The rusty-brown to ochre-buff facial disc is paler around the lemon-yellow eyes, with a thin black border on the eyelids giving the face a 'fierce' expression; blackish ruff and whitish eyebrows are prominent on each side of the greyish bill. The large feet are densely feathered to the end of the greyish-brown toes, which have dark horn-coloured claws. *Juvenile* Downy chick is whitish. Mesoptile is pale brownish to dull yellow and has fluffy feathers on the mantle and back, and indistinct dark barring below; the ear-tufts are smaller or not so apparent.

Call Utters deep booming hoots, *hu-hu hoooooo hoh-hoh*, resonantly repeated fully or in part at intervals of several seconds.

Food and hunting Feeds mainly on mammals up to the size of rabbits, and also on birds, reptiles, frogs, spiders and larger insects. This owl hunts from a perch in open or semi-open areas, or by gliding slowly above the ground.

Habitat Prefers second growth, swamps and open woodlands and scrub with rocky areas, even near human habitations in towns and larger parks. Lives from sea level to mountains, up to 4500m in Andes, but avoids dense cloud forest and primary rainforest.

Status and distribution Has a wide distribution in America, from Alaska and SE Canada south to Uruguay. This species is locally common, but is endangered in some regions owing to habitat destruction and human persecution. Road traffic and power-lines take a heavy

◄ A pair of Great Horned Owls. Race is either *subarcticus* or *wapacuthu*, which are the palest subspecies. Male is whiter than the female, especially on the face. Manitoba, November (*Christian Artuso*).

toll everywhere. Despite these heavy losses, the world population has been estimated at 5.3 million birds, a figure which includes Magellan Horned Owls (111).

Geographical variation Individuals vary in size, tending to become smaller from NE to SW, being small in south of range, and generally darker in more humid regions. Further taxonomic study is required, but 13 subspecies are listed: nominate *virginianus* is found from Canada to Florida; *B. v. saturatus*, from Alaska to California, is mostly very dark; *B. v. subarcticus*, from British Columbia and Mackenzie Valley in east to W. & N. Ontario in the west, is the palest subspecies, white predominating in plumage; *B. v. wapacuthu*, from N. & NE North America, is also very light; *B. v. pacificus*, in SW USA, is fairly similar to the following race; *B. v. heterocnemis*, in E. North America, occurs typically as a dark morph, the pale morph being rare; *B. v. occidentalis*, from C. Alberta to California, typically occurs as a greyish morph, but is possibly a synonym of *wapacuthu*; *B. v. pallescens*, from SE California to N. Mexico, is smaller and paler than nominate; *B. v. elachistus*, confined to S. Baja California south of 30°N, is the smallest subspecies in North America; *B. v. mayensis*, from Mexico to Costa Rica and W. Panama, is smaller than nominate but similar in plumage; *B. v. nigrescens*, from Colombia and Ecuador, is the darkest subspecies; *B. v. deserti*, confined to Bahia, in E. Brazil, has fine barring and vermiculations; *B. v. nacurutu*, from NW Venezuela to Argentina, is smaller than nominate (♂ 300g and ♀ 600g lighter). In captivity, the Eurasian Eagle Owl has hybridised with the Great Horned Owl.

Similar species Snowy Owl (109) is similar in size, but only the subarctic race could be confused with female Snowy; elsewhere, Great Horned Owls are much darker and all subspecies have more prominent ear-tufts. Smaller Magellan Horned Owl (111) hardly overlaps in range and is paler, with a smaller bill and weaker talons; below, it is more finely barred dark and light, and has a very different vocalisation. In northern parts of the American continent, even the Great Horned Owl's distinctive call may be confused with that of the sympatric Great Grey Owl (161), which, however, lacks ear-tufts and has a large rounded head, small yellow eyes, and grey plumage with dark markings.

▼ Great Horned Owl race *nigrescens* is the darkest race. Napo, Ecuador, May (*Roger Ahlman*).

▼ Race *nacurutu* is paler than *nigrescens* and fairly similar to the smaller *Bubo magellanicus*, but ventral barring is more widely spaced. Corrientes, Argentina, November (*James Lowen*).

▲ Race *saturatus* from western North America is much darker than Canadian owls. Washington State, USA, January (*Jim and Deva Burns*).

▲▶ Race *pallescens* is smaller and much paler than the nominate. New Mexico, USA, February.

▶ A large immature bird, still begging food from the female in September. Manitoba, Canada (*Christian Artuso*).

▶▶ Great Horned Owl nominate race has very prominent ear-tufts. Howell, Michigan, USA (*Steve Gettle*).

111. MAGELLANIC HORNED OWL
Bubo magellanicus

L 45cm, **Wt** 830g (1♂); **W** 318–368mm

Other names Magellan Horned Owl, Lesser Horned Owl

Identification A medium-sized owl with neat, rather narrow and pointed, dark brown or blackish ear-tufts. Sexes are alike in plumage; females have slightly longer wings than males, but sexed weights are not available. Light and dark morphs exist; light morph has plain whitish underparts and dark morph is generally darker and browner, with less white below. Above, the plumage is greyish-brown with blackish shaft-streaks, mottled with brownish dots and dark cross-bars. Flight and tail feathers are barred blackish and greyish-brown, and whitish outer webs on scapulars are not prominent. A thin dark line separates the chin from the white throat, the latter being bordered ventrally by a row of dark dots. Below, the plumage is whitish to light brownish-grey with fine dark brown barring. The pale greyish-brown to ashy-grey facial disc, whitish towards the relatively small, bluish-grey

bill, has a blackish ruff. The eyes are bright yellow, the eyelids having a thin blackish rim; the eyebrows paler than crown, but are not distinct. The relatively weak talons and toes are feathered dirty white, with dark horn-coloured claws. *Juvenile* Resembles that of the previous species.

Call Utters two deep hoots, giving the onomatopoeic name of 'tucúquere'; the second hoot is stressed, and followed by a low, quiet, purring sound.

Food and hunting Feeds primarily on mammals up to the size of hares, but also takes birds and reptiles.

Habitat Prefers rocky landscapes, often in mountains and even above timberline, in Andes between 2500m and 4500m, but lives also at sea level and near human settlements.

Status and distribution This owl is locally common from the Andes of C. Peru south to Cape Horn. The introduction of rabbits has increased its numbers in Chile.

Geographical variation Monotypic. DNA-sequence difference between this species and the Great Horned Owl is 1.6 per cent, justifying their separation as two distinct species, which also differ in size and colour.

Similar species Geographically separated Great Horned Owl (110) is darker and larger, with stronger talons and bill; below, the pale bands between the dark bars are much broader, in addition to which the vocalisations are totally different. All other South American owls are much smaller.

Magellanic Horned Owl has smaller and more pointed ear-tufts than Great Horned Owl. Santa Cruz, Argentina, November (*James Lowen*).

Magellanic Horned Owl is smaller than Great Horned Owl *Bubo virginianus nacurutu*, with a relatively smaller bill and talons. It also has denser barring below, yellow eyes and relatively small ear-tufts. Tierra del Fuego, Chile, December (*Arthur Grosset*).

Magellanic Horned Owl also lives near human settlements. Peru (*Fabio Olmos*).

112. EURASIAN EAGLE OWL
Bubo bubo

L 58–75cm, **Wt** 1500–4600g; **W** 405–515mm, **WS** 150–188cm

Identification A large to very large owl with prominent ear-tufts. Female is on average as much as 1 kg heavier than male. The upperparts are dull yellowish to brown with blackish streaks and cross-bars, the flight and tail feathers yellowish-brown with blackish or dark brown bars. The white throat is very prominent when the bird is calling. Below, the buffish orange-brown underparts have blackish streaks on the upper breast, and dark shaft-streaks and fine cross-bars on the lower underparts. The facial disc is greyish-brown with a dull yellowish suffusion, and has a thin and not very prominent slightly darker ruff. The whitish eyebrows accentuate the very large golden-yellow to bright orange-red eyes. An area of whitish feathering surrounds the blackish bill. Has densely feathered tarsi and strong toes, with relatively long and powerful claws blackish-brown with black tips. *Juvenile* Downy chick is whitish. Mesoptile is pale yellowish-brown, with many indistinct darker bars on head, mantle and back and below, fluffy feathers indicating the ear-tufts. Juvenile has milky yellow-orange eyes. *In flight* Shows darker upperwing-coverts and dark tips of greater primary coverts. The nearly noiseless flight is due to soft wingbeats, interrupted by gliding; it sometimes soars.
Call Gives a deep, resonant and booming hoot, *ooo-hu* or *boo-ho*, uttered at intervals of 8–10 seconds, the first syllable emphasised and the second lower.
Food and hunting Feeds mainly on mammals, ranging from shrews to hares, and similarly takes birds of almost all sizes; occasionally eats reptiles, frogs, bats etc. Generally hunts from a perch, but can also use searching flight.
Habitat Prefers open forests with rocky cliffs, but it lives also near human settlements and city buildings. Occurs from sea level to 2000m in Europe, and up to 4500m in C. Asia and the Himalayas.
Status and distribution Human persecution and

▶ Male Eurasian Eagle Owl is a fearsome predator. This is race *hispanus*, which is a little smaller than the nominate. Southern Spain, March (*Vincenzo Penteriani*).

◀ A rabbit is served by the female to the chicks in the nest (race *hispanus*). Southern Spain, March (*Vincenzo Penteriani*).

habitat destruction over a long period of time have endangered populations in Europe, but successful reintroductions have been made in many countries. The European population (excluding Russia) has been estimated to be 12,000 pairs. In Finland, the population has been declining for more than 15 years, the overall decrease from 1982 to 2010 having been 2.2 per cent per year; this decline is connected with the closing of the local refuse sites, especially in rural areas. Unfortunately, human persecution continues still in Finland, often for the claimed reason of protecting game-hunters' interests; e.g. in 2011 more than ten ringed young Eurasian Eagle Owls were killed in different villages. In the UK, the locations of the very few known nest sites are kept secret in order to avoid similar game-related killings to those in Finland. Presence in N. Morocco is unconfirmed.

Geographical variation Races differ mainly in general coloration, strength of dark markings, and size. At least 13 subspecies are recognised: nominate *bubo* occurs from Scandinavia and the Pyrenees east to NW Russia and Moscow; *B. b. hispanus*, from the Iberian Peninsula, is smaller, paler and greyer than nominate; *B. b. ruthenus*, from Moscow to the Ural and Volga Rivers, also is paler and greyer and less buffish; *B. b. interpositus*, from Crimea to Asia Minor and Iran, is darker and more rusty; *B. b. sibiricus*, from W. Siberia to Altai Mountains, is very large and whitish; *B. b. yenisseensis*, from C. Siberia between Ob, Lake Baikal and Altai to N. Mongolia, is darker and greyer and more yellowish than

the previous race; *B. b. jakutensis*, from NE Siberia, is darker and browner than the previous; *B. b. ussuriensis*, from SE Siberia to N, China, also including the Kuril Islands, is darker above than *jakutensis* and with a more ochre wash below; *B. b. kiautschensis*, from Korea and China, is smaller and darker; *B. b. turcomanus*, from the Volga to the Aral Sea and W. Mongolia, is very pale and yellowish; *B. b. omissus*, from Turkmenistan and Iran, is a typical pale ochre desert race; *B. b. nikolskii*, from Iran to Pakistan, is smaller than the previous subspecies and less dark above; *B. b. hemachalana*, from Kyrgyzstan to Baluchistan and the Himalayas, is pale brown in general. In addition, the taxonomy of *Bubo bubo* requires study, as *interpositus* interbreeds freely with Pharaoh Eagle Owl, and *turcomanus* with Rock Eagle Owl, and in both instances the progeny has proven to be fertile. Also, some DNA samples appear to support the separation of *interpositus* as a distinct species.

Similar species During the breeding season allopatric Snowy Owl (109) is largely white, with no clear ear-tufts. Nearly geographically separated Pharaoh Eagle Owl (113) is smaller and more sandy-coloured, with a pale face, dark and light speckled ear-tufts, and black spots on the upper breast. Relatively large fish owls have no or very tousled ear-tufts, and bare tarsi or at least bare toes. Geographically overlapping Great Grey Owl (161) looks a little similar in bulk, but not in weight; it has a large rounded head without ear-tufts and with small, yellow eyes, and the plumage is grey with dark markings.

▲ Adult Eurasian Eagle Owl
race *omissus*. Iran (*Ali Sadr*).

▲ ▶ Adult Eurasian Eagle Owl
race *interpositus*. Russian
Caucasus, September.

◀ A 70-day-old male Eurasian
Eagle Owl of race *hispanus*.
Southern Spain, March
(*Vincenzo Penteriani*).

▶ Eurasian Eagle Owl race
yenisseensis in flight. This
race has only slightly darker
upperwing-coverts, and dark
tips to the greater primary
coverts, but wings are just
as white below as in the very
pale *sibiricus*. Mongolia, June
(*Mathias Putze*).

113. PHARAOH EAGLE OWL
Bubo ascalaphus

L 45–50cm, **Wt** 1900–2300g; **W** 324–430mm, **WS** 100–120cm

Other name Desert Eagle Owl
Identification A fairly large owl with relatively short and pointed ear-tufts, light tan-coloured with dark speckles and sandy-brown edges. Female is on average some 400g heavier than male. Has tawny-rufous upperparts marked with dark and light, giving blotched effect. Flight and tail feathers are barred light and dark. The throat is white and the under-parts pale tawny-brown to sandy-coloured, the upper breast with dark drop-shaped shaft-streaks and a few cross-bars; the abdomen is finely marked with dark, but barring is less prominent than on Eurasian Eagle Owl. The rounded facial disc is plain tawny, rimmed by a row of fine blackish spots. The eyes are yellow to deep orange and the bill black. Tarsi and toes are feathered pale tawny, the tips of the toes sooty-brown with blackish-brown claws. *Juvenile* Downy chick is whitish, with a dull yellowish suffusion on the fore-head, wings and rump. Fluffy mesoptile is barred above, only slightly so on upper breast, and has poorly developed ear-tufts.
Call Utters a short *whu*, downward-inflected and higher in pitch than that of Eurasian Eagle Owl or the Long-eared Owl.
Food and hunting Eats mammals, birds and reptiles, and also scorpions and larger insects. Normally hunts from a perch.
Habitat Inhabits rocky mountain slopes in deserts and semi-deserts, but is found also in dry savannas.

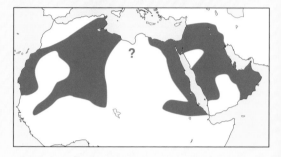

Status and distribution Occurs in N. & NE Africa from Tunisia south to The Gambia, Mali, Sudan and Eritrea, and in Middle East from Syria to W. Iraq and south to Oman. This owl is locally endangered by human persecution, but its status overall is not known.
Geographical variation Monotypic. Paler Saharan and Arabian populations are sometimes separated as race *desertorum*. The biology and behaviour of this owl demand study. It is known to interbreed with Eurasian Eagle Owl (of race *interpositus*) and an intergrade popu-lation exists in the Middle East; in captivity it has inter-bred also with Rock Eagle Owl.
Similar species Largely geographically separated Eurasian Eagle Owl (112) is much bigger and darker, with very prominent, dark ear-tufts. Same-sized African fishing owls (131–133) have no ear-tufts. Partly overlapping Long-eared Owl (243) is smaller and much slimmer, and is boldly striped below.

► Pharaoh Eagle Owl is variable in colour. This owl is a 'typical' morph, with buff-orange ground colour. It has short ear-tufts and blacker markings dorsally than *Bubo bubo*. The tail is tapering and rather narrow. Egypt, February (*Daniele Occhiato*).

► Pharaoh Eagle Owl young; very well camouflaged in its rocky surroundings. Egypt, February (*Daniele Occhiato*).

◄ Adult Pharaoh Eagle Owl: a pale individual. Sweihan, United Arab Emirates, May (*Hanne and Jens Eriksen*).

114. ROCK EAGLE OWL
Bubo bengalensis

L 50–56cm, **Wt** 1100g (1♂); **W** 358–433mm

Other name Indian Eagle Owl

Identification A fairly large owl with brown ear-tufts. Female has longer wings and tail than male, but sexed weights are not available. The crown appears dark, and the yellowish-brown forehead has some small blackish flecks. The tawny-brown upperparts are mottled and streaked with blackish-brown. Flight and tail feathers are tawny to dull yellow, barred blackish-brown; the wingtips are pointed. Underparts are reddish-yellow, more whitish towards the centre; there are small dark streaks on the upper breast, and fine shaft-streaks and faint cross-bars below, and the belly is only faintly barred. The unmarked fulvous-brown to dull yellowish facial disc has a prominent blackish ruff. Whitish eyebrows continue to above the centre of the deep yellow to orange-red eyes. The bill is greenish-horn to slate-black.

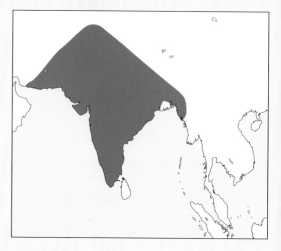

▼ Rock Eagle Owl demonstrating its hunting skills – powerful feet are ready to strike. Karnataka, India, March (*Niranjan Sant*).

Tarsi and toes are feathered reddish-yellow, the outer toe joints bare and the tips of the toes greenish-slate, with dusky-black claws. *Juvenile* Downy chick is whitish with dull yellowish suffusion. Mesoptile has narrow reddish-yellow to brown bars on the head and mantle and below, and the ear-tufts not well developed. *In flight* Shows pointed wings, with 8th primary the longest.

Call Gives a deep, double-noted hoot, *bu-whúoh*, the second note longer and more stressed, repeated at intervals of several seconds. Call is higher than that of the Eurasian Eagle Owl.

Food and hunting Hunts rats and mice, and also birds, reptiles, frogs, crabs and large insects. Hunts from a perch or in a low foraging flight.

Habitat Inhabits rocky semi-deserts, but occurs also in wooded areas and orchards near human habitations. Found mainly in lowlands, but also in mountains at up to 2400m.

Status and distribution Found from W. Himalayas east to W. Burma, and south through Pakistan and India, but not in Sri Lanka. It is not uncommon in suitable habitat but its status is uncertain. Detailed studies are needed for its conservation.

Geographical variation Monotypic. Taxonomic study is required. Eurasian Eagle Owl of subspecies *turcomanus* overlaps in range in Kashmir, and has produced fertile hybrids with the Rock Eagle Owl in captivity. Pharaoh Eagle Owl has also interbred with Rock Eagle Owl in captivity.

Similar species Geographically separated Eurasian Eagle Owl (112) is larger, and has more rounded wings (7th and 8th primaries equal in length), more prominently streaked underparts, and totally feathered toes. Sympatric Brown Fish Owl (128) has yellow eyes, tousled ear-tufts, and bare tarsi and toes.

▼ Rock Eagle Owl has fully feathered tarsi and toes. Eyes are orange-yellow. Karnataka, India, March (*Niranjan Sant*).

▼ Rock Eagle Owl is relatively small but has long ear-tufts. Haryana, India, November (*Amano Samarpan*).

115. CAPE EAGLE OWL
Bubo capensis

L 46–58cm, **Wt** 905–1800g; **W** 330–428mm, **WS** 120–125cm

Identification A fairly large owl with prominent, mostly brown ear-tufts. Female is on average up to 350g heavier than the male. Has dark brown upperparts mottled and spotted with whitish, black and reddish-yellow to brownish-yellow. Large white areas and dark dots on outer webs of scapulars are fairly obvious, and

large white spots on wing-coverts form a whitish bar on the closed wing. Flight and tail feathers are barred light and dark. The throat is white, and the underparts are pale reddish-yellow to brown, the upper breast densely black-blotched and the abdomen with blackish spots and coarse bars. The facial disc is light tawny-brownish, with a black or dark brown ruff. The eyes are orange-yellow to orange and the bill dusky horn-coloured. Tarsi and toes are densely feathered, the outermost parts of toes being brownish, with the underside yellowish and bare, and the claws are dark with blackish tips. *Juvenile* Downy chick is whitish. Mesoptile is dull brownish-white, with dark barring on the head, mantle and back and below. Juvenile has yellow eyes.

Call Gives a powerful disyllabic or trisyllabic *ho hoooo* or *ho hoooo ho*, which it repeats at intervals of several seconds. Sometimes a single prolonged hoot, *hoooooo*.

Food and hunting Feeds on mammals ranging from shrews to hares, and on birds of all sizes, reptiles, frogs, scorpions, crabs and larger insects. Hunts from a prominent perch.

Habitat Inhabits mountainous and hilly regions with rocky cliffs and ravines, but also comes to towns in order to catch feral pigeons. Normally between 2000m and 4200m above sea level in E. Africa, but much lower in S. Africa, where it often occurs at or

◄ Nominate Cape Eagle Owl is the smallest subspecies, with a dark brown crown and long ear-tufts. The dark-blotched breast often has less rufous and a more buffish wash than other races. Western Cape, South Africa (*John Eveson*).

near sea level in flat, dry and open grasslands.

Status and distribution This species is widespread in E. & S. Africa from Eritrea and Ethiopia south through Kenya, Tanzania, Malawi, Mozambique and Zimbabwe to South Africa and Namibia, but distribution is irregular and patchy. It is locally rather common in some areas and totally absent in others.

Geographical variation Three subspecies are listed: nominate *capensis* is found in S. Africa and S. Namibia; *B. c. dilloni*, from Ethiopia and Eritrea, is browner and less coarsely marked below; *B. c. mackinderi*, from Kenya to Mozambique and Zimbabwe, is larger than the others and has more fulvous-tawny plumage. This largest subspecies has often been given full species status, but recent molecular studies show that it hardly differs from the nominate race; it is possible that it may not even be a good subspecies. The taxon *dilloni*, however, is well separated from the nominate race and shows the typical DNA-sequence distance of *Bubo* subspecies.

Similar species Sympatric, yellow-eyed Spotted Eagle Owl (117) is smaller and is spotted and barred below, but without dense blotching on the sides of the upper breast. Only partly overlapping Greyish Eagle Owl (118) is also smaller, greyer and finely barred and vermiculated below, and has dark brown eyes with pink or reddish rims of eyelids. Partly sympatric Milky Eagle Owl (121) is much larger, is pale brownish-grey with dense vermicular markings, and also has dark eyes with flesh-pink eyelids and rather fluffy ear-tufts.

▼ Male of race *mackinderi* at the nest. It is more ochre-blotched above, often with a more rufous tinge below, but the noticeably larger size is the only real distinction from more southerly birds. Nyeri, Kenya, October (*Rick van der Weijde*).

▼ Race *dilloni* is larger than the nominate, less boldly marked above and below, but distinctly barred on the belly. Ethiopia, January (*Rob Hutchinson*).

116. AKUN EAGLE OWL
Bubo leucostictus

L 40–46cm, Wt 486–607g; W 292–338mm

Other name Sooty Eagle Owl

Identification A medium-sized, dark owl having prominent ear-tufts with dusky outer edge. Females are on average only slightly over 50g heavier than males. Crown is dark brown with fine white spots, especially around the base of the ear-tufts. Brown to rufous upperparts have dense dusky vermicular markings and wavy bars, large white areas on outer webs of scapulars forming an indistinct whitish row across the shoulder. Flight and tail feathers are barred light and dark, the rectrices also being white-tipped. Light brownish underparts become whitish on the lower breast and belly, and are dark-barred and densely vermiculated from below the white throat to the upper breast; below, it is less densely barred but with largish white and blackish-brown dots; the white abdomen has small dusky spots. The facial disc is light reddish-brown with fine concentric darker lines and a thin blackish ruff. The eyes are pale yellow to greenish-yellow, and the bill and cere greenish with a yellowish tint. The legs are relatively small and weak; feathered tarsi are dark, and the bare toes pale yellow with blackish claws. *Juvenile* Downy chick is white. Nearly white mesoptile has widely spaced reddish-brown bars; remnants of white mesoptile plumage can be seen up to one year. *In flight* Shows dark appearance and pale-coloured nuchal area. Often flies low down along roads and open trails.

Call Utters a low, accelerating, clucking rattle, *tok tok tok-ok-ok-ok ok*, infrequently.

Food and hunting Believed to eat mainly insects, but this remains to be proven. This owl has been seen to hawk flying cockroaches at dusk; it also seizes prey from foliage or the ground. First it holds the food item in its foot, and then breaks it into pieces by nipping with the bill.

Habitat Prefers lowland rainforests, but locally seen in farmlands with tall trees.

Status and distribution Occurs in W. Africa from Sierra Leone and Guinea to Angola, but reported to be common only in Liberia; its range extends east to as far as Uganda. This species' biology and habits are practically unknown.

Geographical variation Monotypic.

Similar species Geographically overlapping Greyish Eagle Owl (118) has dark brown eyes rimmed with fleshy-red, finely vermiculated underparts, and feathered toes. Sympatric Fraser's Eagle Owl (119) is coarsely barred below and has shorter and fluffier ear-tufts, and dark brown eyes. Partly overlapping Milky (121) and Shelley's Eagle Owls (122) are much larger, with brown eyes.

▶ Akun Eagle Owl is very dark, with light yellow eyes and talons; Fraser's Eagle Owl *Bubo poensis* has dark brown eyes rimmed fleshy-red, and the toes are feathered. Below, Akun Eagle Owl has a white ground colour, but it is almost fully covered with irregular dark blotches and bars, especially on the upper breast. Edo, Nigeria, July (*Tasso Leventis*).

◀ Akun Eagle Owl has considerable variation in the density of markings and of its rufous wash. The bill is yellowish and the head and ear-tufts are dark brown. Facial disc has fine barring. Ghana, May (*Arthur Grosset*).

117. SPOTTED EAGLE OWL
Bubo africanus

L 40–45cm, **Wt** 487–850g; **W** 323–360mm, **WS** 100–113cm

Identification A medium-sized owl with conspicuous ear-tufts. Males are paler and on average 100g lighter than females. A rare brown morph exists. Has dusky brown upperparts with whitish or dull yellow spots, giving a spotted effect (and responsible for this owl's English vernacular name). There are large white areas on the outer webs of the scapulars, but these do not form a distinct row across the shoulder. Flight and tail feathers are barred light and dark. Whitish underparts are finely dark-barred, and several dark greyish-brown blotches are evident on the upper breast below the white chin; the belly is plain white, suffused with dull yellow. The facial disc is whitish to pale ochre with fine dark barring and a blackish ruff. The eyes are bright yellow rimmed by black edges of eyelids, and the bill is black. Tarsi and toes are feathered nearly to the dark horn-coloured tip of the toes, with dark brown to blackish claws. *Juvenile* Downy chick is white. Mesoptile is finely

barred whitish and brown, and has greyish-yellow eyes, the eyes becoming yellow before fledging. *In flight* Shows spotted appearance, as the name suggests.

Call Emits a mellow hoot, *hoo-hoo*, the second note lower, and female duets with a triple hoot, *hoo-whoohoo*. Female often follows male so closely that the calls sound like one.

Food and hunting More than half of the food is invertebrates (arthropods and insects), and the rest consists of small mammals, birds, reptiles and frogs.

Habitat Prefers open or semi-open woodlands, but avoids dense rainforest. It breeds also in large gardens in towns, as in the author's garden in Lilongwe, in Malawi.

Status and distribution This species is widespread from Gabon east to Kenya, and south to the Cape, in South Africa. It breeds locally also in S. Arabia. Human persecution is heavy owing mainly to superstitious beliefs; in Malawi, the author often had to rescue these owls and release them later outside the city.

Geographical variation Three subspecies are listed: nominate *africanus* occurs from Uganda to the Cape; *B. a. tanae*, from the Tana River and SE Kenya, is much smaller and paler; *B. a. milesi*, from S. Saudi Arabia east to Oman, is more tawny-coloured. The Arabian race is a mystery, and requires more taxonomic and ecological studies; even its voice is stated to differ from that of the nominate race, but further details are needed.

Similar species Geographically overlapping Cape Eagle Owl (115) is much larger, with orange-yellow eyes, and is more coarsely marked below, with blackish blotches at the sides of the upper breast. Partly sympatric Akun Eagle Owl (116) is slightly smaller, with bare yellow toes and pale yellow eyes. Largely allopatric Greyish Eagle Owl (118) has dark brown eyes rimmed by black edges of eyelids, is finely vermiculated above, and has dense barring with vermicular markings below. All other African eagle owls have dark brown eyes.

► Spotted Eagle Owl has yellow or orange eyes rimmed by the black edges of the eyelids. Etosha National Park, Namibia (*Michael and Patricia Fogden*).

◄ Spotted Eagle Owl has prominent ear-tufts. The nominate race occurs throughout most of its African range. South Africa, November (*Rick van der Weijde*).

◄◄ Spotted Eagle Owl is noticeably smaller than Cape Eagle Owl *Bubo capensis*, and has less contrasting markings and blotching. The underparts are finely barred. Tanzania, April (*Adam Scott Kennedy*).

118. GREYISH EAGLE OWL
Bubo cinerascens

L 43cm, Wt *c.* 500g; W 284–338mm

Other name Vermiculated Eagle Owl
Identification A medium-sized owl with not very erectile ear-tufts. Females seem only a little heavier than males, but no sexed weights are available. Has greyish-brown upperparts with many dark vermicular markings and a few paler and darker spots. The flight feathers are barred dark and light, and the pale brownish-grey tail feathers have a few dark bars. The light greyish-brown underparts are finely and densely vermiculated dark brown. The pale greyish-brown facial disc is finely dark-vermiculated with concentric lines, and is bordered by a prominent blackish ruff; whitish eyebrows are fairly obvious. The eyes are dark brown, with fleshy-reddish rims of eyelids. The cere is brownish-grey and the pale-tipped bill lead-grey. Tarsi are feathered, but greyish-brown toes are partly bare. *Juvenile* Downy chick is white. Mesoptile is barred whitish and brown, and always has dark brown eyes.

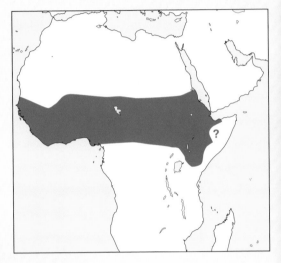

Call A double note, *kuo-wooh*, is uttered at intervals of several seconds, the first syllable much the stronger.
Food and hunting Takes a wide variety of prey, from large insects to birds and mammals, but its diet is not well studied. It is known to hawk flying insects and bats.
Habitat Found in open and semi-open savanna, but also in open parks and gardens. Avoids dense rainforest.
Status and distribution Occurs from Senegal and Cameroon east to Ethiopia and N. & C. Kenya. This owl is common in W. Africa, but its occurrence in E. Africa is not fully clarified. It suffers heavily from human persecution, and is a frequent victim of traffic and powerlines.
Geographical variation Monotypic. Formerly considered to be conspecific with Spotted Eagle Owl *Bubo africanus.*
Similar species Partly sympatric Akun Eagle Owl (116) is darker, with yellowish eyes and bare, yellow toes. Marginally overlapping Spotted Eagle Owl (117) has bright yellow eyes, more erected and longer ear-tufts, and more coarsely barred underparts. Fraser's Eagle Owl (119) overlaps in W. Africa, and also has dark brown eyes and fluffy ear-tufts, but its general coloration is brownish to dull-yellow and coarsely barred below. Other African eagle owls are much larger.

◀ Greyish Eagle Owl has greyish-brown upperparts with many dark vermiculations. Nigeria (*Tasso Leventis*).

▲ The underparts of Greyish Eagle Owl are finely barred with large brownish blotches. Langano, Ethjiopia, January (*Dick Forsman*).

► Greyish or more brownish forms exist of Greyish Eagle Owl. This bird has a very white belly and pale breast without clear brownish blotches. Cameroon (*David Shackleford*).

119. FRASER'S EAGLE OWL
Bubo poensis

L 39–44cm, **Wt** 575–815g; **W** 276–333mm

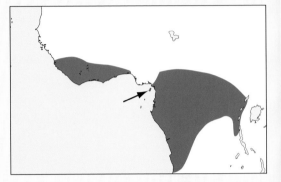

Identification A medium-sized owl with tousled ear-tufts. Sexes alike, but females are on average 200g heavier than males. Has reddish-brown and yellowish-brown upperparts barred with dusky brown. The flight and tail feathers have rather narrow barring of pale brownish to dull yellow and darker brown. The whitish chin is prominent only when the bird is calling. Underparts are light reddish-brown, shading to whitish on the belly and undertail-coverts, with rufous-edged dark wavy markings; the upper breast feathers have broad dusky tips, giving a dark-blotched effect. The facial disc is pale reddish-brown with a distinct dusky ruff, and the light-coloured eyebrows are relatively indistinct. The eyes are dark brown, with bare eyelids rimmed by pale bluish edges. The bill and cere are slate-coloured. Faintly but densely barred tarsi are feathered to the base of the bluish-grey toes, which have blackish-horn claws. *Juvenile* Downy chick is whitish. Mesoptile is light reddish-brown to dull yellow, marked all over with narrow dark brown bars; it has a blackish eyebrow-like zone above the eyes, and poorly developed ear-tufts; the wings and tail are similar to those of the adult.

Call A drawn-out, double hoot, *twow-ooht*, is repeated at intervals of 3–4 seconds, but it utters also a long series of *put put put put* notes resembling the sound of a small engine and continuing for up to 15–20 seconds.

Food and hunting Feeds on small rodents, bats, birds, frogs and reptiles, but also takes large insects and other arthropods.

Habitat Inhabits dense lowland rainforests, sometimes also secondary forests and plantations, from near sea level to 1000m.

Status and distribution Found from Liberia east to SW Uganda and south to NW Angola. This species is very little studied, but is surely threatened by habitat destruction and human persecution.

Geographical variation Monotypic.

Similar species Sympatric Akun Eagle Owl (116) is darker, with pale yellow eyes and pale yellow toes. Also overlapping in range, Greyish Eagle Owl (118) is much paler, with densely vermiculated underparts, but has dark brown eyes. Geographically separated Usambara Eagle Owl (120) has dull orange-brown eyes, and is darker and more coarsely barred below. Other African eagle owls are clearly larger.

◀ Fraser's Eagle Owl has large brown eyes (those of the sympatric Akun Eagle Owl *Bubo leucostictus* are yellow). Ghana, December (*Ian Merrill*).

120. USAMBARA EAGLE OWL
Bubo vosseleri

L 45–48cm, **Wt** 770–1052g; **W** 331–365mm

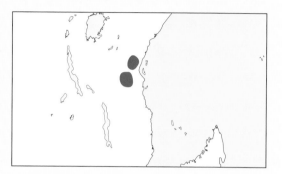

Identification A medium-sized to fairly large owl with short, brown and tousled ear-tufts. Female is on average some 150g heavier than the male. Above, the plumage is dark orange-brown, with blackish-brown barring on the crown, neck, wings and tail. The light tawny-ochre upper breast is densely blotched dark brown, and below becomes paler orange-tawny, mottled whitish to dull yellow, with irregular dark bars; the whitish-buff belly has fine shaft-streaks and darker bars. The facial disc is orange-rufous with a very prominent blackish-brown ruff; the eyebrows are not very distinct. The eyes are dull yellowish-orange to orange-brown, and the bill pale bluish with long dark bristles around the base. The legs are feathered to the base of the dirty yellowish-grey toes, which have dark horn-coloured claws. *Juvenile* Downy chick is whitish. Mesoptile is whitish to dull yellow with fine brown bars on back and below, the flight and tail feathers barred dark on light orange-brown; the blackish-brown ruff and blackish bristles contrast with the pale face, but the sparsely brownish-spotted ear-tufts are hardly visible.

Call Utters a weak, low-pitched *po-a-po-a-po-a-po*, lasting 5–7 seconds, and repeated after 30–60 seconds.

Food and hunting Takes a variety of prey, ranging from larger arthropods to small mammals, birds, reptiles and frogs.

Habitat Inhabits mountain forests between 900m and 1500m above sea level, but is recorded frequently also at forested edges of plantations in the lowlands, down to 200m.

Status and distribution Endemic in the Usambara and Uluguru Mountains, in NE Tanzania. This owl is listed as Vulnerable owing to the extensive level of forest destruction in this part of Africa.

Geographical variation Monotypic. This species was previously considered a subspecies of Fraser's Eagle Owl, but differences in size and colour and its allopatric distribution may justify its separation as a distinct species. More studies on its molecular genetics and vocalisations are needed.

Similar species Geographically separated Akun Eagle Owl (116) has greenish-yellow eyes and long ear-tufts. Sympatric Spotted Eagle Owl (117) has yellow eyes and prominent, pointed ear-tufts. Not overlapping in range, Fraser's Eagle Owl (119) is slightly smaller and paler, with the breast not heavily blotched blackish-brown. Sympatric Milky (121) and allopatric Shelley's Eagle Owls (122) are much larger.

▶ Usambara Eagle Owl is larger than Fraser's Eagle Owl *Bubo poensis*, and the breast is more dark-blotched and less regularly barred. The bill is pale bluish and the blackish ruff is very strong.

121. MILKY EAGLE OWL
Bubo lacteus

L 60–65cm, Wt 1588–3115g; W 420–490mm, WS 140–164cm

Other name Verreaux's Eagle Owl, Giant Eagle Owl
Identification A large to very large owl with short and tousled ear-tufts. It is the biggest owl in Africa. The female is often at least 1kg heavier than the male on average. The light grey-brown upperparts are suffused milky (as the name indicates), with fine whitish vermicular markings above and a row of whitish spots across the shoulder. Flight and tail feathers are barred light and dark. The underparts are pale greyish-brown, finely vermiculated light and dark, darkest on the upper breast and lightest on the flanks. The off-white facial disc has a broad blackish ruff. The eyes are dark brown, rimmed by brown edges of eyelids; the bare pink upper eyelids have ochre 'eyelashes'. The bill is creamy-horn, surrounded by blackish bristles. Tarsi are feathered, feathers covering partly the greyish-horn toes, which have black-tipped dark brown claws.
Juvenile Downy chick is creamy white. Mesoptile is pale greyish with fine vermicular markings and dark barring, a less prominent ruff around the facial disc, and ear-tufts barely evident.
Call Utters a deep, nasal grunting hoot, *gwok-gwok*, repeated in series of one to five with intervals of several

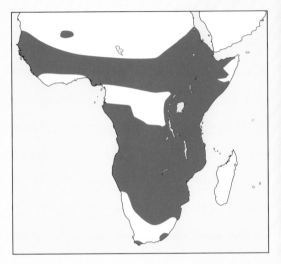

seconds. The call is said to carry for up to as far as 5km, but this is hard to believe. The author kept a male for more than a year at his house in Malawi, and it gave no indication of having so loud and far-carrying a voice.
Food and hunting This species prefers medium-sized mammals and large birds, but it also eats reptiles, frogs, fish and arthropods. It catches prey by gliding down from a perch. It also wades into shallow water after fish, and hawks flying insects in the air.
Habitat This is an owl of tree savannas and riverine forests; it is absent from dense rainforest.
Status and distribution This owl is widespread in Africa south of the Sahara from S. Mauritania and Ethiopia to the Cape in South Africa, where its distribution is very patchy. It is locally rare and endangered, mainly due to human persecution.
Geographical variation Monotypic.
Similar species Largely allopatric Shelley's Eagle Owl (122) is almost the same size but is much darker and coarsely barred below, and does not have pink eyelids. All other African eagle owls are much smaller.

▶ An adult Milky Eagle Owl is a very impressive bird and easily identified by its size and pink eyelids. Kenya (*Tui De Roy*).

◀ Milky Eagle Owl has greyish-brown upperparts with clean whitish spots across the shoulders. Coastal Kenya (*Tasso Leventis*).

122. SHELLEY'S EAGLE OWL
Bubo shelleyi

L 53–61cm; Wt 1257g (1♂); W 420–492mm

Identification A fairly large to large owl with large dusky ear-tufts. Female weights are not known, but believed to be heavier than the male. Light and dark morphs occur. The upperparts are dusky to dark brown with dull yellowish to white bars. Flight and tail feathers are barred dark and light. The yellowish-grey underparts have heavy dark brown barring. The facial disc is off-white to pale tawny with fine, dark concentric lines, and is encircled by a prominent blackish-brown ruff. The eyes are dark brown. The bill is creamy-horn-coloured with a bluish wash near the base, and with brown bristles around the base. Tarsi have dirty-white feathering, leaving only pale creamy tips of the toes unfeathered; the claws are light grey and dark-tipped. *Juvenile* Downy

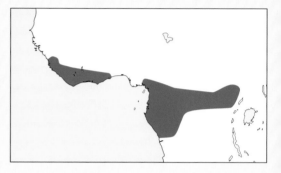

chick undescribed. Mesoptile is yellowish-grey, brown-barred all over; the wings and tail are like those of the adult, and the eyes are dark blue.

Call Utters a loud wailing *kooouw* cry irregularly at intervals of several seconds.

Food and hunting Few data are available. A large flying squirrel has been recorded as prey. Diet and hunting behaviour require study.

Habitat Prefers primary lowland forests near rivers.

Status and distribution Found from Guinea and Sierra Leone east to DR Congo, but only some 20 specimens are known. In C. Africa this owl occurs almost as far east as Uganda. This is obviously an extremely rare species and it is endangered by habitat destruction. It is listed by BirdLife International as Near Threatened.

Geographical variation Monotypic.

Similar species Largely allopatric Milky Eagle Owl (121) is only a little larger, but is very different in coloration, being generally much lighter above and below. All sympatric African owl species are distinctly smaller.

◄ The exceptionally rare Shelley's Eagle Owl is the only large West African eagle owl that is heavily barred below. The much smaller Fraser's Eagle Owl *Bubo poensis* is less densely barred below, and more warm tawny overall. Specimen from BMNH (Tring), collected on Mount Nimba, Liberia in November (*Nigel Redman*).

123. FOREST EAGLE OWL
Bubo nipalensis

L 51– 63cm, **Wt** 1300–1500g; **W** 370–470mm

Other name Spot-bellied Eagle Owl
Identification A fairly large to large owl with very long but almost horizontal ear-tufts. Sexual size or colour differences are not known. Above, the plumage is dark brown with black bars. The blackish-brown primaries have paler greyish-brown bars, and the secondaries are broadly barred dull yellowish to brown. The greyish tail is barred with blackish-brown. The underparts are reddish-yellow to fulvous white with conspicuous blackish V-shaped bars, these becoming broad spots on the belly and vent; the upper breast has a rather prominent pectoral band suffused honey-brown, marked with dark chevrons. There is no dark ruff around the pale facial disc, and the whitish eyebrows are not particularly striking. The eyes are dark brown and the bill wax-yellow or pale yellow. The legs are feathered to the base of the dusky yellowish-grey toes, with dark-tipped horn-coloured claws. *Juvenile* On hatching, the chick looks quite big and has a noticeably large head and beak. Mesoptile is pale yellowish or dull whitish-yellow, the upperparts barred dark brown, the white

underparts indistinctly barred; the wings are like the adult's, but the eyes are bluish-black.
Call Gives an audible, but low and deep double hoot, *hoo hoo*, lasting some two seconds.
Food and hunting Takes fairly large birds and mammals, but also snakes, and lizards. In captivity, it has been known to refuse fish. It hunts large birds by pouncing on them when they are asleep at night-time roosts.
Habitat Inhabits dense primary forests, from lowlands to mountains, up to 2100m in SW India and to 3000m in Himalayas. Despite frequent assertions in the past, it is not always found near water.
Status and distribution Occurs from the Himalayas east to C. Vietnam and separately in SW India and Sri Lanka. It is present at low densities in this vast area, and suffers from habitat destruction and locally from human persecution. In India, these owls are caught by poachers, as they are in demand for use in black-magic.
Geographical variation Two subspecies have been described: nominate *nipalensis* is found from the Himalayas to Vietnam; clearly smaller *B. n. blighi*, from Sri Lanka, may not, in fact, differ from owls of this species in S. India. Recent observations suggest that a small, thinly distributed population might exist in Madhya Pradesh, in C. India, lying between the previously documented populations to the north and south. Taxonomy requires study, as this species is also very similar to the Barred Eagle Owl.
Similar species Allopatric Barred Eagle Owl (124) is much smaller. Sympatric Dusky Eagle Owl (125) is also smaller and has bright yellow eyes.

◀ Forest Eagle Owl of race *blighi* is clearly paler and smaller than nominate. Sri Lanka, June (*Gehan de Silva Wijeyeratne*).

124. BARRED EAGLE OWL
Bubo sumatranus

L 40–46cm, **Wt** 620g (1); **W** 323–417mm

Other name Malay Eagle Owl
Identification A medium-large owl with very long, outward-slanting ear-tufts. Females appear larger and heavier than males, although no sexed weight measurements are available. The crown and upperparts are blackish-brown, vermiculated and mottled with many paler zigzag bars. The dark brown tail has about six tawny-whitish bars and white tips. The underparts are dull yellowish to white with irregular arrow-like dark brown spots, but the upper breast is more densely

marked with broad earth-brown bars, forming a dark breast-band. The facial disc and lores are dirty greyish-white, with no distinct ruff, and the whitish eyebrows are not very conspicuous. The eyes are dark hazel-brown, the eyelid rims yellow to light grey, and the bill and cere pale yellow. Tarsi are feathered a little beyond the base of the yellowish-grey toes, which have dark horn-coloured claws. *Juvenile* Has pure white natal down. Whitish mesoptile is banded with brown, has wings and tail similar to those of adults, but white ear-tufts short and rounded, with fine brown bars; the eyes are dark blue.
Call Utters a deep and loud w*hooa-who* hoot at 2-second intervals.
Food and hunting Feeds on small mammals, birds, snakes, fish and large insects.
Habitat Inhabits primary and secondary evergreen forests with ponds and streams, and is found also in gardens with large trees. From sea level to medium elevations, rarely higher than 1600m.
Status and distribution Occurs from S. Burma and Thailand south to Java and Bali. This species is not uncommon, but is very little studied.

◄ Barred Eagle Owl nominate *sumatranus*. A medium-sized eagle owl with long, outward-pointing ear-tufts. Malaysia, May (*HY Cheng*).

Geographical variation Three subspecies are listed: nominate *sumatranus* is found in the west of the range, from S. Burma south to Sumatra and Bangka; *B. s. strepitans*, from Java and Bali, is much larger than nominate; *B. s. tenuifasciatus*, from Borneo, is somewhat intermediate. Taxonomy requires study, as this owl is very similar to the Forest Eagle Owl, although these two species are fully allopatric.

Similar species Exhibits similarities to larger Forest Eagle Owl (123), which is more boldly spotted below. Somewhat similar to sympatric Brown (143) and Bartels's Wood Owls (146), but those two have no ear-tufts.

▲ Race *tenuifasciatus* from Borneo is similar in size to *sumatranus*. It looks darker, with finer bars on the breast-band and belly, and with more dense, unbroken bars on the flanks. Sarawak, Borneo, September (*Rohan Clarke*).

▶ A female *strepitans* at the nest hole. This race is larger than the nominate, with broad, well-spaced cross-bars on the buffish breast. Java, November (*Willy Ekariyono*).

125. DUSKY EAGLE OWL
Bubo coromandus

L 48–53cm; **W** 380–435mm

Identification A fairly large owl with distinct ear-tufts, which appear quite close together when erect. Females are said to be larger and heavier than a males, although no weights are documented. The crown and upperparts are brownish-grey with blackish shaft-streaks and dark brown and whitish vermicular markings. Primaries and secondaries are barred light and darker greyish-brown, and the pale brownish-grey tail has four or five broad dark greyish-brown bars and white feather tips. Whiter outer webs of scapulars are finely brown-vermiculated, forming only an indistinct row on the scapulars. The underparts are very light yellowish-grey with distinct dark shaft-streaks and brown cross-bars. The whitish facial disc has dark shaft-streaks, and the ruff, although narrow, is dark and visible. The eyes are bright yellow, the cere is bluish-grey, and the bill is bluish-horn with a pale yellowish-horn tip. Tarsi are feathered to or beyond the base of the bluish-grey toes, which have blackish-brown claws. *Juvenile* The short natal down is pure white. Reddish-brown fledgling has a rather mealy appearance caused by longer down tips; the head and neck are dirty grey.

Call Gives a deep, resonant croaking *kro kro kro-krok-rokokokokog*, lasting about three seconds. Calls both during the day and at night.

Food and hunting Feeds on small mammals, birds, reptiles, frogs, fish and large insects. May be seen hunting even in full daylight.

Habitat Inhabits lowland forests with ponds and streams, but occurs also near plantations, roadsides and human habitations. Avoids deserts and arid regions.

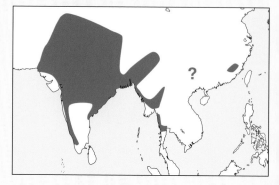

Found from sea level up to 250m.

Status and distribution This owl is not uncommon in the Indian Subcontinent, but is rare in Nepal and Bangladesh. It is extremely rare in W. Burma and S. China.

Geographical variation Two subspecies have been described: nominate *coromandus* occurs from E. Pakistan and India east to Bangladesh; much darker *B. c. klossi*, from W. Burma to SE China, is very poorly known, and only four museum skins are available. A study is required of this owl's biology, behaviour and distribution, especially in the easternmost parts of the range.

Similar species Sympatric Rock Eagle Owl (114) is tawnier in general and has orange eyes. Only partly overlapping Forest Eagle Owl (123) is much larger and is prominently dark-spotted below. Fish owls (127–130) have totally bare tarsi.

▶ Nominate Dusky Eagle Owl from India is a large eagle owl with very whitish underparts bearing long, dark streaks and fine brown cross-bars. Ear-tufts fully erected and close to each other, like twin spires. This is a male. Bharatpur, India, February (*David Behrens*).

◀ Race *klossi* has similar plumage to the nominate, but it is much darker – and extremely rare. There are no clear whitish markings on the wings or scapulars. Kedah, Malaysia, February (*Choy Wai Mun*).

126. PHILIPPINE EAGLE OWL
Bubo philippensis

L 40–43cm; W 341–360mm

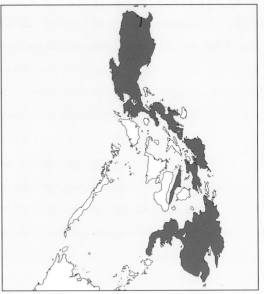

Identification A medium-sized owl with tousled and outward-slanting ear-tufts. No information is available on the sexual size or colour differences. The crown and upperparts are brownish-yellow to reddish-brown with blackish streaks, giving a distinctly striped appearance (crown more spotted-looking). The flight feathers are somewhat paler than the back and have dark bars, and the dull reddish-brown tail is barred dark brown. The chin is light rufous and the throat whitish. The dull yellowish to white underparts have dark brown shaft-streaks, but are almost unmarked on the lower belly. The facial disc is tawny with a very narrow dark ruff. The big yellow eyes are prominently rimmed black;

nictitating membrane and bill are pale bluish. Tarsi are fully feathered to the base of the bare toes, which are light greyish-brown with bluish claws. *Juvenile* Undescribed.

Call Utters a long series of low, deep *hoo-hoo-hoo-hoo* notes, repeated at intervals of some four seconds.

Food and hunting Not studied. This species' large and powerful feet suggest that its diet includes small mammals and birds, and not only insects.

Habitat Occupies lowland forests, often near water, but found also in coconut plantations not far from houses. Occurs obviously only at lower elevations.

Status and distribution Endemic to the Philippines, where it is listed as Vulnerable and rare as a result of extensive habitat destruction and human persecution. Recent records are mainly from Luzon and Mindanao but this owl is very little studied.

Geographical variation Two subspecies have been separated: nominate *philippensis* is confined to Luzon, Cebu and Catanduanes; darker and possibly slightly larger *B. p. mindanensis* is found on Samar, Leyte, Bohol and Mindanao.

Similar species In the Philippines this species could only be confused with Giant Scops Owl (108), but the latter has brown eyes and is far more reddish-brown, with breast and flanks marked with broad drop-shaped streaks.

◄ Philippine Eagle Owl has very powerful legs and large claws (*Michael R. Anton*).

▶ This immature Philippine Eagle Owl has distinct blackish streaks on the upperparts and remnants of natal down on its head (*Michael R. Anton*).

◄ Philippine Eagle Owl nominate *philippensis* from Luzon is a relatively small, short-eared eagle owl with golden-yellow eyes and a large, pale bluish-grey bill. Underparts are pale rufous-brown with dark brown streaks. Luzon, January (*Bram Demeulemeester*).

◄ Philippine Eagle Owl has very short, outward-pointing ear-tufts (*Michael Anton*).

127. BLAKISTON'S FISH OWL
Bubo blakistoni

L 60–72cm, **Wt** 3400–4500g; **W** 498–560mm, **WS** 178–190cm

Identification The second largest owl in the world, this species has almost sideways-slanting, very tousled ear-tufts. Females are on average more than 1kg heavier than males. The upperside is brown with blackish-brown shaft-stripes and dull-yellow feather tips. The deep brown wings have many dull yellowish bars, and seven or eight creamy bars are visible on the tail. The underparts are pale yellowish-brown with blackish-brown shaft-streaks and narrow light brown wavy cross-bars, this pattern petering out on the undertail-coverts, which are creamy with a few dark markings. The tawny-brown facial disc has narrow black shaft-stripes, but no prominent surrounding ruff. The eyes are faded lemon-yellow and the bill greyish. The tarsi are covered with creamy-coloured feathering, and the bare toes are lead-grey with dark horn-coloured claws. *Juvenile* Natal down is white. Juvenile has a bluish bill and toes; from the age of 10–12 days, an almost black mask starts to develop around the pale yellowish-brown eyes. Mesoptile is smoke-brown with a greyish tint, and with slight transversal pattern formed by indistinct elongate spots, especially on the upper head. *In flight* Wing noise can be heard at 50–100m, but the flight is easy, shallow flapping often alternating with gliding. **Call** Utters a powerful, low-frequency *boo-bo-voo* or *khuu-guuuu* lasting about three seconds, and repeated at intervals of 8–10 seconds. Can be heard easily at up to 1.5km.

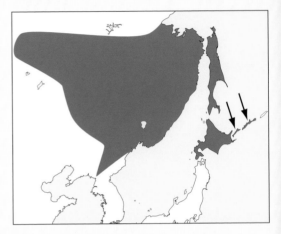

Food and hunting Prefers fish, sometimes large individuals; it also takes crustaceans, frogs, and mammals up to the size of hares. It readily enters the water when catching prey.

Habitat Inhabits moist taiga, thickets and dense coniferous forests along at least partly ice-free rivers, but lives also on rocky sea coasts in the far north. Occurs only in lowlands.

Status and distribution Found in E. Russia (SE Siberia) south to N. Korea and N. Japan (Hokkaido). This owl is rare throughout its range and is listed as Endangered. Its global population is estimated to be not more than 5,000 individuals.

Geographical variation Two subspecies are listed: nominate *blakistoni* is found from Sakhalin and the Kurils to N. Japan; larger and paler *B. b. doerriesi* occurs from SE Siberia south to the Korean border area.

Similar species Slightly sympatric Eurasian Eagle Owl (112) is similar in size but overall darker, with broader streaks on the breast; its eyes are orange, and the ear-tufts more upright and not so tousled. Also partly overlapping, the Ural Owl (160) is paler and much smaller, with dark brown eyes and no ear-tufts. Almost allopatric Great Grey Owl (161) is a little smaller and very much paler, and has a rounded head, a well-developed facial disc with many dark concentric lines, relatively small eyes, and no ear-tufts.

◄ Among owls, Blakiston's Fish Owl is second in size to only Eurasian Eagle Owl. Ussuriland.

▲ Nominate Blakiston's Fish Owl from Japan is smaller than *doerriesi,* with more sombre plumage. Yellow eyes and bluish cere, bill and toes. Hokkaido, Japan, June (*Gaku Tozuka*).

▼ Blakiston's Fish Owl will often enter the water when catching fish. Hokkaido, Japan, January (*Stuart Elsom*).

128. BROWN FISH OWL
Bubo zeylonensis

L 48–58cm, Wt 1105–1308g; W 355–434mm, WS 125–140cm

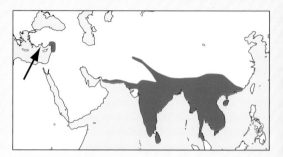

Identification A fairly large to large owl with horizontally oriented, tousled ear-tufts. Females are on average at least 200g heavier than males. The pale chestnut crown and upperparts have broad, black shaft-streaks and brown cross-bars. The scapulars, tertials and wing-coverts are whitish-mottled, and the outer scapulars are mostly white, but there is no white on the forehead. The dark brown flight and tail feathers are barred, vermiculated and tipped dusky to dull yellow. The distinctly white throat and foreneck are marked with dark shaft-streaks, and the light reddish-yellow underparts with fine, wavy reddish-brown cross-bars and bold black shaft-streaks. The facial disc, not prominent, is tawny with black shaft-streaks. The eyes are golden-yellow and the bill pale greenish-grey. The bare legs and feet are dusky to greyish-yellow, with horn-brown claws. *Juvenile* Downy chick is whitish. Mesoptile has upperparts more reddish-brown, with narrower and browner shaft-streaks, and underparts paler and duller, with an indistinct white throat patch and very

narrow shaft-streaks. Even second-year individuals are duller and paler than adults. *In flight* Shows rounded wings and long legs.

Call Emits a deep, hollow and humming *boom-boom* or more muted *tu-whoo-hu*, lasting 1.5–2 seconds and repeated at intervals of several seconds.

Food and hunting Feeds mainly on fish, crabs and frogs, but also on large insects, birds, reptiles and rodents. It locates prey from a perch overlooking water.

Habitat Inhabits lowland forests near streams and lakes, locally near human habitations and plantations, such as rice paddies. Occurs from lowlands up to 1500m.

Status and distribution Distributed from SW Turkey east to SE China and the Malay Peninsula. This species is very common in Sri Lanka and not rare in India; it is probably extinct in Israel since 1975. It is believed that in India and SE Asia more than 1,000 owls, including the Brown Fish Owl, are killed annually during the festival of Diwali by black magicians in the hope of warding off bad luck and gaining magical powers. This is despite the fact that owls are identified with the goddess of wealth, Lakshmi, in whose honour the celebration is held every year.

Geographical variation Four subspecies have been separated: nominate *zeylonensis* is confined to Sri Lanka; *B. z. leschenaultii*, from India, Burma and Thailand, is larger and paler than nominate; very pale *B. z. semenowi* ranges from SW Turkey to Pakistan; darker *B. z. orientalis* occurs from NE Burma to SE China, Hainan, Taiwan and the Malay Peninsula.

Similar species Partly sympatric Malay Fish Owl (129) is a little smaller, and has no shaft-streaks in the facial disc and no cross-bars or vermicular markings below. Also geographically overlapping, the Tawny Fish Owl (130) is the same in size but has a pale face, and the underparts lack bars and are not vermiculated.

◤ Brown Fish Owl has yellow eyes, and the head is flattened over the 'eyebrows'. Kolkata, India, May (*Abhishek Das*).

◀ Brown Fish Owl of the nominate race from Sri Lanka. Smaller than other subspecies, with darker plumage and shorter ear-tufts. The toes are bare and covered with granular scales. Yala, Sri Lanka, November (*Rolf Kunz*).

► Race *leschenaultii* from India is paler than the nominate and has pale fulvous to whitish underparts, with fine wavy brown bars and bold blackish streaks. Bandhavgarh, India, February (*Simon Woolley*).

129. MALAY FISH OWL
Bubo ketupu

L 40–48cm, Wt 1028–2100g; W 295–390mm

Other name Buffy Fish Owl

Identification A medium-sized to fairly large owl with outward-facing, tousled but prominent ear-tufts. Females are always much larger than males; it is likely that sexes differ also in colour, one being paler than the other. The upperparts are rich brown, the wing-coverts with much larger pale spots than the back. Dark brown primaries and secondaries are banded with whitish-yellow or reddish-yellow, and the dark brown tail feathers have whitish tips and three or four dull yellowish to white bars. The rufous-buff or reddish-yellow underparts have narrow dark brown shaft-streaks, but the flanks and thighs are unstreaked. An ill-defined facial disc contrasts the whitish forehead. The eyes are yellow with black-rimmed eyelids, and the bill is greyish-black. Relatively long tarsi are bare, with yellowish-grey toes and dark horn-coloured claws. *Juvenile* Downy chick is whitish. Mesoptile has upperparts paler brownish-yellow than adult, with blackish-brown narrow streaks and fewer or no white spots; the tail has five or six not very distinct narrow dull yellowish to whitish bars.

Call Emits a loud and rattling *kutook, kutook, kutook*, a musical *to-whee to-whee* and a rising *pof pof pof*.

Food and hunting Feeds mainly on fish, but also takes rats, bats, birds, crustaceans, reptiles, frogs and insects. It fishes from a perch, swooping down to catch prey from the surface or in the water. It will also walk in shallow streams.

Habitat Prefers mangroves and other forested areas near water, often close to human habitations and villages, including rice paddies. Occurs mainly in lowlands, but in Sumatra locally up to 1600m.

Status and distribution Found from S. Burma to Bali and Borneo. It is locally rather common, although often persecuted near fish farms.

Geographical variation Four subspecies have been separated: nominate *ketupu* occurs from the Malay Peninsula south to Bali and Borneo; paler *B. k. aagaardi* is found from S. Burma to Thailand; *B. k. pageli*, from N. Borneo, is strongly tinged with red; *B. k. minor*, from Nias Island, off Sumatra, is the smallest of the four races.

Similar species Partly sympatric Brown Fish Owl (128) has some barring and streaks on the underparts, and the only slightly overlapping Tawny Fish Owl (130) is much richer reddish-brown below, the tail is more narrowly barred and the face is more whitish. All eagle owls have feathered legs down to the base of the toes. Wood owls have a rounded head, lack ear-tufts and have dark brown eyes.

▲ Malay Fish Owl, with yellow eyes and greyish-black bill. This is the nominate. Sabah, Malaysia, August (*Christian Artuso*).

▲ ▶ Immature Malay Fish Owl on the ground; it has a very pale yellow appearance. Singapore, October (*HY Cheng*).

◀ The nominate Malay Fish Owl is rich brown above and rufous-buff below. Penang, Malaysia, August (*Choy Wai Mun*).

▶ Race *pageli* is far more reddish than nominate *ketupu*. Kinabatangan, Borneo, May (*Adam Riley*).

130. TAWNY FISH OWL
Bubo flavipes

L 48–58cm; **W** 410–477mm

Identification A fairly large to large owl with rather tousled and horizontal ear-tufts. Females are said to be larger than males, although no weights are known. The orange-rufous to tawny crown and upperparts have broad dark markings on the central part of the feathers and spots of the same colour as the reddish-brown edges. Scapulars are dull yellow, forming a pale band across the shoulder. Flight and tail feathers are strongly barred dark brown and buffish. A prominent white patch on the throat is bordered below by rich orange to reddish-brown underparts with dark brown shaft-streaks, these streaks broader on the breast. The pale facial disc is rather poorly defined, but is accentuated by the white eyebrows, lores and forehead. The eyes are yellow and the bill dark bluish. The legs are feathered to halfway down, the lower parts and the toes bare and dingy greenish-yellow, with greyish-horn

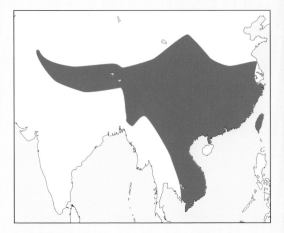

claws. *Juvenile* The upperside has distinct spots and narrower and broader streaks; the underparts are still downy, the chin white, and the streaks finer and paler; tarsi are down-covered to 2.5cm above the base of the middle toe. *In flight* Shows strong banding on the wings and tail.

Call Gives a deep and booming *whoo-hoo* lasting only half a second, and repeated at intervals of several seconds.

Food and hunting Eats mainly fish, crabs, frogs and large insects, but is known to kill also large birds, such as pheasants.

Habitat Inhabits tropical and subtropical old-growth forest near streams, from the plains to 1500m; in India occasionally up to 2450m.

Status and distribution Found from N. India to SE China and Taiwan, as well as in Laos and Vietnam. This owl is locally not uncommon, but it can be threatened by human persecution and habitat destruction.

Geographical variation Monotypic.

Similar species Partly sympatric Brown Fish Owl (128) is much browner and has totally bare legs. Almost allopatric Malay Fish Owl (129) is smaller and generally much less orange-rufous, with bare legs. Voices of these owls also differ, although those of the Tawny Fish Owl need further study.

◀ Tawny Fish Owl has very tousled ear-tufts. Uttarakhand, India, February (*Bill Baston*).

▲ Tawny Fish Owl is similar to Brown Fish Owl *Bubo zeylonensis* and Malay Fish Owl *B. ketupu*, but is more powerful than either, with a distinct pale scapular band, half-feathered tarsi and a white throat-patch. Rajasthan, India, January (*Harri Taavetti*).

► Tawny Fish Owl has yellow eyes, a greenish cere and a dark bluish bill. Uttaranchal, India, January (*Harri Taavetti*).

131. PEL'S FISHING OWL
Bubo peli

L 51–63cm, Wt 2055–2325g; W 407–447mm, WS 150–153cm

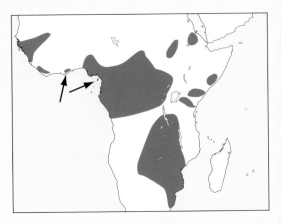

Identification A fairly large to large ginger-brown owl with dusky barring and spotting above, the scapulars with some buff-white. The sexes are alike in colour, but the larger female sometimes looks paler than the male. Considerable individual variation in colour and barring is apparent. It lacks real ear-tufts, but long and loose reddish-brown feathers on the head give impression of a tousled crown; the underparts are pale reddish-brown to dull yellow, with dusky shaft-streaks ending in a rounded spot at the tips. Flight and tail feathers are barred light and dark brown, bars on the primaries being well spaced. The pale throat is prominently inflated when singing. The facial disc is not very prominent, reddish-brown with an indistinct rim. The large eyes are blackish-brown, almost inky black, and seemingly 'bottomless'. The bill is black and the cere grey. Bare straw-coloured legs and toes, the latter with long and curved claws which are grey-horn with darker tips; the scales on the soles of the feet have developed into sharp spicules, and the claws have a sharp inner edge. *Juvenile* Downy chick is white. Mesoptile has head, hindneck and underparts whitish with a dull yellowish wash, and unmarked; head and mantle, as well as wings and tail, are similar to those of the adult, but paler, less reddish-brown. First full plumage is paler than adult, but from the age of ten months similar. *In flight* Looks long-winged and large, but remarkably small-headed as ear-tufts and facial ruff are not well developed. This species does

▶ Pel's Fishing Owl holding a half-eaten fish in the firm grip of its huge, sharp talons. Botswana (*Adam Riley*).

◀ Pel's Fishing Owl is large with variable coloration and markings. Plumage generally rufous; lower breast and belly have dark arrow-like markings. Darker upper breast lacks spotting. Eyes are dark brown and bill colour varies from dark to light grey. Kavango, Botswana, August (*Eric VanderWerf*).

not fly silently, lacking the special sound-suppressing features on the flight feathers.

Call The male's song is a deep and sonorous horn-like boom, *hooommmm-hut*, which carries up to as far as 3km. This owl is said to be vocally most active on moonlit nights, especially towards dawn.

Food and hunting Feeds almost entirely on fish up to the weight of 2kg, but more often between 100g and 250g; it sometimes takes frogs, crabs and mussels. Prefers branches overhanging water as hunting perches. It is not clear how this species locates prey, but it probably detects fish by the ripples which they cause on the water's surface, and then glides down to seize the prey with its powerful talons, before swooping back up to a perch 1–2m above the water; possibly because of its very long, easily wetable flank feathers, it rarely immerses the entire body. (The noisy wingbeats do not seem to be of importance, as this owl takes underwater prey, which cannot hear its approach; sound may not be a vital cue in locating the prey.)

Habitat This species favours riparian habitats, as well as islands in large rivers, swamps or lakes with groups of old trees. Often found in mangrove forests along swamps and estuaries at sea level, but also in some old forests at up to 1700m.

Status and distribution Occurs in 30 African countries south of the Sahara. Although not globally threatened, it is rather sparsely scattered in most areas. Locally fairly common in Congo Basin. Common also in Botswana, where about 100 pairs occur in the 1.5 million-hectare Okavango wetland. In the S. African atlas, the total population was estimated to number fewer than 500 pairs. Locally, a lack of large trees with suitable nest holes within 200m of shallow water with abundant fish may limit populations.

Geographical variation Monotypic.

Similar species Geographically partly overlapping Rufous Fishing Owl (132) is much smaller with honey-brown eyes; it is brighter reddish-brown, and the shaft-streaks on the underparts do not end in rounded spots. More sympatric Vermiculated Fishing Owl (133) is also smaller in size, and has dark brown eyes and yellow bill and cere; its upperparts are densely vermiculated and its whitish underparts are fairly heavily dark-streaked.

▼ This Pel's Fishing Owl has a very pale rufous head and dark spots on the upper breast. Central African Republic (*David Shackleford*).

▼ Pel's Fishing Owl calling. Note how the male is fully puffed up, not only the throat as in other eagle owls. This is a response to a rival male in its territory. KwaZulu Natal, South Africa, January (*Nick Baldwin*).

132. RUFOUS FISHING OWL
Bubo ussheri

L 46–51cm, **Wt** 743–834g; **W** 330–345mm

Identification A fairly large owl without ear-tufts. Female is on average only 90g heavier than male. Has plain brownish-yellow to reddish-brown upperparts, with the flight and tail feathers barred dark and light, and pale yellowish-brown underparts. Forecrown has some indistinct paler mottling; scapulars have whitish outer webs, forming a white row across the shoulder. The upper breast is slightly darker than belly, with narrow dusky to reddish-brown shaft-streaks. The pale cinnamon facial disc is very indistinct, with the ruff tawny-rufous but hardly visible. The eyes are honey-brown to dark brown and the bill blackish-grey. Bare tarsi and toes are light yellow, with pale horn-coloured claws. *Juvenile* Downy chick is completely white. Mesoptile has a distinct rufous wash on the head and reddish-brown streaking on the breast and upper belly; the wings are similar to those of the adult. Immature is paler than adult.

Call Utters a dove-like moaning *whoo*, repeated at one-minute intervals.

Food and hunting Not well documented, but it eats mainly fish. Hunts from branches overhanging water.

Habitat Prefers primary forests along large rivers and lakes; found also in mangroves at coast and in plantations.

Status and distribution Endemic in W. Africa from Sierra Leone and Guinea east to Ghana, but very little studied. Reported observations from Nigeria are unconfirmed. It is listed as Vulnerable by BirdLife International as a consequence of habitat destruction and human persecution.

Geographical variation Monotypic.

Similar species Sympatric Pel's Fishing Owl (131) is larger, and barred and spotted above. Locally overlapping Vermiculated Fishing Owl (133) is the same size, but is densely vermiculated above and heavily streaked below, with a yellowish-horn bill and bright yellow cere.

▼ Rufous Fishing Owl is the plainest and least well-known of the fishing owls. Ghana (*April Conway*).

133. VERMICULATED FISHING OWL
Bubo bouvieri

L 46–51cm, Wt 637g (1♀); W 302–330mm

Identification A fairly large owl with no ear-tufts. Size difference between sexes is not known, as only one sexed weight is available. Large individual variation in colour. The upperparts are yellowish-brown and finely vermiculated, the crown dark brown and more streaked. A pale line across the shoulder is created by large white areas on the outer webs of the scapulars. flight and tail feathers are barred dark and light. The dirty-whitish underparts are heavily streaked dark brown, with plain off-white thighs, undertail-coverts and underwing-coverts. The light reddish-brown facial

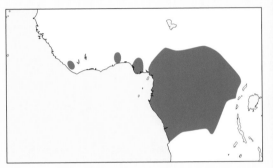

disc is not distinct and the dark brown ruff not prominent. The eyes are dark brown and the bill yellowish-horn. Bare tarsi and toes are chrome-yellow, with blackish claws. *Juvenile* Downy chick is white. Mesoptile is mainly white, often with a faint cinnamon wash on the head and face and below.

Call Gives a low, croaking introductory hoot, *kroohk*, followed by 5–8 short and even notes, *kru-kru-kru-kru-kru*, and ending with one or two hoots as at the beginning.

Food and hunting Feeds on fish, crabs, frogs, small birds and mammals. Hunts from a branch 1–2m above water.

Habitat Occupies gallery forests near rivers and lakes, but presence of water or fish is not an absolute essential.

Status and distribution Occurs in Liberia, Togo and S. Nigeria and in whole of C. Africa north from NW Angola. This owl is obviously not as rare as has been supposed. It requires further investigation.

Geographical variation Monotypic.

Similar species Sympatric Pel's Fishing Owl (131) is larger, is barred and spotted above, has no white outer webs on the scapulars, has narrow streaks and drop-shaped dots on the underparts, and has a dark grey to blackish bill. Only partly overlapping Rufous Fishing Owl (132) is the same in size, but plain reddish-brown to brownish-yellow above and less distinctly streaked below, with a blackish-grey bill.

▶ Vermiculated Fishing Owl is almost the same size as Rufous Fishing Owl but more prominently streaked dark brown below, with yellow cere, bill and legs (*Tasso Leventis*).

◀ Vermiculated Fishing Owl has cinnamon-brown head and back finely vermiculated with dark brown. Flight and tail feathers dark and light barred (*Tasso Leventis*).

134. SPECTACLED OWL
Pulsatrix perspicillata

L 43–52cm, Wt 590–982g; W 305–360mm

Identification A medium-sized to fairly large owl without ear-tufts. Females are on average some 200g heavier than males. Upperparts are uniformly dark brown or blackish-brown. The remiges and rectrices are barred with paler grey-brown. Has a broad dark brown chest-band, below which the plumage is uniformly light yellow to dull yellowish. The facial disc is dark brown, the chin is black, and the white throat forms a semi-collar; white eyebrows, lores and malar streaks form 'spectacles' around the bright orange-yellow eyes. The bill and cere are yellowish-horn, the bill with a greenish tint towards the tip. Creamy to dull yellow tarsi and toes are almost fully feathered, the unfeathered parts of the toes being whitish or pale grey, with dark claws.

Juvenile Downy chick is whitish. Mesoptile is mainly fluffy white, with grey-brown barring on the wing-coverts, and an almost heart-shaped blackish face contrasting strongly with the white body; the wings and tail are brown with paler barring. Can take five years to acquire full adult plumage, but can start to breed before then.

Call Gives a sequence of six accelerating short, dry, rattling hoots, *pum-pum-pum…*, lacking in resonance and lasting between seven and nine seconds, repeated several times after varying intervals. In Colombia, the author wrote into his notes during 1974–76 that the voice is similar to the tapping of a Great Spotted Woodpecker *Dendrocopos major*. It is interesting to note that in Brazil these owls are known locally as 'knocking owls'.

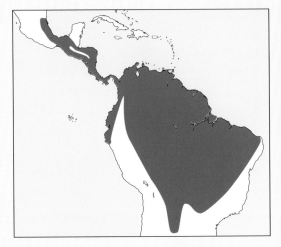

Food and hunting Feeds mainly on small mammals and birds, and also takes crabs, insects and spiders. Hunts from a perch.

Habitat Prefers mature tropical and subtropical forests, but lives also in second-growth woodlands and near plantations and groves. Mainly in lowlands, but occasionally up to 1700m.

Status and distribution Found from S. Mexico to N. Argentina. It is locally rather common, but suffering from deforestation.

Geographical variation Four living subspecies are listed: nominate *perspicillata* occurs from N. Colombia,

Venezuela and the Guianas south to E. Peru and Brazil; *P. p. saturata*, from S. Mexico to Costa Rica and W. Panama, is darker, with uniformly sooty-black head and back, and is finely barred below; both *P. p. chapmani*, from E. Costa Rica to Ecuador, and *P. p. boliviana*, from Bolivia to N. Argentina, are poorly marked. The island race *trinitatis*, confined to Trinidad, is believed to be extinct.

Similar species Only slightly sympatric Short-browed Owl (135) has yellowish-brown to warm brown eyes and short, pale creamy-buff eyebrows, the lower part of the face faintly marked pale creamy or dull yellow, a brown breast-band indistinctly broken at the centre, and only sparsely feathered toes at base. Very marginally overlapping Tawny-browed Owl (136) has dark chestnut eyes and ochre-tawny eyebrows and belly. The similarly sized Band-bellied Owl (137), largely allopatric as it lives in mountain forests of N. Andes at 700–1600m, has dark eyes and white eyebrows, whitish underparts with reddish-brown to dark brown barring, and a brown chest-band broken by dull yellowish or whitish bars.

▶ Race *chapmani* is a dark subspecies, with white crescents on the black face. It has a wide dark brown band across the breast and an unbarred deep-buffish belly. Gamboa, Panama, January (*Yeray Seminario*).

◀ Spectacled Owl of race *chapmani* has a sooty-black crown, lower hindneck and back. It has white 'eyebrows', lores, edges of facial rim and sides of the barred neck. Lower breast and belly are pale yellow to deep buff, without any bars. Tarsi are feathered almost to the toes. Limón, Costa Rica, August (*Daniel Martínez-A*).

◀ This immature Spectacled Owl still has a little white on the head. To reach full adult plumage can take up to five years, so this bird could already be three or four years old. Panama, November (*Mike Danzenbaker*).

135. SHORT-BROWED OWL
Pulsatrix pulsatrix

L 51–53cm, **Wt** 1075–1250g; **W** 363–384mm

Identification A fairly large owl without ear-tufts. Female is some 175g heavier than male. The head and upperparts are plain medium olive-brown to sepia, with flight and tail feathers barred light and dark. The underparts are clear brownish-yellow to dull yellow, with a brown breast-band indistinctly broken in the centre. Warm earth-brown facial disc has an indistinctly paler rim. The rather short creamy-buff eyebrows reach only just behind the pale orange or brownish-yellow eyes, and the lower part of the face is bordered by a rather narrow creamy to dull yellow area from the throat to the lores. Bill and cere are light green. Dirty white toes are sparsely feathered at the base, with greyish-horn claws. *Juvenile* Undescribed.

Call Emits a resonant and rather slow knocking sound, not accelerating as that of the Spectacled Owl.

Food and hunting Little studied but probably eats smaller mammals and birds.

Habitat Inhabits semi-open primary and secondary forests but is found also near human settlements and roads.

Status and distribution E. Brazil from Bahia south to NE Argentina and in SE to the border of Uruguay. This species is believed to be rare and threatened by human persecution and habitat destruction.

Geographical variation Monotypic. This owl is often seen as being a subspecies of Spectacled Owl; only molecular and biological studies can confirm its true taxonomic status.

Similar species Almost allopatric Spectacled Owl (134) is a little smaller and has white 'spectacles' and a more blackish crown and nape, an unbroken chest-band, and bright orange-yellow eyes. Geographically overlapping Tawny-browed Owl (136) is much smaller, with long pale tawny eyebrows and dark chestnut eyes.

136. TAWNY-BROWED OWL
Pulsatrix koeniswaldiana

L 44cm, **Wt** 481g (1 ♀); **W** 300–320mm

Identification A medium-sized owl without ear-tufts. Female is stated to be somewhat larger than male, although no male weights are available. Has crown and upperparts uniformly dark brown. The brown wings are whitish-barred, and the brown tail has four or five narrow white bars and a white terminal band. There is a rather broken brown chest-band above the yellowish-brown remaining underparts. The dark brown facial disc is broken by incomplete 'spectacles' formed by the yellowish-tawny eyebrows and loral streaks. A large white patch on the chin is obvious. The eyes are chestnut-brown and the bill and cere yellowish-horn. The legs are feathered, but the whitish-grey toes bare, with yellowish-grey claws. *Juvenile* Has a browner face and darker eyes compared with the young Spectacled Owl.

Call Said to be a guttural sequence of short *brrr brrr brrr brrr* notes, with emphasis on the second note, the following ones becoming weaker.

Food and hunting Not sufficiently known.

Habitat Occupies mature tropical and subtropical forests but also seen in degraded and marginal forests. From lowlands up to 1500m.

Status and distribution E. Brazil to E. Paraguay and NE Argentina. This species is uncommon to rare. It is locally endangered as it depends to some extent on intact mountain forests. It is too little known for its conservation needs to be identified.

Geographical variation Monotypic.

Similar species Partly overlapping Spectacled (134) and Short-browed Owls (135) are both much larger. Geographically separated Band-bellied Owl (137) is a little larger, with a whitish belly heavily barred reddish-brown, and white eyebrows.

◀ Tawny-browed Owl has brown eyes and distinctly tawny to fulvous 'eyebrows'. Belly is cinnamon-yellow and the broken chest-band is brown. This is the smallest of the spectacled owls. North-east Brazil, May (*Lee Dingain*)

137. BAND-BELLIED OWL
Pulsatrix melanota

L 44–48cm, Wt 420–500g; W 275–325mm

Identification A medium-sized to fairly large owl without ear-tufts. Females are said to be somewhat larger than males, although only two unsexed weights are known. The crown and upperparts are dark brown with some paler mottling, the dark wings narrowly white-banded and the dark brown tail with some six narrow white bars and a white terminal band. The underparts are dull yellowish to white with distinct reddish-brown to dark brown barring; a broad brown band on the upper breast, indistinctly broken in the centre, is mottled with whitish and yellowish tones. The facial disc is dark brown with prominently wide, white eyebrows and loral streaks, these forming white 'spectacles'; a dark area borders a white patch on the throat, with a white half-collar below. The eyes are reddish-brown to blackish-brown, the bill pale yellow and the cere greyish. The whitish to dull yellow legs are feathered, the light greyish-brown toes bare, with blackish-tipped horn-coloured claws. *Juvenile* Unknown.

Call Said to give a short purring call followed by a sequence of four or five popping notes, with emphasis on the third note. Several deep, muffled hoots have been recorded in Peru.

Food and hunting Not studied.

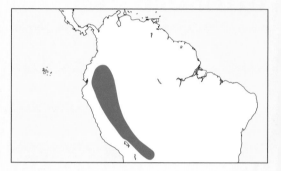

Habitat Inhabits dense and humid mountain forests, preferably between 700m and 1600m. Sometimes found in more open woodland and at lower elevations.

Status and distribution Patchily distributed from Colombia south to Bolivia, but mainly east of the Andes. A specimen presumed to be from Colombia lacks locality and date. This species could be at risk through deforestation but it is very little known.

Geographical variation Monotypic.

Similar species Partly overlapping Spectacled Owl (134) has yellow eyes and no rusty bars on the belly. Other *Pulsatrix* owls are geographically separated.

▼ Band-bellied Owl is a relatively large, dark brown spectacled owl, with white 'eyebrows', lores and throat. Wildsumaco, Ecuador, October (*János Oláh*).

▼ This young bird has a black face and dark brown eyes on a white downy head. It also has a strong white nuchal collar. Manu, Peru, November (*Santiago David-R*).

138. TAWNY OWL
Strix aluco

L 36–46cm, Wt 325–800g; W 248–323mm, WS 94–105cm

Other name Eurasian Tawny Owl

Identification A medium-sized, round-headed owl without ear-tufts. Females are on average over 100g heavier than males. It occurs in three morphs, brown, grey, and rufous, also with intermediates. First morph is generally brown above, the feathers with dusky streaks and some cross-bars. The forehead and fore-crown are darker; the scapulars have whitish outer webs, forming a whitish row across the shoulder. Flight feathers are barred lighter and darker brown, and the tail is brownish above with darker mottling, the outer feathers with some indistinct bars. Underparts are pale ochre to brownish-white with dark brown streaks and fainter cross-bars. The light brownish facial disc has some dusky concentric lines and a narrow dark brown ruff, with off-white eyebrows. The eyes are blackish-brown, the bluish-grey eyelids having pale fleshy edges. The bill is ivory-yellowish, with bristles around the base. The tarsi and most of the grey toes are feathered, with horn-coloured claws. In other morphs, grey and reddish-brown, respectively, replace the normal background. *Juvenile* Downy chick is white. Mesoptile is light brownish-white or greyish-white, densely barred with diffuse brown, grey or reddish. Juvenile has pink-rimmed eyes, an opaque pupil, and pale fleshy cere, eyelids and toes. *In flight* Rarely seen by day. The short, broad wings and short tail enable it to manoeuvre easily in wooded areas.

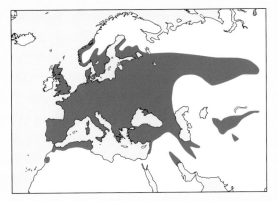

Call This species' well-known musical hoot consists of a long drawn-out *hoóo*, a pause of 2–6 seconds, an abrupt and subdued *hū*, followed at once by a prolonged resonant and tremulous *huhuhuhooo*. In addition, it frequently gives a loud and penetrating *ke-wik* call. Directors of horror and action films very often use the sound of a Tawny Owl in night scenes, indirectly giving the impression that this Old World owl has a global distribution.

Food and hunting Takes a great variety of small mammals, birds, reptiles, frogs, fish and some insects. Usually hunts from a perch, but catches bats and insects in flight.

Habitat Occupies mixed and deciduous forests, often

▶ This flying nominate Tawny Owl demonstrates well the dark alula spots. Tuscany, Italy, February (*Daniele Occhiato*).

near rivers, clearings or open farmland; common also in parks in cities and in gardens near housing. From lowlands to 2800m, and in Pakistan and NW India even up to 3800m.

Status and distribution Widely distributed from Britain south to N. Africa, north to C. Finland and east to N. Pakistan and Kyrgyzstan. It is rather common in Britain and C. Europe, but included in the Red Data list in Kyrgyzstan. In C. Europe the population is estimated to be 198,000 pairs. In Finland there are some 1,500 pairs annually. Ural and Tawny Owls are capable of hybridisation. In captivity two hybrids showed maternal and paternal characters, as well as intermediate ones; one hybrid proved fertile in backcrosses with both parental species. Interbreeding has not been noted in the wild, and in Finland the continuous annual increase of 1.6 per cent in Ural Owl numbers has not caused a noticeable reduction in Tawny Owl numbers; on the contrary, Tawny Owl has also increased its population by 0.8 per cent annually between 1982 and 2008. Competition is therefore very different from that between close relatives (Spotted and Barred Owls) in North America.

Geographical variation Eight subspecies have been listed: nominate *aluco* occurs from Scandinavia to the Mediterranean and Black Sea and W. Russia; *S. a. mauretanica*, from NW Africa to E. Syria, is larger, with wingspan more than 20 per cent greater than nominate; *S. a. sylvatica*, from Britain and W. Europe to Iberian Peninsula, is often grey-brown and more boldly patterned than nominate; *S. a. siberiae*, from west of the Urals to Irtysh, in C. Russia, is larger and distinctly paler, with much white; *S. a. sanctinicolai*, from Iran to NE Iraq, is a pale desert form; *S. a. wilkonskii*, from Asia Minor and Palestine to N. Iran and the Caucasus, has a coffee-brown morph; *S. a. harmsi*, from Turkestan, and *S. a. biddulphi*, from NW India and Pakistan, occur predominantly as grey morphs.

Similar species Smaller Hume's Owl (139) is partly sympatric, but has orange-yellow to pale ochre-yellow eyes and is generally more sandy-coloured. Very similar wood owls – Mottled (142), Brown (143), Himalayan (146) and Mountain (147) – and Spotted Owl (156) are all allopatric. A close relative, the Ural Owl (160), is partly overlapping in range, but is much larger, has a long, distinctly barred tail and smaller dark brown eyes, and is boldly dusky-streaked and without cross-bars below.

▼ A rufous-morph *sylvatica* Tawny Owl. Devon, England, March (*Chris Townend*).

▲ Race *sylvatica* is more boldly patterned than the nominate. This owl has taken a Grey Squirrel *Sciurus carolinensis,* which makes up to 1% of the Tawny Owl's food in Great Britain. Bedfordshire, England, September (*Lee Dingain*).

▲▶ A greyish-brown morph Tawny Owl, of race *sylvatica*. Cadíz, Spain, December (*Andrés Miguel Domínguez*).

▶ A rufous-morph bird of the nominate race. Utrecht, The Netherlands, April (*Lesley van Loo*).

▼ Race *mauritanica* has medium-grey upperparts with dark grey bars and vermiculations. Facial disc has fine grey barring and a distinct rim. Atlas Mountains, Morocco, February (*Augusto Faustino*).

▼▶ A juvenile Tawny Owl. Somerset, England, June (*Gary Thoburn*).

139. HUME'S OWL
Strix butleri

L 30–34cm, **Wt** 162–225g; **W** 243–256mm, **WS** 95–98cm

Other name Hume's Tawny Owl

Identification A small to medium-sized, round-headed owl without ear-tufts. Females are a little larger than males. Crown and nape have brown and blackish spotting, and often some pale lines. The upperparts are greyish-ochre or sandy, with dusky streaks and fine vermicular markings on mantle and back. Whitish outer webs and dark shaft-streaks on scapulars, the forming creating a whitish row across the shoulder, and there are several whitish and yellowish spots on wing-coverts. Wing and tail feathers are boldly barred, the primaries with whitish and dusky brown bars, the secondaries and tail with light and dark brown bars. Underparts are creamy white to pale ochre, becoming white on lower breast and belly; the breast and flanks are barred and spotted with orange-buff, with darker shaft-streaks. Light yellowish-grey facial disc is nearly circular, with a thin darkish ruff. The eyes are brownish-

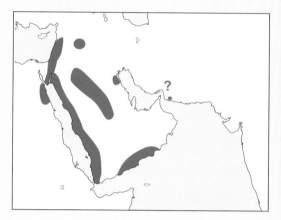

orange or bright yellow, rimmed by black edges of the eyelids. The bill is yellowish-horn and the cere pale ochre. Whitish-creamy tarsi are feathered to the base

of the yellowish-grey toes, which have horn-coloured claws. *Juvenile* Downy chick is white. Mesoptile is like a very pale Tawny Owl (138), but with yellowish eyes; talons are much weaker than those of the Tawny Owl.
Call Utters a clear, rhythmic, hooting *hoo oo, hoo-u hoo-u,* repeated after a few seconds. This is higher-pitched than that of the Tawny Owl, and without latter's tremulous character.
Food and hunting Hunts small mammals, birds, reptiles, insects and scorpions. It sometimes hawks insects in the air, but hunts mainly from a perch.
Habitat Inhabits semi-deserts and rocky ravines with water, and palm groves in oases, sometimes even near human habitations. One of the least forest-dependent or tree-dependent of all owls.

Status and distribution Occurs from Syria and Israel south to both sides of the Red Sea and patchily through the Arabian Peninsula, and locally in S. Iran. DNA tests have shown differences of 9–12 per cent between *Strix aluco, S. butleri* and *S. woodfordii,* proving all of them to be distinct species.
Geographical variation Monotypic.
Similar species Slightly overlapping Tawny Owl (138) is larger and much darker, with dark brown to blackish eyes; it normally perches upright and has partly feathered toes, whereas Hume's Owl perches rather more at an angle and has bare toes. Geographically well-separated African Wood Owl (140) is fairly similar in size, but has a very brown body and face, with dark brown eyes.

◄ Hume's Owl is a pale-faced, brownish-eyed owl, which has buffish-white and faintly marked underparts and well-feathered tarsi. The bill is greyish-blue. Dead Sea, Israel, December (*Amir Ben Dov*).

▼ Hume's Owl has brown and grey-mottled upperparts, and a less clearly rimmed facial disc. Side of the head is light brownish. Tawny Owl *Strix aluco* is larger and darker, and typically perches upright. Dead Sea, Israel, December (*Amir Ben Dov*).

140. AFRICAN WOOD OWL
Strix woodfordii

L 30–35cm, Wt 240–350g; W 222–273mm, WS 79–80cm

Identification A small to medium-sized, round-headed owl with no ear-tufts. Females are on average more than 50g heavier than males. The dark brown head and neck are spotted with white, and the dusky reddish-brown mantle, back and uppertail-coverts with narrow white barring and shaft-streaks and dull yellow vermicular markings. White outer webs of the scapulars form a pale row across the shoulder. Flight and tail feathers are barred light and dark. The dull rusty-coloured underparts are barred densely with whitish and brown. The pale yellowish-brown facial disc has darker concentric lines. The eyes are dark brown, surrounded by a dusky ring, and with fleshy-pink eyelids. The bill and cere are yellowish. Dull yellow tarsi are feathered, with light brown bars, and the yellow-horn toes are bare, with greyish-brown claws. *Juvenile* Downy chick is white and has pink skin. Mesoptile is pale reddish-brown, with white feather tips above and white and brown bars below.

Call Utters a rhythmic and loud hoot, *whoo-whu whu-uh-uh whu-och*, repeated at intervals of several seconds. The first part is louder, the rest given in varying rhythm.

Food and hunting Feeds mostly on large insects, such as crickets, cicadas, moths and beetles, but it also eats frogs, reptiles, small birds and small mammals. Often snatches prey from vegetation while in flight.

Habitat Occupies wooded areas, ranging from primary forest to dense woodland, and occurs also in city parks and plantations. From sea level to 3700m.

Status and distribution Occurs south of the Sahel in W. Africa from Casamance, in Senegal, east to Ethiopia, and south to the Cape in South Africa. This owl is resident in 33 African countries, and in many of them is fairly common, but it could be locally endangered by forest destruction.

Geographical variation Four subspecies are listed: nominate *woodfordii* is found from S. Angola to the SE coast of Africa and south to the Cape; *S. w. umbrina*, from Ethiopia and SE Sudan, is more brown in colour; more blackish *S. w. nigricantior* occurs from S. Somalia to Zanzibar and E. DR Congo; *S. w. nuchalis*, from Senegal to W. DR Congo and including Bioko Island, is brightest russet with the broadest white chest-bars.

Similar species Geographically fully separated Tawny Owl (138) is heavier and also allopatric Hume's Owl (139) has orange-yellow eyes and sandy-coloured plumage. Sympatric Maned Owl (162) likewise has yellow eyes, but also rather long, bushy ear-tufts.

African Wood Owl of race *uchalis* has very broad chest barring. (*Tasso Leventis*)

Female African Wood Owl of the race *nigricantior* with two juveniles, sleeping in a large tree. The wings and tail of the juveniles resemble those of the adult, but the underparts remain partly downy. Tanzania, February (*Martin Goodey*).

African Wood Owl of the nominate race; a very round-headed owl, with dark eyes and a yellow bill. Shakawe, Botswana, December (*Ralph Martin*).

141. SPOTTED WOOD OWL
Strix seloputo

L 44–48cm, Wt 1011g (1♂); W 297–376mm

Identification A medium-sized to fairly large owl without ear-tufts. Size difference between the sexes is not known, as only one male weighed. Above, the plumage is coffee-brown overall, with black-edged white spots, the paler brown mantle, back and upper-tail-coverts with black-margined white bars and spots. White or dull yellowish outer webs of scapulars have dark bars and blackish edges. Dull yellow underparts are barred black and white, the white bands broader. It has an orange-buff facial disc on a chocolate-brown head. The eyes are dark brown, and the bill and cere greenish-black. The legs and toes are feathered, visible parts of the toes being dark olive, with horn-coloured claws. *Juvenile* Natal down is whitish. Mesoptile has the upperparts predominantly barred white and dark brown. Subadult has broadly white-tipped upperwing-coverts, and more white on the scapulars.

Call The song consists of a rolling staccato *huhuhu* followed by a deep drawn-out *who*.

Food and hunting Feeds mainly on rats and mice, small birds and large insects.

Habitat Occupies unpopulated mangrove swamps and forests near the coast, but also occurs in plantations

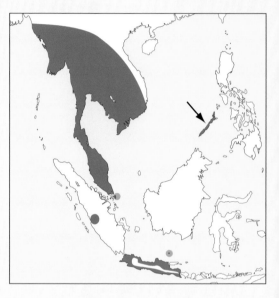

and cleared forests, and even in parks in towns and villages.

Status and distribution Found from S. Burma to Indonesia and SW Philippines. This species is sparsely distributed at low density, but could be partly over-looked as a result of its cryptic appearance.

Geographical variation Three subspecies are listed: nominate *seloputo* from S. Burma to Java; *S. s. baweana*, from Bawean Island, off N. Java, is much smaller and paler, with more narrow bars below; *S. s. wiepkeni*, from Calamian Islands and Palawan, in SW Philippines, has yellowish-rufous ground colour below.

Similar species Partly sympatric Brown Wood Owl (143) has more narrowly barred, yellowish-buff under-parts, unspotted dark brown forehead, crown and nape, dark rufous breast, eight distinct bars on tail, and a reddish-brown facial disc with rather prominent black ruff and whitish-russet eyebrows. Geographically separated Nias Wood Owl (144) is smaller and gener-ally deep reddish-brown, with a prominent rufous nuchal collar. Also smaller, Bartels's Wood Owl (146) partly overlaps in range in Java, but is dark greyish-brown above and densely barred reddish-brown below, with a distinct ochre nuchal collar.

◄ Nominate Spotted Wood Owl from Malaysia. This bird has a yellowish facial disc and distinct yellowish throat band. Penang, Malaysia, May (*Choy Wai Mun*).

► Nominate Spotted Wood Owl with a typical orange-buff facial disc. White outer webs of the scapulars have dark bars and blackish edges. This owl has very little yellow below and very black claws. Singapore (*Ong Kiem Sian*).

▼ Nominate Spotted Wood Owl showing white throat patch. Singapore, April (*HY Cheng*).

▼ ► The fledgling is very pale – the face and primaries already resemble those of the adult but the back is very distinctive. Singapore, April (*HY Cheng*).

142. MOTTLED WOOD OWL
Strix ocellata

L 41–48cm; **W** 320–372mm

Identification A medium-sized to fairly large owl without ear-tufts. Females are presumably larger than males, although no weights are available. Above, the plumage is mottled and vermiculated with reddish-brown, black, white and buff, the nape white and black mixed with coffee-brown (appearing distinctly black-spotted). Flight and tail feathers are barred light and dark. The chestnut-and-black throat is stippled with white, and the underparts are barred narrowly blackish on a background of white mixed with golden-buff to orange-buff. The whitish facial disc is finely barred concentrically with blackish-brown. The eyes are dark brown, with dusky-pink or dull coral-red eyelids, and the bill is horn-black. Feathered tarsi and toes are pale reddish-buff, the soles yellow and the claws blackish. *Juvenile* Mesoptile has a whiter crown than adults, and the whitish nape, mantle and wing-coverts are finely blackish-barred. *In flight* If disturbed, it can fly long

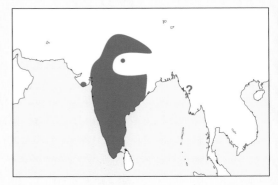

distances even in bright sunlight without any apparent discomfort, settling again within the seclusion of the foliage canopy.

Call A loud, shivering, hollow *chuhuawaarrrrr* song is heard during the breeding season. During other seasons, the calls consist of single, metallic hoots and occasional screeches.

Food and hunting Feeds primarily on rats, mice, other rodents, lizards, crabs and birds; in addition, it takes scorpions and large insects.

Habitat Inhabits open woodlands and lightly wooded plains, often on the outskirts of villages and plantations.

Status and distribution Occurs in the Indian Subcontinent eastwards from N. & E. Pakistan and south to s. India, and in W. Burma. Although locally common, this species is not well studied.

Geographical variation Three subspecies are known: nominate *ocellata* is found in C. & S. India and Bangladesh; *S. o. grandis*, from the Kathiawar Peninsula (S. Gujarat), in NW India, is larger and greyer above; *S. o. grisescens*, from N. India and the S. Himalayas in Pakistan south to Rajasthan and east to Bihar and W. Burma, is paler above and a little larger than the nominate.

Similar species Allopatric Spotted Wood Owl (141) has no darker concentric lines on its plain orange-buff facial disc. Partly sympatric Brown Wood Owl (143) is more yellowish-buff below, with a tawny face, black ruff and conspicuous blackish rings around the eyes.

◄ Races *grisescens* and *ocellata* meet in southern Gujarat. This individual is an intergrade. June (*Arpit Deomurari*).

▲ Mottled Wood Owl race *grandis* is less brown below than the nominate. Gujarat, India, November (*Alain Pascua*).

▲ ▶ Race *grandis* is larger than the nominate, with a greyer and less black-spotted appearance above. Gujarat, March (*Arpit Deomurari*).

▶ Nominate Mottled Wood Owl is a smart owl with a rounded head. Its facial disc has black concentric rings; the ruff is a blend of white, black and dark brown. Gujarat, India, December (*Ian Merrill*).

143. BROWN WOOD OWL
Strix leptogrammica

L 34–45cm, **Wt** 500–1100g; **W** 286–400mm

Identification A medium-sized owl without ear-tufts. Females are clearly larger and at least 200g heavier than males. The dark, blackish-brown head has a rufous tint, and is separated from the chestnut upperparts by a distinct cinnamon-buff or reddish-brown nuchal collar; the mantle and back are densely barred dark brown to blackish. The primaries are barred deep reddish and dark brown and the secondaries and wing-coverts are barred fulvous and dark tawny-brown; the similarly barred tail has whitish feather tips. The neck is brown, but the throat has a narrow, horizontal white band; there is a rufous or chestnut pectoral band on the upper breast, below which the underparts are light reddish-buff with thin blackish or dark brown barring. The fulvous or reddish-brown facial disc has a blackish

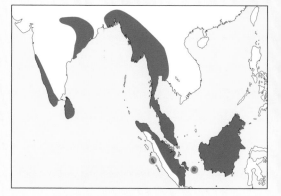

area around the dark brown eyes, and a narrow but distinct black ruff, and the eyebrows are whitish-buff or pale orange-buff. The bill is greenish-horn and the cere bluish-grey. The tarsi and basal parts of the toes are feathered, the distal two phalanges of the toes being bare and pale bluish-grey, with dusky lead-coloured claws. *Juvenile* Natal down is light rufous-buff, gradually replaced by the barred juvenile plumage. Mesoptile has buff facial disc rimmed with a narrow dark brown or black line, and a black zone around the eyes; very pale rufous or ochre-buff crown, mantle and underside, faintly barred with russet, reddish-buff wings with dark bands, and a yellowish-white tail banded dark reddish-brown and white-tipped.

Call Utters a single hoot and a vibrating *who whooow-wwwooh* lasting less than one second, and repeated at intervals of several seconds.

Food and hunting Feeds on rats, mice and shrews, but also takes birds, fish and reptiles.

Habitat Occupies dense tropical forests from along the sea coast to the lower hills; mainly in lowlands up to 500m.

Status and distribution Found from India to W. Burma and W. Thailand in the east and to Sumatra and Borneo in the south. It is generally fairly rare, but rather elusive.

Geographical variation Five subspecies are listed: nominate *leptogrammica* is found only in C. & S. Borneo; *S. l. vaga*, from N. Borneo, is larger than the nominate; *S. l. maingayi*, from Burma and Thailand and south through the Malay Peninsula, is deeper and richer in colour; *S. l. indranee*, from S. & C. India and Sri Lanka, is much larger than the nominate; *S. l. myrtha*, from

Sumatra, Mentawai Islands and Belitung, is much smaller than the previous race.

Similar species Clearly smaller Nias Wood Owl (144) is allopatric and generally rufous, with a deep reddish-brown nuchal collar and facial disc, the latter indistinctly blackish-rimmed. Geographically separated Bartels's Wood Owl (146) lacks a prominent rufous band on the upper breast, but has a more ochre-tawny face, a broad ochre-coloured nuchal collar, densely barred tail, and fully feathered toes. Larger Mountain Wood Owl (147), also allopatric, has no dark pectoral band and the brown parts are less reddish-brown.

▶ Race *vaga* from N. Borneo is larger than the nominate. Duller, more greyish-brown and less rufous. Borneo, November (*James Eaton*).

◀ Race *maingayi* is deeper and more richly coloured than the nominate. Facial disc is rich rufous, and the bill is bluish. Collar is almost black. Pahang, Malaysia, November (*Choy Wai Mun*).

◄ A pair of *indranee* Brown Wood Owls. This race is more variable in colour than the nominate, and the chest-band is not distinct. There is little difference between the sexes; the female is slightly larger. Karnataka, India, April (*Niranjan Sant*).

144. NIAS WOOD OWL
Strix niasensis

L *c.* 35cm; **W** 273–286mm

Identification A medium-sized owl with no ear-tufts. There is no sexual difference in wing length, but no sexed weights are available. The head is mostly warm maroon-chestnut, separated from the deep reddish-brown upperparts by a broad rufous nuchal collar. The mantle and back are barred blackish, and the outer webs of the scapulars are light buffish, these forming a pale line across the shoulder. The primaries are barred black and chestnut, and the fulvous secondaries are barred darker brown; the rufous-buff tail has a number of darker reddish-brown bars. A narrow dark line borders the lower edge of the whitish throat, below which is a deep rufous pectoral band; the rest of the underparts are paler reddish-brown, densely barred by the very fine, darker brown distal edges of the feathers. The rich rufous facial disc is bordered by an indistinct, narrow blackish ruff, with a blackish zone around the dark brown eyes. The bill is bluish-horn and the cere bluish-grey. The feathered tarsi are light reddish-brown with darker bars, and the unfeathered parts of the toes bluish-grey, with dusky-grey claws. *Juvenile* Said to be more rufous than that of the Brown Wood Owl.
Call This species' voice is not well known, but it possibly utters a disyllabic *whoo-hooh*.

Food and hunting Unknown.
Habitat Inhabits tropical forests at lower altitudes.
Status and distribution Endemic to Nias Island, off Sumatra. No precise information on this owl's status is available.
Geographical variation Monotypic. This owl was previously considered to be a subspecies of Brown Wood Owl (143). Genetic study is required in order to confirm its taxonomic status.
Similar species All other wood owls are larger, and none of them is known to exist on Nias Island. The Nias subspecies of the Malay Fish Owl (129) is much larger and has prominent, outward-pointing ear-tufts.

145. HIMALAYAN WOOD OWL
Strix nivicola

L 35–40cm, Wt 375–392g; W 280–320mm

Other name Chinese Tawny Owl

Identification A medium-sized, round-headed owl without ear-tufts. Females are slightly larger than males. Grey and rufous morphs exist. The grey morph is dark greyish-brown above, with light and dark mottling, barring and vermiculations; rufous morph is reddish-brown above, with pale and whitish mottling, and tawny-rufous below. Outer webs of scapulars and tips of greater wing-coverts are whitish, forming two distinctive pale rows across the wing area. The wing and tail feathers are broadly banded dark brown. The chin is dark greyish-brown and the throat whitish; below this, the pale greyish-brown or dirty-whitish to light tawny-buff underparts have heavy blackish streaks and cross-bars. The greyish-brown or brownish-yellow facial disc is emphasised by distinct whitish eyebrows, these forming a white 'X' between the blackish-brown eyes, which have pink eyelids. The bill is yellowish-grey. Tarsi are feathered to the tips of the toes, which have dark horn-coloured claws with blackish tips. *Juvenile* Mesoptile is greyish-brown, evenly and heavily barred, the white-frosted crown less barred.

Call Utters two or three clear hoots, repeated rapidly, with very short intervals.

Food and hunting Diet consists of small mammals, birds and large insects.

Habitat Occupies oak and coniferous forests between 1000m and 2650m.

Status and distribution Found from the Himalayas east to E. China and Taiwan. This species is said to be fairly common, but is not well studied.

Geographical variation Three subspecies are separated: nominate *nivicola* occurs from Nepal to SE China and NW Burma; *S. n. ma*, from NE China and Korea, is paler than the nominate, and *S. n. yamadae*, from Taiwan, is darker than the nominate. Formerly, all these owls were treated as subspecies of the Tawny Owl, but are now separated on the grounds of their different vocalisations and geographical isolation; they differ further from that species in their plumage colours.

Similar species Allopatric Tawny Owl (138) is distinctly

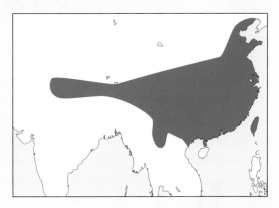

dark-streaked above and less coarsely patterned below, its tail is less barred, and a second pale row across wings is not visible. Sympatric Mountain Wood Owl (147) is larger, with a dark-rimmed pale face, and the underparts dirty whitish and brown-barred, with no dark shaft-streaks.

▶ Himalayan Wood Owl is similar to Tawny Owl *Strix aluco*, but mottled and unstreaked above, with two pale wing-bars and broadly barred tertials. Prominent white patch on the upper breast. A grey-brown morph (as here) and a very rufous morph occur. Emei Shan, China, June (*James Eaton*).

◄ Rufous-brown morph of Himalayan Wood Owl showing a very dark brown face, and a rufous tinge on the flanks. Note clear black ruff, yellow bill and strong white line on the scapulars. Hunan, China, May (*Jonathan Martinez*).

146. BARTELS'S WOOD OWL
Strix bartelsi

L 39–43cm, Wt 500–700g; W 335–376mm

Identification A medium-sized owl without ear-tufts. Females seem to be clearly larger than males, although no sexed measurements are available. The crown and upperparts are sepia-brown, often with an indistinct yellow-buff or ochre nuchal collar. The primaries and secondaries are barred reddish-buff and dark brown, with dull yellow edges and tips; the scapulars are paler than the remiges. The light rufous uppertail has many dark brown bars. The chin and throat are pale whitish-buff, and the underparts reddish-brown with numerous fine dark rufous bars; it does not have a darker breast-band. The facial disc is tawny-fulvous, with a blackish area around the dark brown eyes, and with contrasting whitish-yellow eyebrows. The bill is bright bluish-plumbeous. Tarsi are feathered nearly to the tips of the plumbeous toes, which have dusty blackish-grey claws. *Juvenile* Said to have reddish-brown down with fine dark rufous bars.

Call Utters a loud and explosive *whooh*, repeated at long intervals. This is audible to several hundred metres.

Food and hunting Feeds on small mammals up to the size of fruit-eating bats; it also takes large insects.

Habitat Inhabits primary mountain forests and forest edges at elevations of between 700m and 2000m.

Status and distribution This species is known for

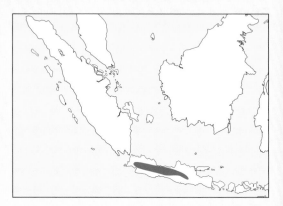

certain only from W. Java, but some inaccurately labelled museum skins seem to indicate that it could occur also in Sumatra and N. Borneo. It is rare, and is seriously threatened by the cutting of mountain forests in Indonesia.

Geographical variation Monotypic. This owl was previously considered to be a subspecies of Brown Wood Owl (143). Genetic study is required in order to confirm its taxonomic status.

Similar species Similar-sized Spotted Wood Owl (141) partly overlaps in range in Java, but has no blackish zone around the eyes and no ochre nuchal collar.

147. MOUNTAIN WOOD OWL
Strix newarensis

L 46–55cm, Wt 970g (1♂); W 377–442mm

Identification A fairly large owl with no ear-tufts. Females appear larger than males, although only one male weighed. Has the crown to mantle and back dark sepia-brown. The scapulars, wings and tail have pale or white barring, but the wing-coverts are uniformly brown. Off-white underparts are densely barred with dark brown; no distinct dark pectoral band is visible on the upper breast. The very light ochre facial disc and the broad white eyebrows contrast the dark brown eyes. The disc has a narrow ruff formed by small blackish-brown feathers. Feathered tarsi and toes, with dusky-grey claws. *Juvenile* Undescribed.

Call Utters a double hoot, *to-hooh*, in which the second note is emphasised and slightly vibrating. The voice resembles that of the Rock Dove *Columba livia*.

Food and hunting Hunts small mammals, and birds up to the size of pheasants; it also takes reptiles, including small monitor lizards.

Habitat Inhabits primary mountain forests with some open areas for hunting, between 1000m and 2500m, but locally ascends to 4000m.

Status and distribution Occurs from Pakistan and

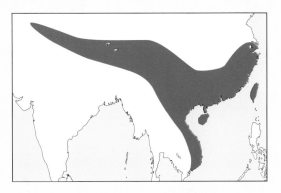

Nepal east to China and Taiwan and south to Indochina. This species is rarely seen and very little studied, but it must be threatened by deforestation.

Geographical variation Four subspecies: nominate *newarensis* from Pakistan to Sikkim; darker *S. n. laotiana*, from S. Laos to C. Vietnam; smaller *S. n. ticehursti* from N. Burma to SE China; and larger *S. n. caligata*, from Hainan and Taiwan.

Similar species Sympatric Himalayan Wood Owl (145) is smaller and distinctly streaked below.

▼ Race *laotiana* is darker on the upperside than the nominate, and more buff below. Coloration similar to Brown Wood Owl *Strix leptogrammica vaga*, but Mountain Wood Owl is larger. Cambodia, January (*James Eaton*).

▼ Young Mountain Wood Owl is very white. Hong Kong, June (*Martin Hale*).

148. MOTTLED OWL
Strix virgata

L 30–38cm, Wt 235–307g; W 230–274mm

Identification A small to medium-sized, round-headed owl without ear-tufts. Females are over 50g heavier than males. Pale and dark morphs exist. Pale morph has crown, nape and upperparts dark brown with a rufous tint, with whitish to pale buff flecking and sparse barring, and a row of whitish spots across the shoulder. Wing feathers have very broad dark bars and rather narrow light bars. Whitish or dull yellow underparts are streaked dark brown. The brown facial disc, with white eyebrows and whiskers, is bordered by a whitish or buffish ruff. The eyes are dark brown and the bill yellowish. Tarsi are feathered, and the bare yellowish-grey to brownish-grey toes have pale horn-coloured claws. Dark morph is almost blackish-brown above, darker ochre-buff below, the breast sides heavily mottled dark brown. *Juvenile* Downy chick is whitish. Mesoptile is dull yellow, barred above indistinctly, with whitish facial disc and pale pink bill.

Call Emits a clear and resonant frog-like *gwho*, which after four or five such hoots becomes lower-pitched *gwho-gwho-gwho-gwhóho*, repeated at intervals of several seconds.

Food and hunting Feeds on small mammals and birds, but also takes snakes and frogs.

Habitat Inhabits primary and secondary forests, mostly humid lowland forests, but also dry and thorny forests, from sea level to 800m, sometimes to 2500m.

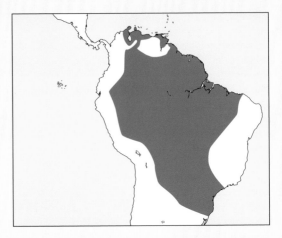

Status and distribution Occurs from N. Colombia to NE Argentina. This species is locally common, but could be threatened by habitat destruction. Further studies of its ecology, vocalisations and breeding biology are needed.

Geographical variation Four subspecies: nominate *virgata* is found from N. Colombia to E. Ecuador and Trinidad; *S. v. macconnellii*, from the Guianas is barred below; larger and more rufous *S. v. superciliaris* from Amazonian Peru to NW Argentina; and dark-winged *S. v. borelliana* from SE Brazil to NE Argentina.

◄◄ Dark morph Mottled Owl of race *borelliana* is similar to the dark morph nominate. Minas Gerais, Brazil, March (*Luiz Gabriel Mazzoni*).

◄ Pale morph nominate Mottled Owl has finely streaked breast. Mexican Wood Owl (150) is similar in size, but more coarsely streaked below. Arrierito, Colombia, June (*Frank Lambert*).

149. RUFOUS-LEGGED OWL
Strix rufipes

L 33–38cm, **Wt** *c.* 350g; **W** 250–275mm

Identification A medium-sized, round-headed owl without ear-tufts. Females have longer wings and are obviously heavier than males, although no sexed weights are available. Pale and dark morphs are known, but males are also generally darker than females. The plumage is sepia-brown above, with dense, fine whitish bars on the head, crown and hindneck, and not very prominent white and pale buff spots on the scapulars. Wings and tail are barred dull yellow and dusky. The throat is whitish, otherwise the plumage is bright cinnamon-buff below, densely barred white and blackish. The facial disc is light ochre to orange-brown with darker concentric lines, the eyebrows and lores whitish-buff, and the eyes dark brown. The cere is pale yellowish and the bill wax-yellow. Tarsi and toes are feathered, with brownish-horn claws. Dark morph has the entire plumage much darker, with a tawny ground colour, and with bold bars merging together. *Juvenile* Downy chick is whitish. Mesoptile is warm buff, indistinctly dusky-barred, with a tawny facial disc and whitish flecks on the head.
Call Utters a fast guttural *kokoko-kwowkwowkwówk-wowkwow-kwowkwok*, which is different from Chaco Owl.

Food and hunting Feeds on small mammals, birds, reptiles, frogs and insects. It hunts mainly from a perch.
Habitat Prefers mountain slopes with moist forests, from sea level up to 2000m, locally perhaps higher.
Status and distribution Found from SC Chile and SW Argentina (Patagonia) south to Tierra del Fuego. Occurs on Chiloé Island, and migrants occasionally turn up in the Falkland Islands as well.
Geographical variation Monotypic.
Similar species Allopatric Chaco Owl (151) is much paler, has no prominent eyebrows, and its off-white facial disc has more distinct, dark concentric lines.

▼ A pale morph male calling with the throat fully puffed up. It has sepia-brown upperparts, barred and spotted whitish and orange-buff. Scapulars and secondaries have large white spots. Tierra del Fuego, Argentina, December (*James Lowen*).

▼ Pale morph Rufous-legged Owl is densely barred below with a light ochre facial disc. This is an elusive species. La Campana, Chile, November (*Rob Hutchinson*).

150. MEXICAN WOOD OWL
Strix squamulata

L 29–33cm, Wt 177–356g; W 221–265mm

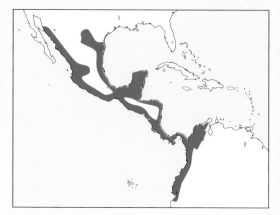

Identification A small to medium-sized, round-headed owl without ear-tufts. Females are on average some 80g heavier than males. The crown, nape and upperparts are mainly greyish-brown, mottled and vermiculated dark and light; the slightly darker head and nape have a number of whitish spots. Large whitish parts on outer webs of scapulars form a distinct row across the shoulder. The greyish-brown flight and tail feathers have finely blackish-edged paler and darker bars. The throat is whitish and there is a narrow brownish collar between the upper and lower breast, below which the underparts are whitish to pale buff with strong dark brown shaft-streaks, and a nearly unmarked whitish-yellow belly. The dark brownish-grey facial disc has fine whitish shaft-streaks running from around the blackish-brown eyes to the whitish ruff bordering the disc. The eyelids are light greyish-flesh in colour and the eyebrows yellowish. Dirty whitish to pale grey whiskers are visible around the base of the greenish-yellow bill. The tarsi are feathered to the base of the light grey toes, which have yellow, black-tipped claws. ***Juvenile*** Mesoptile is buffish to light cinnamon, dusky-barred above, with a whitish facial disc and pinkish bill. In common with *Pulsatrix* owls, it has several immature plumages.

Call Gives frog-like, short guttural hoots, *kwow-kwow-kwow-gwot*, repeated at intervals of *c.* 13 seconds.

Food and hunting Hunts small mammals, birds, reptiles, frogs and large insects. This owl forages around clearings, catching bats and insects in flight.

Habitat Occupies humid primary and secondary forests, as well as gallery forests and plantations, not far from human habitations. From sea level to 2200m.

Status and distribution Distributed from Mexico south to SW Ecuador. It is fairly common, but could be affected by forest destruction.

Geographical variation Three subspecies have been described: nominate *squamulata* occurs in W. Mexico; *S. s. tamaulipensis* is found in NE Mexico and *S. s. centralis* from E. & S. Mexico to SW Ecuador; this last subspecies is the heaviest of all. Further studies are needed in order to confirm the taxonomic status of this owl, as it has been separated from Mottled Owl only recently and without DNA evidence.

Similar species Allopatric Mottled Owl (148) is very similar, but generally darker, and has a less prominent pale line across the shoulder, and on the flight and tail feathers the pale bars are very narrow and the dark ones very broad. Sympatric Black-and-white Owl (154) is distinctly barred blackish and whitish below, the head and mantle are blackish and separated by a broad dark-barred pale collar, and the bill and toes are yellow. Partly overlapping Spotted (156), Fulvous (157) and Barred Owls (158) are much larger.

▶ Race *centralis* is darker than nominate *squamulata*, but paler above than the light morph of the nominate race of Mottled Owl, *Strix virgata*. Outer webs of the scapulars have relatively large whitish areas, forming a distinct row across the shoulder. Guatemala, December (*Knut Eisermann*).

◀◀ Mexican Wood Owl is much smaller than the sympatric Spotted *Strix occidentalis*, Fulvous *S. fulvescens* and Barred Owls *S. varia*, and paler than its southern relative, the Mottled Owl *S. virgata*. This is race *centralis*. Guatemala, March (*Knut Eisermann*).

◀ A female Mexican Wood Owl of race *centralis* on a nest. Note very large dark brown eyes and a greenish-yellow bill. Costa Rica, April (*Daniel Martínez-A*).

151. CHACO OWL
Strix chacoensis

L 35–38cm, Wt 360–500g; W 251–291mm

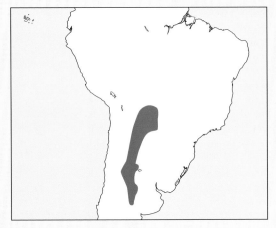

Identification A medium-sized owl with no ear-tufts. Females are some 70g heavier than males. The crown and nape are finely barred light and dark. The upperparts are dark greyish-brown. The wing-coverts have coarsely white and light orange-buff blotching. Flight feathers are boldly barred with dark greyish-brown and orange-buff, and the dark greyish-brown tail has a few narrow pale buffish bars. Below, the throat is white and foreneck and upper breast are finely barred dark greyish-brown and greyish-white, the very light orange-yellow lower breast and belly having bold but fewer bars. The pale greyish-white facial disc has narrow darker concentric lines, but the darker ruff is not prominent. The eyes are blackish-brown, and the bill is horn-coloured, becoming yellowish at the tip. Tarsi

and upper parts of toes are feathered, the bare parts of the toes being greyish-brown, with reddish-horn claws. *Juvenile* Downy chick is whitish. Light greyish-brown mesoptile is very fluffy, with buffish suffusion and a pale greyish facial disc.

Call Utters a deep, frog-like croaking *cru cru cru craw craw*, the first three notes quiet, but the following two very loud. The voice is similar to that of Rusty-barred Owl (152), but clearly different from that of Rufous-legged Owl (149).

Food and hunting Hunts small mammals and other small vertebrates, such as birds; it also takes insects and other arthropods. Hunts mainly from a perch.

Habitat Prefers dry, semi-open landscapes and thorny scrub in hilly places or mountain slopes up to 1300m.

Status and distribution Occurs from S. Bolivia south to the Argentine provinces of Córdoba and Buenos Aires. It is not uncommon, but is little studied.

Geographical variation Monotypic. Formerly considered a subspecies of Rufous-legged Owl *Strix rufipes*, although it may be more closely related to Rusty-barred Owl *Strix hylophila*.

Similar species Allopatric Rufous-legged Owl (149) has a shorter tail and rufous feathering on the legs. Slightly sympatric Rusty-barred Owl (152) has russet upperparts, and its yellowish toes are not feathered.

◄ Face of Chaco Owl is off-white with darker concentric rings; tarsi are whitish with rufous bars (Rufous-legged has rufous face and feathering on tarsi and toes). Central Chaco, Paraguay, October *(Paul Smith)*.

152. RUSTY-BARRED OWL
Strix hylophila

L 35–36cm, Wt 285–395g; W 280mm (1)

Identification A medium-sized, round-headed owl without ear-tufts. Females are over 50g heavier than males. Above, the plumage is dark brown to warm brown, densely barred rusty-orange, with buffish and whitish areas on the outer webs of the scapulars forming an indistinct pale row across the shoulder. Flight feathers are dull yellow or rusty and barred brown, and the warm brown tail has narrow whitish-buff bars. The throat and underparts are whitish, coarsely brown-barred on upper chest and flanks; buffish-whitish undertail-coverts have dark barring. The rounded facial disc is rusty with dusky brown concentric lines, the brown ruff not very prominent. The eyes are dark brown, offset by greyish eyebrows and whiskers. The bill, cere and toes are yellowish-horn. Tarsi are feathered, and the bare toes have horn-coloured claws. *Juvenile* Downy chick is white. Mesoptile lacks concentric lines on the face and is more buffish than that of Rufous-legged Owl (149).

Call Utters a deep grunting *grugruu-grugruugrugru*, similar to that of Chaco Owl (151) but with a different rhythm. More studies are required in order to elucidate the differences.

Food and hunting Varied diet contains small mammals, birds, reptiles, insects and maybe also amphibians.

Habitat Prefers primary and secondary forests with dense undergrowth, but sometimes settles near human habitations. From lowlands up to 2000m.

Status and distribution Occurs in SE Brazil from Rio de Janeiro south to Rio Grande do Sul, as well as to E. & S. Paraguay and NE Argentina. It is locally not rare, but is vulnerable to deforestation.

Geographical variation Monotypic.

Similar species Sympatric Tawny-browed Owl (136) has chestnut eyes and a broken brown breast-band, and is pale yellowish-cinnamon below, plain or with faintly darker barring. Partly sympatric Mottled Owl (148) is streaked below. In eastern Chaco it may overlap with similar Chaco Owl (151), but the latter has greyish-brown (never rusty) upperparts, and fully feathered toes. More geographically overlapping Black-banded Owl (155) has yellowish-orange eyes, and is entirely blackish with fine whitish barring.

► Rusty-barred Owl resembles Rufous-legged Owl *Strix rufipes* but is only irregularly dark-barred below, and belly is almost white; much browner below than Chaco Owl *Strix chacoensis* with a darker face and distinct dark ruff. Minas Gerais, Brazil, September (*Guilherme Gallo-Ortiz*).

153. RUFOUS-BANDED OWL
Strix albitarsis

L 30–35cm; W 274mm (1)

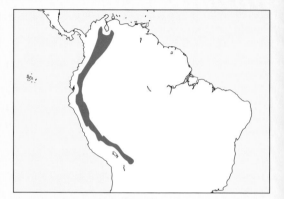

Identification A small to medium-sized, round-headed owl with no ear tufts. No data on size differences between the sexes. Above, the plumage is brown, heavily barred and spotted orange-buff, the flight and tail feathers barred blackish-brown and buffish. There is no whitish scapular row across the shoulder. Dull yellowish-white to pale ochre breast has dark brown mottling and barring, looking somewhat like an indistinct pectoral band on upper breast; whitish below, the feathers having a dark central streak and one terminal cross-bar, giving a chequered or ocellated appearance; large, squarish silvery-white spots cover lower breast and belly. The tawny facial disc has light buffish concentric rings and a dusky ruff, with a blackish area around the orange to deep yellowish-orange eyes, and buffish-white eyebrows and lores. The bill and cere are yellowish-horn. Off-white tarsi are feathered, the dirty to creamy-white toes are bare, with pale horn-coloured

claws. *Juvenile* Downy chick is whitish. Dull yellowish mesoptile has a paler head, with few dark spots and bars, a dark brown facial disc with brown eyes, and some ocellated feathers on the underparts.

Call Utters four or five deep, guttural, subdued *gwo* notes, followed after a short pause of 0.7 seconds by a very loud, higher-pitched hoot, *gwóoh*, the whole phrase lasting about two seconds. In Colombia, the author noted in 1974 that the rhythm is different from that of Black-and-white (154) and Black-banded Owls (155).

Food and hunting Unknown.

Habitat Inhabits cloud forests and mountain forests with dense undergrowth, mainly from 1700m to 3700m. In Venezuela, it is also found in open areas with scattered trees between forested areas.

Status and distribution Occurs patchily from Venezuela and Colombia south in Andean region to C. & E. Bolivia. This owl is little studied, but is believed to be rare and threatened by forest destruction.

Geographical variation Monotypic.

Similar species Partly geographically overlapping Band-bellied Owl (137), Mottled Wood Owl (142) and Black-and-white Owl (154) all have dark brown eyes, the first and last of those also being larger, and the second streaked without cross-bars. Almost allopatric Black-banded Owl (155) lives mainly in lowland forests, and is blackish overall, finely white-barred, with orange-yellow eyes.

◄ Rufous-banded Owl is a relatively short-tailed, compact wood owl. Band-bellied *Pulsatrix melanota*, Mottled *Strix virgata* and Black-and-white Owls *S. nigrolineata* have dark brown eyes, while those of Rufous-banded are deep orange. Napo, Ecuador, November (*Roger Ahlman*).

154. BLACK-AND-WHITE OWL
Strix nigrolineata

L 35–40cm, Wt 404–535g; W 255–293mm

Identification A medium-sized, round-headed owl without ear-tufts. Females are some 80g heavier than males, but the sexes are almost equal in wing length. Has sooty blackish-brown crown and nape, and a prominent collar of dusky and whitish bars around the hindneck. Otherwise it is rather uniformly dark brownish-black above, with two narrow whitish bars on the primaries, and the blackish-brown tail has four or five whitish bars and a white terminal band. There is a black bib on the throat, and the whitish underparts are densely blackish-barred. The facial disc is very dark, with the ruff and eyebrows densely speckled whitish and black. The eyes are dark reddish-brown to blackish-brown, the bill pale orange-yellow and the cere yellowish. Feathered tarsi are barred dusky and whitish, the bare toes dirty yellow to orange-yellow and with yellowish-horn claws. *Juvenile* Downy chick is whitish. Mesoptile is overall dirty whitish, above narrowly barred blackish-brown, below creamy white with dusky bars; the eyes are dark brown, but with an oily-bluish suffusion.

Call Utters a rapid, low, guttural *wobobobobobo* followed after a short pause by a loud *wów* and, lastly, a faint, short *ho*, these three parts repeated at intervals of several seconds. Sometimes leaves out the introductory series and abbreviates the call to a two-note *ho-hóoe*.

Food and hunting Feeds mainly on insects, such as large beetles and orthopterans; it also takes small mammals and bats, birds and tree-frogs. Catches food items often from a perch, but also hawks prey in the air.

Habitat Inhabits gallery forest and rainforest, and also swampy or flooded deciduous woodlands and mangroves, but also found not far from human habitations. From sea level up to 2400m.

Status and distribution Found from C. Mexico south to NW Peru. This species is fairly common in twelve countries. It clearly overlaps in range with Black-banded Owl (155) in Colombia, where some individuals thought to be hybrids have been found, so these species are likely to be very close relatives (perhaps conspecific); even the Ecuadorian 'mystery owl' was initially believed to be intermediate between the two (see next species).

Geographical variation Monotypic.

Similar species Clearly geographically separated Rusty-barred Owl (152) has similarly dark brown eyes, but is rusty all over and brown-barred. Partly sympatric Rufous-banded Owl (153) has yellowish-orange eyes and a chequered or ocellated lower breast and belly. Narrowly sympatric Black-banded Owl (155) is blackish-brown above and distinctly whitish-barred below, and its bill is yellow and the bare toes bright ochre.

▶ Black-and-white Owl has a blackish crown, nape and upperparts; tail has four to five narrow white bars. Bill and feet are orange-yellow. Belize, December (*Christian Artuso*).

155. BLACK-BANDED OWL
Strix huhula

L 31–35cm, Wt 397g (1); W 243–280mm

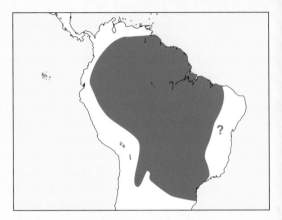

Identification A small to medium-sized, round-headed owl with no ear-tufts. There are no data on sexual size differences. This species is extremely variable in appearance, the white bars ranging from very thin to nearly as wide as the black bars. The head and upperparts are sooty-brown to blackish, densely barred from forehead to lower back with narrow, wavy whitish lines; it has no pale scapular row. Flight and tail feathers are brownish-black, the primaries very dark, the wing with some paler bars, and the tail with four or five narrow whitish bars and a white terminal band. A blackish bib is visible on the chin, and below this the underparts are barred blackish and white, the white bars below clearly broader than those above. The blackish facial disc is densely marked with whitish concentric lines, and the ruff and eyebrows are finely speckled whitish and black. The eyes are brown to dark brown, with pinkish-fleshy rims of the eyelids. Blackish bristles surround the yellow bill. Feathered tarsi are mottled white and black, and the bright ochre toes are bare, with pale horn-coloured claws. *Juvenile* Downy chick is whitish. Light brown mesoptile is heavily whitish-barred, with wing and tail feathers similar to those of the adult, and the dark eyes have an oily or bluish appearance.

Call Utters three or four deep, guttural *wobobo* followed, after a pause of 0.6 seconds, by a louder *whúo*, the phrase being repeated after a while. Generally seems to sing very little. Its calls sometimes sound like those

of Black-and-white Owl (154) and the two will, indeed, respond to one another's calls, which explains the interbreeding in Colombia.

Food and hunting Poorly known, but said to eat mainly insects; also takes small mammals and other small vertebrates.

Habitat Inhabits rainforests, but also plantations in forested areas. Said to prefer lowlands up to 500m in the eastern lowlands and slopes of Andes. Recently, however, it has been found at much higher elevations in the W. Andean subtropics at Baeza (Napo), in Ecuador, more than 1000m above the regular range of this owl.

Status and distribution Colombia, east of Andes, south to N. Argentina and SE Brazil. Suffers from forest destruction, but is often also overlooked. Further studies are needed.

Geographical variation Two subspecies are listed: nominate *huhula* occurs from E. Colombia to N. Argentina; *S. h. albomarginata*, from E. Paraguay and NE Argentina to E. Brazil, is more blackish and has more prominent white barring than nominate. A 'mystery owl' recorded at Baeza, in N. Ecuador, has now been identified as belonging to the present species, although it sounds more like Black-and-white Owl; its voice is somewhat lower-pitched and could afford it at least new subspecies status, but DNA studies have so far been inconclusive owing to the small number of blood samples. (Some have already rushed to name it as '*Strix sanisidroensis*'.)

Similar species Sympatric Mottled Owl (148) is dark brown above and streaked below. Also geographically overlapping, Rusty-barred Owl (152) is barred rusty and brown above, and coarsely barred brown below.

Black-banded Owl from San Isidro, Ecuador, lives 1000m higher than any other population – for this reason, this owl was originally known as the 'San Isidro Mystery Owl'. However, it does not have a fully black face, as seen in Black-and-white Owl populations. October (*Dušan M. Brinkhuizen*).

Black-banded Owl of the nominate race. Smaller than Black-and-white Owl and lacks the black face; its tail has a broad white tip (*Mike Danzenbaker*).

Black-banded Owl from San Isidro has a white tip to the tail just like nominate *huhula*, but it could be a new subspecies due to its range and minor colour and vocal differences. November (*Mike Danzenbaker*).

156. SPOTTED OWL
Strix occidentalis

L 41–48cm, **Wt** 520–760g; **W** 301–328mm

Identification A medium-sized to fairly large, round-headed owl with no ear-tufts. Females are on average at least 80g heavier than males. Has slightly rufous-tinged dark brown crown with whitish flecks and small spots, and dark brown upperparts with numerous transverse and arrow-shaped white spots. Relatively large white areas on scapulars, and dark-barred outer webs. Flight feathers are barred light and dark, with fine vermiculations on light bars, and dark brown tail feathers have a number of narrow whitish bars and pale terminal band. Below, it has a whitish throat, upper breast barred dark brown and whitish, and bold pattern of dark brown and white markings on lower breast and belly. The pale buffish-brown facial disc has some darker concentric lines and a not very prominent dark ruff. The eyes are blackish-brown, the bill pale greyish-brown and the cere yellowish-horn. Densely feathered tarsi to toes, with distal parts of light greyish-brown toes bristled, soles yellowish, and claws dusky horn-coloured. *Juvenile* Downy chick is whitish. Dull yellowish-white mesoptile is indistinctly barred above and below.

Call Gives four strong hoots, *whoop hu-hu-hooo*, uttered in typical rhythm. Believed to be able to make fine adjustments to its own call to imitate that of the nearest neighbour.

Food and hunting Feeds on all suitable prey that it can catch: small mammals, birds, lizards, frogs, insects. Most prey is taken from the ground, but bats and flying insects are often captured on the wing.

Habitat Inhabits well-shaded, mature coniferous and mixed forests, from sea level to 2700m. Prefers the vicinity of water.

Status and distribution Found from British Columbia, in SW Canada, south in USA to California, and from Arizona to C. Mexico. Suffers badly from intense logging of old-growth forests; this not only because of the habitat loss, but also because logging seems to favour the westward expansion of its close relative, the more adaptable and powerful Barred Owl (158). Increasing predation and hybridisation with the latter constitute a real risk and more and more cases have been recorded in which Spotted Owls have interbred with the more common Barred Owl.

Geographical variation Three subspecies have been separated: nominate *occidentalis* is found from Nevada to C. & S. California; *S. o. caurina*, from British Columbia to N. California, is darker brown; *S. o. lucida*,

from Arizona to C. Mexico, is paler and profusely white-spotted. The last-mentioned subspecies has been a candidate for recognition as a full species, 'Mountain Spotted Owl' or 'Mexican Spotted Owl', but more detailed studies are required before confirming its new taxonomic status.

Similar species Geographically separated Fulvous Owl (157) is pale-faced, and fulvous-streaked below. Partly sympatric Barred Owl (158) is larger, paler, with upper breast distinctly barred, lower breast and belly boldly streaked but not barred. Spotted × Barred Owl hybrids ('Sparred Owls') have markings on the back of the nape and head similar to those of Barred Owl, but the breast looks more like that of Spotted Owl; rectangular bars on the head and facial coloration are intermediate between the two; bars on the tail resemble those of Spotted Owl, but are farther apart.

▶ Race *lucida* from Arizona has a paler appearance than the nominate, with much yellowish-buff suffusion; also, its markings are larger and more distinct. Arizona, USA, May (*Eric VanderWerf*).

▲ ◄ Nominate Spotted Owl. This species is rather smaller than Barred Owl *Strix varia*. It has a large head with white spotting, and a dark appearance. California, USA, February (*Mike Danzenbaker*).

▲ Spotted Owl female at a nest with young. California, USA, May (*Paul Bannick*).

◄ Race *caurina* is darker than the nominate and only has a dark morph. Washington State, USA, July (*Paul Bannick*).

157. FULVOUS OWL
Strix fulvescens

L 41–44cm, **Wt** *c.* 600g; **W** 300–333mm

Identification A medium-sized, round-headed owl without ear-tufts. Females are some 100g heavier than males. Has dark rufous crown, nape and upperparts with whitish and dull yellowish flecks and short ochre scallops. Flight feathers are barred dark and light, and tail has three to five broader pale and dark bars. Plumage is fulvous-ochre below, with brown barring on neck, sides of head and upper breast; lower breast and belly are broadly streaked reddish-brown. The light ochre facial disc, darker around the blackish-brown eyes, has a narrow dark brown ruff, with whitish eyebrows. The bill is corn-yellow and the cere yellowish. Tarsi are feathered reddish-yellow down to near the base of the toes, the bare parts of toes yellowish, with dusky horn-coloured claws with darker tips. *Juvenile* Downy chick is whitish. Mesoptile is cinnamon-brown, barred with pale yellowish-orange and white.

Call Utters loud, barking hoots or rhythmic *who-wuhú-woot-woot* song, the number of single notes varying, as do inter-phrase intervals.

Food and hunting Feeds on small mammals, birds, reptiles, amphibians and insects. Normally hunts from a perch.

Habitat Occupies mountain pine forests and humid evergreen pine–oak forests at 1200–3100m.

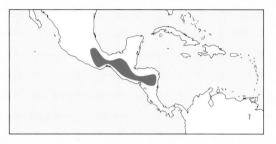

Status and distribution Occurs from Chiapas, in S. Mexico, to Honduras. This species is little studied, but forest-logging may represent a threat.

Geographical variation Monotypic. Recently separated from Barred Owl (158), but the voice resembles more that of Spotted Owl (156); possibly forms a superspecies with those two, but comparative studies, molecular and biological, are required to verify this speculation.

Similar species Sympatric Black-and-white Owl (154) has a dark facial disc and blackish head. Geographically separated Spotted Owl (156) is not streaked, but boldly spotted below. Also allopatric, Barred Owl (158) is paler, greyer and larger, with densely feathered toes; Mexican race *sartorii* of Barred lives only 100km farther north than Fulvous Owl (which is 20 per cent smaller than it).

◄◄ Fulvous Owl has dark rufous upperparts with whitish and dull yellowish flecks. Three to five broader pale and dark bars occur on the tail. Pale facial disc is darker around the blackish-brown eyes. Guatemala, March (*Knut Eisermann*).

◄ Fulvous Owl resembles the more northerly and clearly larger Barred Owl *Strix varia* but it is pale-(almost white-) faced, and fulvous-streaked below. Guatemala, March (*Knut Eisermann*).

158. BARRED OWL
Strix varia

L 48–55cm, **Wt** 468–1051g; **W** 312–380mm, **WS** 107–111cm

Identification A fairly large, round-headed owl with no ear-tufts. Females are up to 200g heavier than males. The plumage is brown to greyish-brown above, scalloped with whitish bars on crown, mantle and back, and with dense light and dark barring on the side of the head and on the hindneck; whitish spots on wing-coverts. Flight feathers are barred whitish-buff and brown, and the brown or greyish-brown tail has four or five whitish bars. Below, the plumage is pale greyish-brown to dirty whitish, densely barred light and dark on the foreneck and upper breast, and with lower breast and belly boldly streaked rufous-brown to dark brown. Pale greyish-brown facial disc has darker concentric lines, but no prominent ruff and no dark area around blackish-brown eyes. The bill is yellowish and the cere pale greyish. Tarsi feathered to near tips of toes, bare parts of which are yellowish-grey, with dark horn-coloured claws with blackish tips. *Juvenile* Downy chick is whitish. Fluffy brownish-white mesoptile has indistinct darker barring above and below. Juvenile has pinkish skin and a light bluish-green cere.

Call Emits a highly distinctive and rhythmic *whohú-buhóoh whohú-buhóoh* (often transliterated as 'you cook today, I cook tomorrow'), repeated at intervals of several seconds.

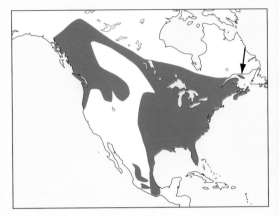

Food and hunting Feeds mainly on small mammals, but also takes birds, frogs, lizards and fish. It also catches insects and other arthropods, snails and slugs. It can catch fish by wading in shallow water.

Habitat Inhabits mixed and old coniferous forests, but also swamps and open wooded areas, including logged old-growth forest; occurs locally in large parks with old trees. From lowlands up to 2500m.

Status and distribution Wide distribution in North America south to S. Mexico. Fragmentation of old-growth forest has facilitated the spread from east to west, as the Barred Owl seems to adapt easily to logged areas, especially if supported by the provision of large nest-boxes. It is unfortunate that this larger and more aggressive owl is able to displace the Spotted Owl. Increased opportunities for cross-breeding may threaten locally to swamp the pure Spotted Owl gene pool.

Geographical variation Four subspecies are separated: nominate *varia* is found from SE Alaska and SW Canada to N. California, N. Texas and North Carolina; *S. v. georgica*, from SE USA south of North Carolina to Georgia and Florida, is paler and a little smaller than nominate; *S. v. helveola*, from Texas and adjacent lowlands of Mexico, has a pale cinnamon ground coloration; *S. v. sartorii*, from C. & S. Mexico's mountain regions, is the darkest subspecies of all.

Similar species Partly sympatric Spotted Owl (156) is smaller and boldly spotted, not streaked, below and is darker above. Narrowly allopatric Fulvous Owl (157) is also smaller and tan-coloured, with a pale facial disc lacking darker concentric lines and having a darker area around the eyes.

◀ Barred Owl nominate *varia* has greyish-brown plumage with distinct white spots above and below, and feathered toes. The visible bare parts are yellow. Ontario, Canada, January (*Chris van Rijswijk*).

◀▶ Race *georgica* is smaller and paler than nominate *varia*. Florida, December (*Deborah Allen*).

◀ This Barred Owl pair shows little sexual size dimorphism but the face of the male (left) looks a little darker than that of the female. Florida, December (*Jim and Deva Burns*).

▶ A Barred Owl swoops to capture a vole, which is just in front of it. Canada, January (*Chris van Rijswijk*).

159. SICHUAN WOOD OWL
Strix davidi

L 58–59cm; W 371–372mm

Other name Pere David's Owl
Identification A large owl with no ear-tufts. Females are always much larger than males, although no weights are available. The plumage of the crown and upperparts is very brown with pale or whitish streaks and spots, these especially prominent on the nape and mantle. Has finely dark-barred whitish outer webs of scapulars. The dark brown primaries and secondaries are barred whitish, and the rather long tail is barred light and dark and white-tipped. The underparts are greyish-white, boldly dark-streaked, and with a suggestion of cross-bars; the upperbreast is washed light brownish. The pale buff facial disc has some concentric dark lines and a ruff of light and dark spots, with much white on the forehead. The relatively small eyes are dark brown with pink eyelid rims, and the bill yellow. Totally feathered legs are light greyish-brown, with pale horn-coloured claws and yellow soles. *Juvenile* Downy chick is white. Mesoptile is dirty whitish, barred greyish-brown on head, nape and mantle and below.

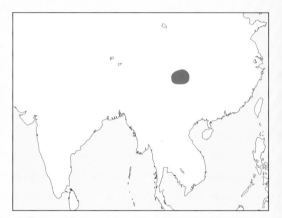

Call Utters a deep *whoo bububub*, similar to that of Ural Owl (160), but call is described also as a long, quivering hoot rising in pitch.
Food and hunting Not well documented, but seems to feed mainly on mammals, such as pikas *Ochotona* spp.
Habitat Inhabits mixed conifer–deciduous or old coniferous forests, usually from 2900m to 3300m, but sometimes as high as 5000m.
Status and distribution Endemic to W. Sichuan in C. China where it is probably endangered owing to extensive deforestation in the region. This species is listed as Vulnerable by BirdLife International.
Geographical variation Monotypic. This owl was formerly regarded as an isolated subspecies of the Ural Owl, and has now been separated because the two are geographically isolated from each other (they are not likely to encounter each other without crossing the Gobi Desert). Studies of the biology and DNA of the Sichuan Wood Owl are urgently needed in order to confirm its taxonomic status and conservation needs.
Similar species Smaller Himalayan Wood Owl (145) normally lives at lower altitudes (1000–2650m), and has very coarse and rather broad, blackish shaft-streaks and heavy cross-bars below. Partly sympatric Mountain Wood Owl (147) lives at the same altitudes and is of the same size as the Sichuan species, but is densely barred, not streaked, below. Geographically separated Ural Owl (160) is very similar, but generally much paler, with the central tail feathers more broadly barred.

◄ Sichuan Wood Owl has scapulars with small white bars on the outer webs. Wings are dark grey-brown, with pale spots and bands. Sichuan, June (*James Eaton*).

▲ Sichuan Wood Owl is generally considered close to Ural Owl *Strix uralensis*, but it has similar concentric rings on the facial disc to Great Grey *S. nebulosa* and Barred Owls *S. varia*. Eyes are dark brown (again like Barred Owl). Sichuan, June (*James Eaton*).

► Sichuan Wood Owl taking off from a tree, showing its large wings. Sichuan, June (*James Eaton*).

160. URAL OWL
Strix uralensis

L 50–62cm, Wt 451–1307g; W 267–400mm, WS 115–135cm

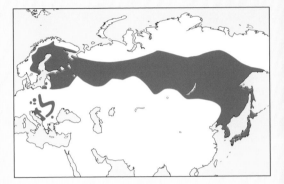

Identification A fairly large to large, round-headed owl with no ear-tufts. Females are up to 300g heavier than males. Light and dark morphs exist. Light morph has pale greyish-brown crown and upperparts with whitish and dusky mottling, spotting and streaking. Large white areas on the scapulars form a row across the shoulder. The flight feathers are conspicuously barred light and dark, and the wedge-shaped, long dark brown tail has five to seven broad greyish-white bars. The whitish throat and very pale greyish-brown or dirty-white underparts are heavily brown-streaked. The round and prominent facial disc is uniformly brown-tinged whitish to pale ochre-grey, with a ruff consisting of rows of small dark and light pearl-like spots. The fairly small eyes are very dark brown, with pink to reddish rims of the eyelids. The bill is yellow-horn. Thickly feathered legs are pale greyish-brown, with dark-tipped yellowish-brown claws.
Juvenile Downy chick is less white than that of Tawny Owl (138), but bill and claws of the Ural Owl are much larger. At the age of four weeks, Ural Owl chick has a much more developed facial disc. *In flight* Not unlike a Common Buzzard *Buteo buteo*, but with much deeper wingstrokes, and appearing very greyish to brownish-white, streaked dark brown, most clearly on underparts.
Call Emits a deep hooting *wóhu wohu-huwóhu*, with a pause of about 3–4 seconds after the first two syllables. This song is repeated at intervals of 10–15 seconds.
Food and hunting Takes a large variety of mammals and birds, but voles alone make up 60–90 per cent of the diet. It occasionally eats frogs, lizards and large insects.

Habitat Occupies mature but not too dense coniferous and mixed deciduous forests; often near bogs and some open areas. This species is well adapted to sustainable logging of the forest, especially when aided by the provision of large nestboxes near forest clearings. In Finland, it has bred in nestboxes left on the ground after an entire forest had been felled during the winter.

Status and distribution Distributed from Norway south to the Balkans, and from the Baltic countries eastwards through Russia and Siberia to Sakhalin and Japan. In C. Europe the average estimated population is 2,700 pairs and in Finland 3,000 pairs. Numbers have increased since the supply of large nestboxes; in Finland, the annual increase in numbers of Ural Owl nests between 1982 and 2008 was 1.6 per cent. Hybridisation between Ural and Tawny Owls in captivity has demonstrated that there is no genetic barrier preventing fertile cross-breeding; the hybrids

◄ Ural Owl of the Japanese race *hondoensis* in flight. Note very long tail and large rounded wings. May (*Kazuyasu Kisaichi*).

showed maternal and paternal characters, as well as intermediate ones, and their vocabulary was more varied than that of either parental species, adding new 'inventions' to the original repertoire of the parents. Despite intensive studies of these species, however, there are as yet no records of wild hybrids.

Geographical variation Eight subspecies have been named: nominate *uralensis* occurs from the S. Urals east to Siberia and E. Russia; *S. u. liturata*, from Swedish Lapland south to eastern Alps and east to Volga River in Russia, is darker than the nominate; *S. u. macroura*, from NW Carpathians to W. Balkans, is darker still and larger; *S. u. yenisseensis*, from the C. Siberian plateau, is darker above and more heavily streaked below; *S. u. nikolskii*, from Transbaikalia and Sakhalin south to NE China and Korea, is brownish-tinged; *S. u. fuscescens*,

from W. & S. Honsu to Kyushu, in Japan, is much smaller and rufous-brown; *S. u. hondoensis*, from N. & C. Honsu, in Japan, is more rusty-brown; smaller and very pale *S. u. japonica* is confined to Hokkaido, in N. Japan. At least *liturata* intergrades with nominate *uralensis*.

Similar species Partly sympatric Tawny Owl (138) is much smaller, with a large head and short tail, its underparts not heavily streaked but with dark shaft-streaks and cross-bars. Allopatric Sichuan Wood Owl (159) has some darker concentric lines on its facial disc and is generally darker, and its central tail feathers are not barred, but densely vermiculated and marked with fine scribbles. Geographically overlapping Great Grey Owl (161) is much larger, with small, yellow eyes on a huge rounded head, and its facial disc has very distinct concentric lines.

▼ Race *macroura* from the Carpathians to the western Balkans is larger and darker than nominate Ural Owl, but it also has an unusual 'melanistic' morph, shown here, that is entirely deep brown, including the face. Krakow, Poland, May (*Chris van Rijswijk*).

▼ Race *macroura* 'melanistic' morph from behind. It is not known for sure whether these very dark owls are real morphs or just represent abnormal melanistic colour mutations. This owl has clearly banded flight and tail feathers, like a normal morph. Krakow, Poland, August (*Chris van Rijswijk*).

▲ Race *japonica* of Ural Owl is similar to *nikolskii* but smaller; pale areas are whiter than in *liturata*. Hokkaido, Japan (*Jan Vermeer*).

▲◄ Race *liturata* has a greyish-white face without any markings. Flight feathers and tail have broad, transverse dark-brown bars on a buffish or buffish-white ground colour. Finland, April (*Harri Taavetti*).

◄ A young Ural Owl peeking from its nestbox in Finland – the bird is almost ready to jump out from the nest. May (*Hugh Harrop*).

161. GREAT GREY OWL
Strix nebulosa

L 57–70cm, **Wt** 568–1900g; **W** 387–483mm, **WS** 130–160cm

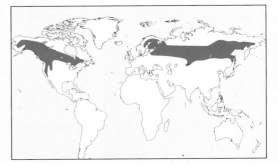

Identification A large to very large, round-headed owl with no ear tufts. The massive head may be as much as 510mm in circumference. Females are on average more than 300g heavier than males. The dark grey crown and upperparts, with a brownish tint, have indistinct dusky streaks and are densely mottled and vermiculated darker. The flight feathers are barred dark and light, and the relatively long, wedge-shaped tail is barred and mottled grey and dusky. Pale greyish underparts are dark-vermiculated, with mottling and diffuse longitudinal dark streaks; the belly is barred dusky. The circular, whitish facial disc has six or more brown concentric rings, and the thin dark ruff is bordered below by very prominent white patches in the middle of the foreneck. The eyes (which appear relatively small on the large head) are yellow, edged on the inside with a touch of black and two large outward-facing comma-shaped white marks. The ivory-coloured to yellow bill is surrounded by a black beard, and has a white moustache above. Densely feathered tarsi and toes are grey with dusky mottling, the relatively weak and only slightly curved claws being blackish. *Juvenile* Downy chick is whitish, with pink skin. Light greyish mesoptile has dusky bars and whitish mottling above and below, and a dusky face with pale yellowish-grey eyes and greyish-yellow cere and bill. *In flight* Appears heavy-headed and long-tailed, a shape unlike that of any other bird. It flies easily with slow, airy wingbeats, and is very agile in passing between trees.

Call Utters a series of regularly spaced, deep *ho* notes of equal duration and interval, up to twelve in succession. The whole call lasts for *c.* 6–8 seconds with a mean interval of 33 seconds between calls.

Food and hunting This is a vole specialist, but it occasionally takes other small mammals, such as shrews and lemmings; birds, frogs and invertebrates feature only rarely in the diet. In California it eats mainly pocket gophers, which are much larger than voles. Prefers semi-open areas for hunting, and is able to locate prey by hearing only and to catch with its 'clenched feet' prey hiding beneath snow cover. It can break snow crust hard enough to carry an 80kg person, and penetrate as deep as 45cm.

► Nominate Great Grey Owl from North America is darker than Eurasian *lapponica* and has less distinctly defined grey streaking above, less prominent white 'eyebrows' and a vertical dark throat-band. Alberta, Canada, June (*Anne Elliott*).

Habitat Inhabits spruce, pine and mixed forests near clearings and swampy areas, from lowlands up to 3200m above sea level. During winter it often comes into the vicinity of farmhouses and to areas with open fields.

Status and distribution One of the few owls living right across the globe in the Holarctic forest belt, in Europe the most northerly nests are in N. Norway and the most southerly in Poland, and in North America it ranges from Alaska south to N. California and NE Minnesota. This species appears to be not as rare as was previously thought, but it is difficult to study owing to its nomadic movements. The average population in Europe (including Russia east to Ural Mountains) is estimated to be 4,400 pairs. More than 10,000 Great Grey Owls were reported in Minnesota during the winter of 2004–2005, so the North American population far exceeds that of Europe.

Geographical variation Only three subspecies are listed: nominate *nebulosa*, from C. Alaska eastwards through much of S. Canada and south in USA to Idaho, W. Montana, Wyoming and NE Minnesota, has a grey face, more suffused streaks on the underparts, and is generally browner with white markings; *S. n. yosemitensis*, from Sierra Nevada, in California, recently separated from *nebulosa* on the basis of DNA studies; *S. n. lapponica*, from N. Eurasia to Kyrgyzstan, NE China and Sakhalin, has a less grey face and more prominent streaking, without barring, below.

Similar species Smaller Barred Owl (158), sympatric in Canada and N. USA, has much less well-marked concentric lines on the facial disc, blackish-brown eyes and a whitish-barred greyish-brown back. Also smaller, the Ural Owl (160), which overlaps in range in Eurasia, has dark brown eyes, is boldly streaked below, and lacks dark concentric rings on the pale face.

► The Eurasian race *lapponica* of Great Grey Owl is clearly paler than the nominate. Finland, February (*Harri Taavetti*).

◄ This Great Grey Owl is in the 'tall-thin' position. Finland, April (*Harri Taavetti*).

▼ The fledgling Great Grey Owl leaves the nest well before it is able to fly. Note the very pale, almost whitish eyes of this owl. Vestmanland, Sweden, June (*Stefan Hage*).

▲ In flight, Great Grey Owl has a shape like a flying 'broken tree stump'. Skanör, Finland, March (*Harri Taavetti*).

▶ Male Great Grey Owl feeding the female on the nest. This occurs during incubation and when the young are small. Finland, June (*Jari Peltomäki*).

162. MANED OWL
Jubula lettii

L 34–40cm, Wt 183g (1♂); W 241–285mm

Identification A medium-sized owl with rather long, bushy ear-tufts, these, together with elongated feathers on sides of head and nape, giving a maned appearance. No data are available on sexual size differences, if any. The reddish-brown crown feathers have white borders, giving a scaly appearance, and the rufous or chestnut-brown upperparts have dark and light markings, the mantle and back also with some barring. Pale buffish scapulars have whitish but dark-edged outer webs, forming a light band across the shoulder, and the deep reddish-brown wing-coverts have dark vermiculations and shaft-streaks. The flight feathers have about four dark bars, and the tail is barred rufous and dark. The throat is white, and the reddish-brown underparts become more buff and paler towards the belly; there are many fine whitish vermiculations on the upper breast, and very prominent dark shaft-streaks and vertically elongated white spots on upper and lower breast feathers. The light rufous-buff facial disc is finely vermiculated dusky brown, and has a prominent blackish-brown ruff, and white eyebrows and forehead. The eyes are deep yellow to orange-yellow, the bill ivory-coloured to yellow, and the cere yellowish-green. Dull yellowish feathering covers the tarsi; unfeathered

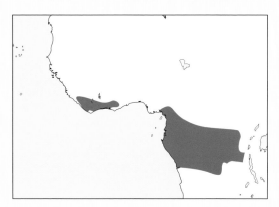

yellow toes have some grey patches on upperside, and the brownish-horn claws have dark tips. *Juvenile* Downy chick unknown. Rufous mesoptile has faint barring below and a less developed mane, but the wings and tail are similar to those of the adult.

Call Reported to be a mellow hooting *who*, followed after ten seconds by another *who*, slightly higher-pitched than the first, the two notes repeated in the same succession for some time. This hoot, however, could belong to the Rufous Fishing Owl (132), also found in W. Africa, and the real voice of Maned Owl remains undescribed.

Food and hunting Feeds mainly on large insects, but also takes small vertebrates. Food and hunting behaviour not yet studied.

Habitat Occupies lowland forest and gallery forest, often near rivers or lakes. This owl is not known to occur anywhere outside the forest.

Status and distribution Patchily distributed from Liberia east to Ghana and mainly from S. Cameroon to River Congo and NC, SC and EC DR Congo. This species is insufficiently studied to allow assessments of whether it is common or endangered, but it certainly suffers from increasing destruction of the tropical forest in W. and C. Africa.

Geographical variation Monotypic.

Similar species No similar species with a maned head is found in Africa.

◄ This poorly known African owl is similar to Crested Owl *Lophostrix cristata*. However, this is not likely to be anything more than convergence; both birds have whitish 'eyebrows' extending to sideways-slanting ear-tufts. Gabon, June (*Karen Hargreave*).

163. CRESTED OWL
Lophostrix cristata

L 38–43cm, **Wt** 425–620g; **W** 280–325mm

Identification A medium-sized owl with large, white ear-tufts. Females can be much as 100g heavier than males. Three colour morphs have been distinguished, but one is also regarded as a different subspecies. Brown morph is plain dark chocolate-brown above, with whitish dots on the wing-coverts and outer webs of primaries. The wings are barred light and dark, and the rather uniformly dark brown tail has very fine darker mottling. It has a dull yellow throat and dark chocolate-brown neck and upper breast, below which it is pale brownish with a number of faint brown vermiculations. The facial disc is uniformly deep dark brown, like the crown, and the white of the ear-tufts extends inwards to meet the white eyebrows and forehead. The eyes are dark orange-brown and the bill dark yellowish-horn. Tarsi are feathered to the base of the light greyish-brown toes, which have dark horn-coloured claws. Rufous morph is generally light rufous instead of dark chocolate, and has a darker brown collar on the upper breast. Grey morph is in general pale greyish-brown,

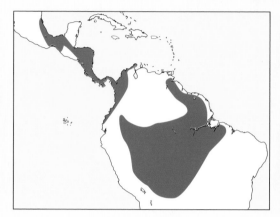

with a blackish facial disc, and yellow or orange eyes. *Juvenile* Mesoptile has a whitish head and body, a dark facial disc, short ear-tufts, and wings and tail as those of the adult. *In flight* Shows rounded wings with 7th primary the longest.

Call Has a frog-like croak, *k-k-kkk-krrrrrrrao*, uttered at intervals of several seconds. Subspecies *stricklandi* has a shorter call, *gurrr* or *kwarrr*, given at irregular intervals.

Food and hunting Not well studied, but feeds chiefly on large insects, and also some vertebrates. Its hunting behaviour is unknown.

Habitat Found mainly in primary forests, such as lowland rainforest, but also in second-growth woodland, from sea level up to 1950m.

Status and distribution Ranges from S. Mexico to Bolivia and Amazonian Brazil, although it is absent from large parts of Colombia and Venezuela. It is said to be rather common, but is so little studied that its true status is not known.

Geographical variation Three subspecies are listed: nominate *cristata* occurs in South America east of the Andes from E. Venezuela and Suriname south to N. Bolivia; *L. c. wedeli*, from E. Panama to NW Venezuela, has yellow eyes; *L. c. stricklandi*, from S. Mexico to W. Colombia, is in general greyish-brown, and has yellow eyes and a slightly different voice. This last subspecies may even be specifically distinct, but further studies of its DNA and vocalisations are required.

Similar species This is the only Central and South American owl having such long, white ear-tufts and eyebrows as well as unstreaked underparts.

◣ Nominate Crested Owl looks much paler than *stricklandi*. Ecuador, August (*János Oláh*).

◣ ▶ This Crested Owl from Mexico looks more reddish than the Costa Rica bird opposite. Chiapas, Mexico. January (*Christian Artuso*).

◀ Crested Owl's northernmost race *stricklandi* has a tawny to chestnut dark-rimmed face. Eyes are yellow to orange-brown. Densely barred below, with bars wider on belly and undertail-coverts. Costa Rica, March (*Robin Chittenden*).

Race *stricklandi* has a dark back with much paler scapulars, unspotted greater coverts and secondaries with fine vermiculations. Outermost coverts sometimes have pale patches; the primaries have a few bars. Chiapas, Mexico, January (*Christian Artuso*).

164. NORTHERN HAWK OWL
Surnia ulula

L 36–41cm, **Wt** 215–450g; **W** 218–258mm, **WS** 72–81cm

Identification A medium-sized owl without ear-tufts. Females are on average some 50–60g heavier than males. Has a densely whitish-spotted crown, and dark grey to dark greyish-brown upperparts, the nape with an indistinct 'occipital face' and the dusky-grey mantle and back with some whitish dots. The mainly white scapulars form a fairly broad white band across the shoulder. The dark grey-brown flight feathers are marked with rows of white spots, and the long and graduated tail is dark greyish-brown with a number of narrow whitish bars. The blackish chin is bordered with whitish, and the whitish underparts are fully barred greyish-brown. The facial disc is whitish with a broad blackish ruff at the sides. The eyes are yellow, with blackish eyelid rims and white eyebrows. The bill is yellowish-green and the cere pale greyish-brown. Tarsi and toes are fully feathered, the toes with dirty yellow soles and black-tipped dark brown claws. *Juvenile* Downy chick is whitish. Mesoptile has the back similar to that of adults, but light grey crown and darker-mottled underparts, and also has blackish facial disc with whitish lower part. Juvenile has more golden-yellow eyes. Facial disc of the fledgling becomes whitish, and dark barring begins to develop on underparts. *In flight* This species has a swift and direct flight, in which the pointed wings and long

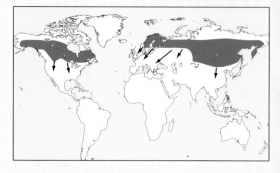

tail give a hawk-like effect. Sometimes known to hover like a Kestrel *Falco tinnunculus*.

Call Gives a trilling but melodious whistle, *ululululu-lulululululu*, lasting up to ten seconds and containing 12–14 notes per second; the whole phrase is repeated at intervals of several seconds.

Food and hunting Voles form 93–98 per cent of the diet. In addition, this species takes some shrews, birds and occasionally frogs, fish and larger insects.

Habitat Found mainly in fairly open coniferous and mixed forests, often near edges of marshes and areas cleared by felling. Sometimes it is seen even near human habitations.

▶ Northern Hawk Owl subspecies *caparoch* is darker than the nominate. Michigan, USA (*Steve Gettle*).

◀ A day-active nominate Northern Hawk Owl, from Finnish Lapland in June – where there is no night in summer (*Daniele Occhiato*).

Status and distribution Has an holarctic distribution in the boreal zones of Eurasia and North America. The population is declining and, except in years with extraordinary abundance of voles, is currently retreating northwards. This obviously has something to do with climate change. Recent estimate of European population (including Russia east to Ural Mountains) was 23,600 pairs. In 2011, an estimated 5,000 pairs of Northern Hawk Owls were breeding in Finnish Lapland out of a total Finnish population varying between 2,000 and 6,000. The species winters irregularly south to C. & SE Europe (including, rarely, the Shetlands and Orkneys), the Kuril Islands, and N. USA south to Oregon, Nebraska and New Jersey.

Geographical variation Three subspecies have been listed: nominate *ulula* occurs from Scandinavia through Siberia to Kamchatka and Sakhalin, and south to Tarbagatay; *S. u. tianschanica*, from the Tien Shan of C. Asia, N. China and perhaps N. Mongolia, has dark parts of the plumage more blackish and white purer than in nominate; *S. u. caparoch*, from Alaska through S. Canada to extreme N. United States, is darker than the other two.

Similar species Globally sympatric Tengmalm's Owl (209) is short-tailed and smaller, with underparts not boldly barred but blotched and mottled. Northern Saw-whet Owl (210), overlapping in range in North America, is much smaller, with a short tail, its facial disc without a broad black ruff, and is diffusely brown-streaked below.

▼ Hunting over snow proves successful, with Bank Vole on the menu. Finland, March (*Harri Taavetti*).

▲ Northern Hawk Owls at the nest stump. The female (left) is protecting the young when the male arrives at the nest. Canada (*Jon Groves*).

A young Northern Hawk Owl sitting on a burnt-stump nest. Montana, USA, June (*Paul Bannick*).

▶ Northern Hawk Owl in flight looks more like a hawk than an owl. Finland, May (*Harri Taavetti*).

165. EURASIAN PYGMY OWL
Glaucidium passerinum

L 15–19cm, Wt 47–100g; W 92–112mm, WS 32–39cm

Identification A tiny to very small round-headed owl without ear-tufts. Females are on average 10–15g heavier than males, but just before breeding female can be 40g heavier than male. The crown and upperparts are dark brown or greyish-brown, the crown finely marked with creamy-whitish spots, the mantle back with small whitish dots near lower edge of each feather; two large blackish spots with whitish surrounds on nape, suggesting false eyes (known as 'occipital face'). Flight feathers are barred pale and dusky, and brown tail feathers have about five narrow whitish bars. Has a whitish throat and off-white underparts, with brown mottling on sides of breast and flanks, as well as brown streaks from throat to belly. Laterally indistinctly defined facial disc is light greyish-brown with several darker concentric lines formed by minute dark spots. Relatively small eyes are yellow, with blackish-edged eyelids. The bill is yellowish-horn and the cere grey. Whitish-brown tarsi are feathered to the base of the toes, the unfeathered parts of which are yellowish, with blackish-tipped dark horn-coloured claws. *Juvenile* Downy chick white. Mesoptile resembles the adult, and is not so fluffy as many young owls; has a rather dark facial disc with very prominent eyebrows and white chin spot, and plain, unspotted dark brown crown. *In flight* The wings appear very rounded, and the flight is noisy and undulating.

Call Utters a rather long sequence of well-spaced, monotonous, flute-like *deu* notes at about two-second intervals; bouts of calling may last for several minutes. It resembles the voice of a Bullfinch *Pyrrhula pyrrhula*.

Food and hunting Voles, mice and shrews form normally more than half of the diet, and small birds some 40 per cent. Occasionally takes lizards, fish and large insects. It

often catches small birds in dashing flight from a bush, and locates ground prey from a perch.

Habitat Prefers coniferous forests, but occurs also in mixed forests, especially at higher elevations. Occurs from lowlands to mountains, up to 2150m in Alps.

Status and distribution Distributed from C. & N. Europe eastwards through Siberia to Sakhalin, Manchuria and N. China. In Europe it is found from N. Finland down to the Spanish Pyrenees and Greece. Locally common in Finland, and even in Germany expanding its range towards west. In C. Europe the total population is estimated to be 10,200 pairs and in Finland 10,000 pairs.

Geographical variation Only two subspecies are listed: nominate *passerinum* is found from C. & N. Europe east to Yenisei in Siberia; *G. p. orientale*, from E. Siberia, Manchuria, Sakhalin and N. China, has paler upperparts, showing purer white and more sharply defined spots, and the breast and flanks more strongly brown-marked.

Similar species Geographically separated Collared Pygmy Owl (168) is similar in size, but with large whitish areas on the throat and foreneck, as well as on the belly; it is barred orange-buff above, and more barred than streaked below.

▶ A nest hole full of nestlings – eight! Kuortane, Finland, May (*Esko Rajala*).

▶▶ Female Eurasian Pygmy Owl peeking from the nest hole. Brabant, The Netherlands, February (*Lesley van Loo*).

◀ Five recently fledged Eurasian Pygmy Owl young near a nest in Finland. June (*Matti Suopajärvi*).

The small head of the Eurasian Pygmy Owl often looks flat-crowned. It lacks a full facial disc but has clear white 'eyebrows' and lores. Finland, March (*Harri Taavetti*).

Nominate Eurasian Pygmy Owl is mainly dark brown and spotted buffish-white all over. Finland, September (*Han Bouwmeester*).

▼ Eurasian Pygmy Owl lacks bold white scapular spots. Flight feathers are dark brown, with narrow buffish-white bars. Poland, May (*Menno van Duijn*).

166. PEARL-SPOTTED OWL
Glaucidium perlatum

L 17–20cm, Wt 61–147g; W 100–118mm, WS 38–40cm

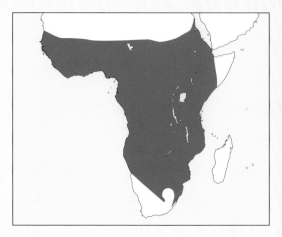

Other name Pearl-spotted Owlet

Identification A very small, round-headed owl without ear-tufts. Females average up to 40g heavier than males, depending on body condition. Has top of head and upperparts deep cinnamon-brown, finely whitish-spotted on forehead and crown; mantle and back sometimes paler, almost sand-coloured, with blackish-bordered white spots on mantle resembling pearls; two large sooty-brown spots, diffusely whitish-bordered, on nape create a very prominent 'occipital face'. Whitish outer webs of scapulars with thin dusky edges make a clear white row across the shoulder. Flight feathers are barred light and dark, and the brown tail feathers have about six rows of black-rimmed whitish spots. It has a whitish throat, and off-white underparts streaked dusky brown. Pale greyish-brown, ill-defined facial disc has some diffuse concentric lines, and fairly prominent, whitish eyebrows. The eyes are yellow, the bill yellowish-horn and the cere brown. Feathered tarsi are off-white with brown spots, and the sparsely bristled brownish-yellow toes have dark-tipped horn-coloured claws; talons are fairly large and powerful. *Juvenile* Completely white downy chick has pinkish skin. Mesoptile resembles adult, but has deep yellow eyes, an unspotted mantle and crown, and is less clearly streaked below. Juvenile has pink tongue and gape, not blackish (as in African Barred Owlet).

Call Gives a clearly fluted whistle, *feu-feu-feu-feu-feu-feu*, rising gradually in volume and in pitch to a climax of long, loud notes. Calls mainly after dark, male often duetting with female.

Food and hunting Prefers arthropods, but also takes reptiles, birds and small mammals, and even snails.

Habitat Lives mainly in open savannas with short grass, scattered trees and thorny scrub, but also in mopane woodlands and riverside forests.

Status and distribution Occurs south of Saharan Africa from S. Mauritania across to Ethiopia and south to N. Cape Province, in South Africa. Absent only from deserts and rainforests. A rather common and resident breeder in 35 African countries.

Geographical variation Nominate *perlatum* occurs from S. Mauritania east to Sudan in the east and R Congo and DR Congo in the west; *G. p. licua*, from Ethiopia and Uganda to Angola, Namibia and South Africa, is greyer and more heavily spotted.

Similar species Partly overlapping Red-chested Pygmy Owl (167) has a strong orange-buff suffusion on the sides of the chest and flanks; it has a pale nuchal band but no occipital face, and the dark upperparts lack white on the scapulars. Sympatric African Barred Owlet (196) is barred on the crown, forehead and upper breast, has no pearl-like spots, and no occipital face. Allopatric Etchécopar's Owlet (195) and only narrowly sympatric Chestnut (198) and Albertine Owlets (197) all have a barred head without an occipital face.

◄ Nominate Pearl-spotted Owl. The Gambia, January (*Nik Borrow*).

▼ Race *licua* is paler than the nominate. Head cinnamon-brown, nape ochre to buff, wing-coverts less rufous. Belly is white with fine dark shaft-streaks. Namibia, November (*Wil Leurs*).

▶ The white face and undersides of race *licua* mean this bird can be confused with any pale *Athene*. Namibia, November (*Adam Scott Kennedy*).

▲ A pair of the nominate race; *perlatum* is brighter and more contrastingly coloured than *licua*, with larger white spots on the flanks and heavy cross-barred stripes on the belly. Sierra Leone, December (*Jon Hornbuckle*).

▶ This southern *licua* has a very dark face. It is likely that white-faced owls are older than dark-faced ones. South Africa, August (*Niall Perrins*).

167. RED-CHESTED PYGMY OWL
Glaucidium tephronotum

L 17–18cm, Wt 80–103g; W 99–127mm

Other name Red-chested Owlet

Identification A very small, round-headed owl without ear-tufts. Nominate race females are on average less than 1g heavier than males, but *elgonense* females are 10g heavier. Has a plain greyish-brown crown, white-flecked head sides, and an indistinct 'occipital face' created by not very prominent blackish 'false eyes' on nape bordered above and below by broad white bars; dusky dark grey-brown mantle, back and rump, and no whitish outer webs on scapulars. The primaries are plain dusky brown, and there is inconspicuous barring on the secondaries, and the darkish brown tail feathers have with three large, rounded white spots on inner webs of central rectrices. The throat is white, and the whitish breast has visible dark brown spots, the rufous-washed breast sides shading into extensive reddish-brown on the flanks; rest of underparts are light rufous-buff. Pale grey facial disc has an indistinct ruff. The eyes are yellow, the bill greenish-yellow, and the cere wax-yellow. Plain rufous-buffish tarsi are feathered, and the yellow toes are sparsely bristled, with dark-tipped yellowish claws. *Juvenile* Undescribed.

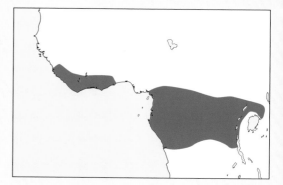

Call Utters a long series of up to 20 hollow whistles, *füfü-fü-fü-fü-fü-fü-fü*, with interval of 0.5–0.7 seconds between notes. These phrases are repeated at intervals of several seconds.

Food and hunting Feeds mainly on insects, but also takes small birds and small mammals.

Habitat Found in rainforests and forest–scrub mosaic, as well as forest edges and clearings, from lowlands up to 2150m.

Status and distribution Occurs from Sierra Leone, in W. Africa, east to Uganda and W. Kenya, and from Cameroon to R Congo and DR Congo and east to Uganda and Rwanda. Its status is not well known, but it is mostly fairly rare. Suffers from forest destruction. Research is required on this species' ecology and conservation needs.

Geographical variation Four subspecies are listed: nominate *tephronotum* occurs in Sierra Leone, Liberia, Ivory Coast and Ghana; *G. t. pycrafti*, from Cameroon to Gabon, has underparts less rufous than nominate; *G. t. medje*, from DR Congo and Rwanda to SW Uganda, is more slate-coloured above, with less chestnut; *G. t. elgonense*, from E. Uganda and W. Kenya, is darker and browner above than *pycrafti* and larger, with clear sexual size dimorphism. Research is needed in order to confirm the taxonomy of this owl.

Similar species Partly sympatric Pearl-spotted Owl (166) has a white-spotted crown, whitish scapular line, and streaking below; its relatively long tail is marked with rows of dusky-edged whitish spots. Geographically overlapping African Barred Owlet (196) has a barred crown and upper chest, a white scapular line, and no occipital face.

▲ Race *medje* Red-chested Pygmy Owl is slaty above, with less chestnut than the nominate. Nyungwe, Rwanda, March (*Ron Hoff*).

◄ Nominate Red-chested Pygmy Owl is more rufous below than *pycrafti* and *elgonense*. Ankasa, Ghana, May (*Nik Borrow*).

► Red-chested Pygmy Owl has a reddish-brown chest and white throat. Head is bluish-grey. Race *medje* is larger than the nominate. Nyungwe, Rwanda, March (*Ron Hoff*).

168. COLLARED PYGMY OWL
Glaucidium brodiei

L 15–17cm, Wt 52–63g; W 80–101mm

Other name Collared Owlet

Identification A tiny round-headed owl without ear-tufts. Females are at least 10g heavier than males. Very variable in coloration irrespective of provenance; occurs also as a rufous or deep reddish-brown morph. Has dull grey-brown or reddish-brown crown, nape, ear-coverts and sides of neck with broken bars and whitish-fulvous or variously russet spots; light collar on hindneck together with a black spot on each side of the nape form an 'occipital face'. Brown or greyish-brown upperparts are barred with whitish-fulvous. White outer webs of scapulars create a bold streak across the shoulder. Primaries are blackish-brown, barred with white and buff, and the dark reddish-brown to greyish-brown tail is fairly densely barred with whitish to reddish-buff. The underparts are white, fulvous-white or rufous-white, sides of breast and flanks marked with broad dark brown bars, the lower flanks with fewer and drop-like markings. The light brown face has ill-defined white eyebrows. The eyes are pale lemon-yellow or more straw-yellow or golden-yellow. The bill is greenish-yellow, darker at the base, and the cere is greenish to bluish. Feathered greenish-yellow to olive-grey legs and toes have paler and more yellow soles, and dark horn-coloured claws. *Juvenile* Has the back fully barred, as on the adult, but with a more streaked head. *In flight* Typically flies with a series of rapid wingbeats alternating with gliding.

Call Gives a three-note whistle, *wüp-wüwü-wüp*, with intervals of 0.5–0.6 seconds between the notes, repeated at intervals of several seconds. When singing, it turns its head in all directions, creating a ventriloquial effect.

Food and hunting Takes mainly small birds, but also insects, lizards and small mammals.

Habitat Inhabits mountain forests, often with clearings, and open woodlands with scrub. Usually from 700m to 2750m, but sometimes up to 3200m.

Status and distribution Found from the Himalayas in N. Pakistan east to China and Taiwan, and south through the Malay Peninsula to Sumatra and Borneo. It is locally not rare, but more research on its biology, behaviour and ecology is needed.

Geographical variation Four subspecies are known: nominate *brodiei* occurs from the Himalayas to SE China, Hainan, the Malay Peninsula and N. Vietnam; *G. b. pardalotum*, from Taiwan, has the head olive-brown with ochre spots and other markings, and a broad buff

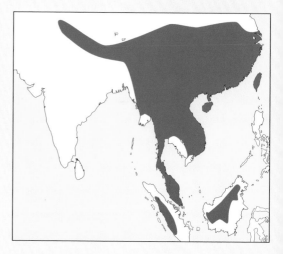

collar; *G. p. sylvaticum*, from Sumatra, and *G. p. borneense*, from Borneo, are very little known.

Similar species Sympatric Asian Barred Owlet (193) is larger, with no occipital face. All geographically overlapping scops owls have ear-tufts.

▼ Collared Pygmy Owl of the race *sylvaticum* from Sumatra has a very white belly and centre of breast. Flanks have brown spots, and the upper breast has a reddish-brown band. Sumatra, Indonesia, July (*Jon Hornbuckle*).

The nominate race from front and back. Eyes are lemon-yellow; head has prominent whitish spots on the crown. Back is clearly barred; the rear view shows the strong, buffish-rufous occipital 'face'. Uttaranchal, India, March (*Subharghya Das*).

▼ Race *pardalotum* is brownish above and below, with drop-shaped shaft-streaks on a white belly. Taiwan (*Pei-Wen Chang*).

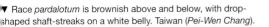

▼ An immature nominate Collared Pygmy Owl; similar to the adult with very well-feathered legs. China, May (*Aurélien Audevard*).

169. NORTHERN PYGMY OWL
Glaucidium californicum

L 17–19cm, Wt 62–73g; W 87–105mm

Identification A very small, round-headed owl with no ear-tufts. Females have longer wings and are some 10g heavier than males. Grey, red and brown morphs and intermediates are known. Brown morph is brown above, the crown with fairly dense, small whitish spots; two large diffuse whitish-rimmed blackish dots on the nape form and 'occipital face'. Partly whitish outer webs of scapulars are rather indistict. Flight feathers are barred light and dark, and the brown tail feathers have about six incomplete whitish bars on both webs, hardly reaching the central shaft. Has a whitish throat and foreneck, and whitish underparts with distinct dark brown streaks, the chest sides and upper flanks being brown with several small white spots. Rather flat facial disc is pale greyish-brown with some concentric lines, a less than prominent ruff and more obvious whitish eyebrows. The eyes are bright yellow. The bill is yellowish-horn. The greenish-yellow cere is distinctly swollen around the nostrils, where it looks more like a cone-shaped buckle. Off-white tarsi are feathered, and the greyish-yellow toes sparsely bristled, with dark-tipped greyish-horn claws. *Juvenile* Downy chick is whitish. Mesoptile has an unspotted crown and less fluffy plumage than in many non-*Glaucidium* owls.

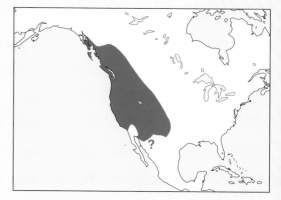

In flight Shows more rounded wings compared with Mountain Pygmy Owl (171).

Call Utters a series of somewhat ventriloquial evenly spaced *toot* notes at intervals of about two seconds, and for periods of up to 20 seconds.

Food and hunting It regularly takes insects, frogs and reptiles, but also birds and small mammals. Prey can sometimes be twice the size of the owl.

Habitat Occupies mixed and coniferous forests with mature trees, from lowlands up to more than 3000m.

Status and distribution Found from southern Alaska, USA, south through W. Canada to N. Mexico. This species is locally not rare but it suffers from the effects of logging.

Geographical variation Three subspecies are listed: nominate *californicum* occurs from SE Alaska south to California; very dark *G. c. swarthi* is confined to Vancouver Island, Canada; *G. c. pinicola*, from Idaho and Montana, USA, south to N. Mexico, is more spotted than the nominate and occurs in grey and red morphs. This owl was earlier treated as a subspecies of Mountain Pygmy Owl; some DNA evidence exists to support its new classification as a distinct species, but more research on its biology, ecology and taxonomy is needed.

Similar species Partly sympatric Mountain Pygmy Owl (171) is smaller and short-tailed, with pointed wingtips, is less streaked below, and has more extensive white on the throat, neck and upper breast.

◄ Northern Pygmy Owl has a brown tail with six incomplete whitish bars on both webs, hardly reaching the central shaft. Vancouver, Canada, January (*Mike Danzenbaker*).

▲ ► Northern Pygmy Owl in the 'tall-thin' position. Note how this owl can erect the head feathers sideways, as if it had ear-tufts. Many if not all pygmy owls can do this. Canada, January (*Mike Danzenbaker*).

▲ Northern Pygmy Owls mating. Washington State, USA, April (*Paul Bannick*).

▼ Female Northern Pygmy Owl at the nest hole. Colorado, USA, June (*Paul Bannick*).

▼ ► Fledgling Northern Pygmy Owl soon after leaving the nest. Colorado, USA, June (*Paul Bannick*).

170. BAJA PYGMY OWL
Glaucidium hoskinsii

L 15–17cm, Wt 50–65g; W 86–89mm

Other name Cape Pygmy Owl

Identification A tiny, round-headed owl without ear-tufts. Females are said to be heavier than males. Females are often more reddish than males, but no grey or red morphs are known. The plumage is sandy grey with a reddish tint above, with finely whitish-spotted crown; two blackish or dark brown spots, rimmed whitish and pale buff, on the nape appear as 'false eyes'; irregularly light-spotted mantle and back. Dull yellow spots on the scapulars are not very prominent. Flight feathers are barred light and dark, and the dark brown tail feathers have six incomplete thin whitish bars, not reaching the central shaft of the feathers. A clear white patch on the throat and foreneck is surrounded by greyish-brown mottling and streaking; below, the plumage is mostly off-white, the upper breast sides and some of the flanks with greyish-brown mottling, the remaining underparts fairly densely dusky-streaked. The facial disc is rather poorly developed, with whitish eyebrows. The eyes are yellow, rimmed by blackish edge of eyelids. The bill

is greenish-yellow. The cere is greenish-grey, swollen around nostrils in a cone-shaped buckle. Feathered tarsi, with yellowish-grey, bristled toes and dark-tipped horn-coloured claws. *Juvenile* Undescribed.

Call Utters a series of relatively high-pitched *kwiu kwiu kwiu* notes, these separated by intervals of about one second.

Food and hunting Preys on insects and other arthropods, as well as reptiles, and also takes small mammals and birds.

Habitat Inhabits pine and pine–oak forests, mainly from 1500m to 2100m, wintering often in deciduous forests at lower elevations (500m).

Status and distribution Endemic in Sierra Victoria and Sierra de la Giganta, in S. Baja California, Mexico. This is a rare and vulnerable species on account of its very limited range and the degree of habitat destruction in that area.

Geographical variation Monotypic.

Similar species The only pygmy owl in S. Baja California. Sympatric Elf Owl (200) is even smaller, and short-tailed, with unstreaked underparts and without 'occipital face'.

◄ Baja Pygmy Owl is smaller than Northern Pygmy Owl *Glaucidium californicum*; the plumage pattern is very similar but the tail is noticeably shorter. Baja Mexico, March (*Jonathan Newman*).

171. MOUNTAIN PYGMY OWL
Glaucidium gnoma

L 15–17cm, **Wt** 48–73g; **W** 82–98mm

Identification A tiny, round-headed owl without ear-tufts. Females are larger and on average 15g heavier than males. The crown is finely speckled with whitish or dull yellow, and two large blackish or dark brownish nuchal spots, edged buffish-white above and pale cinnamon-buff below, form an 'occipital face'. The greyish-brown to dark brown upperparts have whitish and dull yellowish spots on the mantle and back. Outer webs of scapulars are irregularly spotted with whitish or pale cinnamon-buff. Flight feathers are barred light and dark, and the brown to rufous tail has five or six narrow whitish bars. Below, the plumage is off-white to whitish-buff, with a fairly plain area from the throat to the lower breast, and reddish-brown and whitish mottling on the sides of the upper breast and part of the flanks, the remaining underparts having dusky streaks. The light brownish facial disc is speckled paler and darker, with narrow whitish eyebrows. The eyes are yellow. The bill is horn-yellow. The yellowish-grey cere is typically swollen around the nostrils. Tarsi are feathered, and yellowish-grey toes bristled, with brownish-horn claws with darker tips. *Juvenile* Downy chick is whitish. Mesoptile has an unspotted, greyish crown and more greyish cere than adults. *In flight* Shows pointed wings and a shorter tail than Northern Pygmy Owl (169).
Call Utters a prolonged series of short staccato notes,

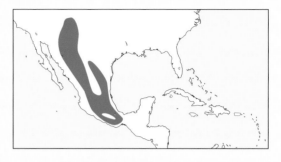

gewgew-gewgew-gew-gew, given in double and single hoots.
Food and hunting Feeds chiefly on insects, but also takes small mammals, birds and reptiles.
Habitat Occupies humid pine or pine–oak evergreen forests in mountains from 1500m to 3500m.
Status and distribution Highlands from S. Arizona, USA, south to C. Mexico, where it does not seem to be rare, but suffers from forest destruction.
Geographical variation Monotypic. Formerly, three other species were included with this owl; they have now been separated as Northern (169), Baja (170) and Guatemalan Pygmy Owls (175), although the taxonomy of the whole group still requires intensive study.
Similar species Partly sympatric Northern Pygmy Owl (169) is larger and long-tailed, with rounded wingtips.

▼ Mountain Pygmy Owl from rear and front showing well the occipital 'face', white spots on wings and almost uniform greyish-brown back. Tail has at least five very clear spot-bands. Arizona, USA, August (*Jim and Deva Burns*).

172. RIDGWAY'S PYGMY OWL
Glaucidium ridgwayi

L 17–19cm, **Wt** 46–102g; **W** 81–113mm

Identification A very small, round-headed owl without ear-tufts. Females are larger and 20g heavier than males. A highly polymorphic species with grey-brown and red morphs, as well as intermediates. The first of these morphs is greyish-brown above, fairly densely streaked dull yellow to whitish on forehead and crown, the nape with an obvious 'occipital face' formed by very clear black 'false eyes' with pale buff borders; the mantle, which is slightly darker than crown, is irregularly marked with whitish-buff spots. The wing-coverts are barred light and dark, and large whitish areas on the outer webs of the scapulars form a distinct row of spots across the shoulder. Flight feathers are dark-greyish brown with whitish-buff spots on both webs, forming incomplete bars, and the dark brown tail has

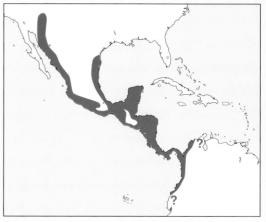

six to eight rufous, ochre or whitish-yellow bars. The throat is off-white, and the greyish-brown sides of the upper breast have dull yellowish streaks and spots; remaining underparts are whitish with bold brown or greyish-brown streaks. Light brown facial disc is flecked with whitish, and the eyebrows are whitish. The eyes are yellow, and the bill and cere greenish-yellow. Tarsi are feathered, and the yellowish toes bristled, with dusky horn-coloured claws with blackish tips. Red morph is similar in pattern, but in general rufous to orange-brown, with the tail barred reddish-brown. **Juvenile** Downy chick is whitish. Mesoptile has a less clear pattern than the adult, the crown often greyer than the back, and the plain forehead with fine pale shaft-streaks.

Call Gives a long series of hollow *poip-poip-poip-poip* notes, a little like those of Ferruginous Pygmy Owl, but slower in sequence, some 2.5–3 notes per second.

Food and hunting Insects make up almost 60 per cent of the diet, reptiles more than 20 per cent and birds 10 per cent; the rest consists of small mammals. This species normally catches its prey from a perch, or in dashing flight into dense foliage or thickets. Prey is sometimes heavier than the owl itself.

Habitat Prefers open woodlands and semi-open areas with giant cacti and thorny scrub, generally in lowlands but sometimes up to 1500m or slightly higher.

◄ Ridgway's Pygmy Owl is often considered a race of Ferruginous Pygmy Owl *Glaucidium brasilianum*, but there are vocal and DNA differences. Texas, February (*Jim and Deva Burns*).

Status and distribution Occurs from S. Arizona and Texas, USA, south to NW Colombia. It is locally common, but has decreased significantly in the northern parts of its range (S. USA).

Geographical variation Two subspecies are known: nominate *ridgwayi* is found from Texas and Mexico south to NW Colombia; *G. r. cactorum*, from S. Arizona to W. Mexico, is greyer in general, and the pale tail bars are always rufous or buff (never white). This owl was for a long time treated as a subspecies of Ferruginous Pygmy Owl (183), but has recently been separated on the basis of DNA data and vocal differences. The *G. brasilianum* complex, however, obviously represents a superspecies, with various related allospecies and paraspecies. Further studies are needed in order to resolve conclusively which of these have already reached species level.

Similar species All other *Glaucidium* species in the area have a spotted, rather than streaked, crown, and a less densely barred tail. Sympatric Elf Owl (200) is much smaller and short-tailed, without an occipital face, and is densely vermiculated below.

▶ Ridgway's Pygmy Owl has a clear occipital 'face'. Texas, February (*Jim and Deva Burns*).

▼ This red morph has scapulars with roundish spots; greater wing-coverts are also variably spotted. Tail has 7–8 reddish-brown bars. Guatemala, June (*Knut Eisermann*).

▼▶ A female looking out from a typical nest hole. Texas, April (*Jim and Deva Burns*).

173. CLOUD-FOREST PYGMY OWL
Glaucidium nubicola

L 16cm, Wt 73–80g; W 90–96mm

Identification A very small, round-headed owl without ear-tufts. Females have slightly longer wings but are not heavier than males. It has two large whitish-rimmed dark spots on nape, forming an 'occipital face'. Dark warm brown back and mantle with a slight russet tinge are practically unspotted, but large, bold white spots, often tinged light reddish, are apparent on the upper-wing-coverts and scapulars. Primaries and secondaries are barred whitish to dull yellowish and dark, and the dark sepia-coloured tail is relatively short and has five incomplete bars of irregularly shaped white spots. The chin, sides of upper throat and centre of breast are white, and the centre of belly and undertail-coverts are similar but with a buffish-brown wash; reddish-brown sides of breast have a few very small white spots, and the flanks are rufous, paler or more whitish towards centre of underparts, with dark brown streaks. The indistinct brown facial disc has some whitish concentric rings, but no ruff, and whitish eyebrows. The relatively large eyes are yellow, the eyelids with a blackish rim. The bill is greenish-yellow, and the cere dirty greenish-yellow and somewhat swollen. Tarsi feathered, yellowish toes slightly bristled and with dark horn-coloured claws with blackish tips. *Juvenile* Undescribed.

Call Utters a long sequence of soft couplets, *deewdeew-deewdeew*, but single notes and intervals between couplets are longer than those of Mountain (171) and Costa Rican Pygmy Owls (174).
Food and hunting Preys mainly on insects and other invertebrates, but also takes small vertebrates, lizards and even frogs.
Habitat Inhabits humid cloud forests on steep Andean slopes from 1400m to 2000m. In Ecuador it occurs also in young secondary forest and forest edges with dense undergrowth.
Status and distribution Found from Colombia to W. Ecuador, and maybe south to N. Peru. This species is locally not so rare, but it suffers from forest destruction on the W. Andean slopes. In Ecuador it should be ranked at least as vulnerable, because the amount of suitable habitat is less than 5000km² in extent, and this owl is known from fewer than ten localities.
Geographical variation Monotypic. Probably a member of the Mountain Pygmy Owl superspecies group, as suggested by vocalisations and DNA evidence; a study of this whole group of closely related species is needed.
Similar species Geographically separated Costa Rican Pygmy Owl (174) has rufous-bordered 'false eyes' on the hindneck, and is clearly less white on the belly and upper breast. Sympatric Andean Pygmy Owl (185) normally lives at higher elevations, and has a longer tail, more pointed wings, and a slightly barred or spotted mantle.

◄ Cloud-forest Pygmy Owl is similar to Mountain Pygmy Owl *Glaucidium gnoma* but differs in having a shorter tail and less spotted mantle and back. Eyes are yellow and relatively large for a pygmy owl. Ecuador, November (*Winnie Poon*).

174. COSTA RICAN PYGMY OWL
Glaucidium costaricanum

L *c.* 15cm, **Wt** 53–99g; **W** 90–99mm

Identification A tiny owl without ear-tufts. Females are 35g heavier than males. Brown morph is earth-brown above, densely marked with more or less rounded whitish dots and pale flecks; the head is a little paler than the back, and on the nape has two large, rounded blackish 'false eyes' bordered with white above and dull yellowish below; the earth-brown mantle and back are spotted irregularly with whitish and dark brown. Scapulars have rather indistinct pale cinnamon-buff edges. Primaries and secondaries have narrow dirty whitish or pale buffish barring, and the dark brown to blackish tail has five to seven bands of whitish spots. Shows some white on central upper breast, merging with the white belly; lower flanks somewhat streaked. The brownish facial disc is mottled lighter and darker, with no distinct ruff. The eyes are bright yellow and the bill yellowish-horn. The dirty-yellowish cere is typically swollen around the nostrils. Tarsi are feathered to the base of the greyish-yellow, sparsely bristled toes, which have dark horn-coloured claws with blackish tips. Rufous morph is in general reddish-brown or chestnut, with buffish-white underparts, the whitish markings replaced by dull yellow or fulvous, and light bars on tail tawny-buff or pale cinnamon-buff. *Juvenile* Undescribed.

Call Utters a long series of couplets, *dewdew-dewdew-dewdew*, sometimes even triple notes, given at fairly long intervals. Double notes are similar to the song of Mountain Pygmy Owl (171), but totally different from that of Andean Pygmy Owl (185).

Food and hunting Hunts mainly insects and other arthropods, but also lizards, small mammals and small birds. It strikes prey in short, swift dashing flight from perches in dense foliage.

Habitat Inhabits cloud forest and mountain forests with clearings, from 900m up to timberline.

Status and distribution Occurs from C. Costa Rica to W. Panama, possibly also to E. Panama; the distribution is not yet fully known. Considered to be fairly common in Costa Rica, but less so in Panama.

Geographical variation Monotypic. Closely related to Mountain Pygmy Owl, and not a subspecies of Andean Pygmy Owl as was formerly believed.

Similar species Sympatric Ridgway's Pygmy Owl (172) is large with long tail and Central American Pygmy Owl (179) is smaller with a greyish-brown head.

◄◄ This rufous morph has caught a large thrush – not much smaller than the owl itself. Costa Rica, April (*Mike Danzenbaker*).

◄ Costa Rican Pygmy Owl's brown morph. Brown head is somewhat paler than earth-brown upperparts. Sides of breast are brown with a row of mottling towards centre of the underparts. Dark horn claws with blackish tips. Cuanacaste, Costa Rica, April (*Daniel Martínez-A*).

175. GUATEMALAN PYGMY OWL
Glaucidium cobanense

L 16–18cm; W 82–98mm

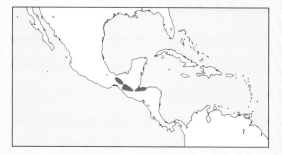

Identification A very small owl without ear-tufts. No data on sexual size differences. Brown and red morphs occur. Red morph predominates. It has a bright rufous or chestnut crown, forehead, sides of head and nape with subdued paler spots; blackish 'false eyes' are bordered dull yellowish to white, above a buffish nuchal collar. Reddish-brown back and mantle have indistinct lighter and darker markings, and there are slightly paler rufous edges on the scapulars. Bright reddish-brown wings have narrow paler, dark-edged bars, and the rufous tail has five to eight dull yellowish spot-bars, each spot finely edged blackish or dark brown. Below, it has a narrow buffish-rufous band across the throat, and is whitish centrally from neck to belly, the cinnamon-buff or rufous sides of breast and flanks having broad but somewhat diffuse reddish-brown streaks. Pale reddish-yellow facial disc has darker streaks from the eyes to the border of the disc, and whitish eyebrows, but no distinct ruff; lores and chin are whitish. The eyes are yellow, the bill yellowish-horn, and the cere black-tinged pale yellow. Tarsi are feathered to the base of the dirty yellow, bristled toes, with yellow soles and dark horn-coloured claws with blackish tips. Brown morph has a similar plumage pattern, but is in general brown to dark brown, with the pale markings whiter. *Juvenile* Downy chick is whitish. Mesoptile resembles adult, but is less distinctly marked, with more diffuse pattern, and with more greyish unmarked crown and mottled forehead.
Call Utters a long series of couplets, *püpühp-püpühp-*

püpühp..., with emphasis on the second whistled note. The voice is different from that of Mountain (171) and Costa Rican Pygmy Owls (174).
Food and hunting Not studied.
Habitat Found in mountain forests at higher elevations.
Status and distribution Found from Chiapas, S. Mexico, to Guatemala and Honduras. Its status is not known.
Geographical variation Monotypic. This owl was formerly regarded as a subspecies of Mountain Pygmy Owl, but separation suggested on grounds of vocal differences; molecular studies are required in order to to confirm the new taxonomic status.
Similar species Allopatric Mountain Pygmy Owl (171) is darker, with brown breast sides finely buff-spotted. Sympatric Ridgway's Pygmy Owl (172) has distinct pale shaft-streaks on the crown and forehead. Geographically separated Costa Rican Pygmy Owl (174) has the breast sides and flanks brown with whitish-buff mottling, and a large whitish area on central breast and belly.

◄ Typical rufous morph of Guatemalan Pygmy Owl. Guatemala, January (*Knut Eisermann*).

▲ Guatemalan Pygmy Owl has pale buff to cinnamon tail-bars. Guatemala, May (*Knut Eisermann*).

▲▶ Guatemalan Pygmy Owl is bright rufous to chestnut above, with subdued paler spots on the head. Guatemala, May (*Knut Eisermann*).

▶ Guatemalan Pygmy Owl has bright rufous chest-sides, and flanks with diffuse stripes. Guatemala, May (*Knut Eisermann*).

176. CUBAN PYGMY OWL
Glaucidium siju

L 17cm, **Wt** 55–92g; **W** 87–110mm

Identification A very small, round-headed owl without ear-tufts. Females are on average more than 20g heavier than males. Two morphs occur. Grey-brown morph is greyish-brown above, irregularly spotted whitish and buff, and with mantle indistinctly barred; the crown has whitish spots, and the nape very clear blackish 'false eyes' narrowly bordered whitish above and ochre-buff below. Scapulars lack prominent pale markings. Flight feathers are barred light and dark, and the brownish-grey tail has five or six narrow dusky-edged whitish bars. Below, it is off-white, densely barred brownish-ochre on sides of upper breast, with an unmarked longitudinal central area

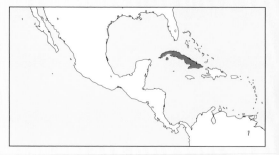

from throat to lower breast, the rest of the underparts spotted and streaked with brown. The pale greyish-brown facial disc has indistinct dark speckling and narrow whitish eyebrows. The eyes are yellow, rimmed by black edges of the eyelids; the bill is yellow-horn, and the cere yellowish-grey and only slightly swollen. Tarsi are feathered and bristled toes yellowish with dusky horn-coloured claws with darker tips. Rufous morph has reddish-brown general coloration. *Juvenile* Downy chick is white. Mesoptile has unspotted crown, but otherwise resembles adult.

Call Emits a whistle-like *jiu* at short intervals, or equally spaced sequence of *jiu, jiu, jiu...* starting softly and increasing in frequency and tone. Notes are given at intervals of some four seconds.

Food and hunting Preys mainly on insects and small reptiles, but also takes small mammals and birds. Hunts normally from a perch.

Habitat Inhabits open and semi-open woodlands, forest edges, plantations and large parks with mature trees, from sea level up to *c.* 1500m.

Status and distribution Endemic in Cuba and the Isle of Pines (Isla de la Juventud). It is locally rather common, but not well studied.

Geographical variation Two subspecies are separated: nominate *siju* is confined to Cuba; *G. s. vittatum*, from the Isle of Pines, is much larger and heavier, and more distinctly barred above, than the nominate. It is remarkable that *vittatum*, living very close to Cuba, is so different; research on the ecology and biology of both subspecies is needed.

Similar species This is the only *Glaucidium* owl in Cuba and Isle of Pines. Sympatric Cuban Bare-legged Owl (105) is longer, although not much heavier, and has long, bare tarsi, prominent whitish eyebrows and, most commonly, brown eyes.

▲ Female Cuban Pygmy Owl looking out from the nest hole. Zapata, Cuba, March (*Oliver Smart*).

▲▶ Cuban Pygmy Owl has a distinct occipital 'face' on its nape. Zapata, Cuba, February (*Arthur Grosset*).

◀ Cuban Pygmy Owl is off-white below with an unmarked white longitudinal central area from throat to belly. Zapata, Cuba, November (*Adam Riley*).

▶ Cuban Pygmy Owl has brown or rufous-brown flanks and yellow eyes. Zapata, Cuba, March (*Oliver Smart*).

177. TAMAULIPAS PYGMY OWL
Glaucidium sanchezi

L 13–16cm; **Wt** 52–56g; **W** 86–94mm

Identification A tiny owl without ear-tufts. Females have somewhat longer wings but are not even 5g heavier than males. Sexes differ also in plumage. Adult male has rich olive-brown crown, nape and upperparts, the crown often greyer; the forecrown has fine pale cinnamon to whitish spotting, spots extending along the sides of the crown to the nape, and hindneck has striking 'false eyes' creating an 'occipital face'. Upperwing-coverts are pale-spotted, and paler areas on outer webs of scapulars are rather indistinct. Flight feathers are barred light and dark, and the relatively long brown tail has five or six broken whitish bars. Below, the plumage is whitish with reddish-brown streaking and mottling; pale buff spots are visible on sides of upper breast. The brownish facial disc has white to dull yellow flecks and is bordered by short whitish eyebrows. The eyes are yellow, the bill horn-yellow and the cere yellowish-grey. Tarsi are feathered, and the bristled yellowish-grey toes have horn-coloured claws with darker tips. Adult female differs from male in having rufous crown, nape and upperparts. *Juvenile* Downy chick is white. Mesoptile has greyer and unspotted crown, and pale cinnamon tail-bars, as well as a more greyish cere than adults.
Call Utters a sequence of two or three hollow, high-pitched and drawn-out notes, *phew-phew-phew*, given at regular intervals and repeated after several seconds.

The call differs from that of others of genus in having fewer and longer notes, with long intervals between the notes.
Food and hunting Not studied.
Habitat Occupies subtropical evergreen and semi-deciduous cloud and mountain forests from 900m to 2100m.
Status and distribution Endemic in NE Mexican mountains. Its status is not fully known, but it is regularly reported in Tamaulipas and San Luis Potosí and in extreme N. Hidalgo.
Geographical variation Monotypic. This species and the Colima (178), Central American (179), Sick's (180) and Pernambuco Pygmy Owls (181) were earlier treated as subspecies of what was called the 'Least Pygmy Owl *Glaucidium minutissimum*', but any similarity to *G. minutissimum* is obviously due only to convergence and not to a close relationship. However, further studies of the ecology, taxonomy and biology are badly needed.
Similar species Sympatric Mountain Pygmy Owl (171) is larger, with a spotted mantle and relatively short tail. Allopatric Ridgway's Pygmy Owl (172) is also larger, with a streaked crown. Also geographically separated, Colima Pygmy Owl (178) is similar in size but much paler.

◀ Tamaulipas Pygmy Owl is tiny with a finely spotted forecrown. It has dark rufous-brown (female) or dark tawny (male) chest-sides and underpart-streaks. Female upperparts are distinctly more red than the male in this photo. Tamaulipas, Mexico, March (*Dominic Mitchell*).

178. COLIMA PYGMY OWL
Glaucidium palmarum

L 13–15cm, **Wt** 43–50g; **W** 81–88mm

Identification A tiny, round-headed owl without ear-tufts. Females are a little larger and 5g heavier than males. Has greyish tawny-brown upperparts, and the crown a little less greyish than the mantle, with the forehead to nape fairly densely spotted whitish or dull yellow, and 'false eyes' on the nape below a cinnamon band across the hindneck. Flight feathers are barred light and dark, and the greyish-brown tail has six or seven buffish-white bars of which only four are visible (two are concealed by the uppertail-coverts). It has off-white underparts, with cinnamon-brown streaking on sides of upper breast and below. The light ochre facial disc has indistinct darker concentric lines, with short whitish eyebrows above. The eyes are yellow, and the bill and cere yellowish-horn. Tarsi are feathered, the pale yellowish toes bristled, with horn-coloured claws with darker tips. *Juvenile* Downy chick is whitish. Mesoptile has unspotted grey crown contrasting with brown upperparts, the forehead with some paler flecks and the nape with false eyes, but no clear cinnamon band below the 'occipital face'.

Call Utters a series of hollow, short *whew-whew-whew* whistles, about three notes per second and up to 24 in one phrase; often successive increase in number of notes.

Food and hunting Takes small birds, reptiles and small vertebrates, but also larger insects and other invertebrates.

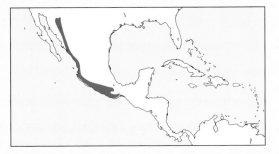

Habitat Found in arid to semi-humid tropical decid-uous thorn forests and oak and pine–oak woodlands, palm groves, dry oak forests and swampy forests, from foothills up to 1500m.

Status and distribution Occurs along the Pacific coast in W. Mexico. It is locally rather common.

Geographical variation Monotypic. This species and the Tamaulipas (177), Central American (179), Sick's (180) and Pernambuco Pygmy Owls (181) were earlier treated as subspecies of what was called the 'Least Pygmy Owl *Glaucidium minutissimum*', but now sepa-rated on grounds of different vocalisations.

Similar species Sympatric Mountain Pygmy Owl (171) is larger, with spotted mantle and more visible pale bars on tail. Also overlapping Ridgway's Pygmy Owl (172) is larger and long-tailed, with streaked crown.

▼ Colima Pygmy Owl is generally greyish tawny-brown, paler on the crown, which is peppered with white. Flight feathers are barred light and dark, and tail has 3 or 4 visible pale bars. Toes are pale yellowish. Mexico, March (*Gary Thoburn*).

179. CENTRAL AMERICAN PYGMY OWL
Glaucidium griseiceps

L 14–16cm, Wt 50–57g; W 85–90mm

Identification A tiny owl with no ear-tufts. Females are a little larger and 5g heavier than males. Has a brownish-grey crown and nape, the latter with a prominent 'occipital face'; minute whitish spots on the forecrown, spotting sometimes extending to hindcrown, but a plain rich brown mantle and back. Primaries are greyish-brown with rows of light spots, and the rufous secondaries have pale buffish bars. The brown tail has two or three visible broken whitish bars, a further two bars being concealed normally by uppertail-coverts. Below, the plumage is off-white, with a large whitish area from throat to central breast; sides of upper breast are mottled reddish-brown, and flanks and below are rufous-streaked. Light grey-brown facial disc has whitish flecks and indistinct concentric lines, and is bordered above by short whitish eyebrows. The eyes are yellow, and the bill and cere yellowish-horn with a slight greenish tint. Tarsi are feathered to the base of the pale yellowish toes, which have dark-tipped horn-coloured claws. *Juvenile* Downy chick is whitish. Mesoptile has nearly unspotted grey crown, contrasting with rich brown upperparts; tail-bars vary from whitish to pale cinnamon. *In flight* Shows contrasting secondaries.
Call Utters a series of hollow, ringing *pew-pew-pew* whistles, three per second and up to 18 notes in a phrase, the phrases repeated after variable intervals.

Food and hunting Not well documented, but feeds mainly on insects and spiders, and takes some small mammals and birds.
Habitat Inhabits tropical humid evergreen forests and bushlands, and also plantations and semi-open areas, from sea level to hills; up to 1300m in Guatemala.
Status and distribution Occurs from SE Mexico to Panama. It is locally not rare, but is too little known for any true estimate of its status to be made. Detailed observations on this owl are lacking.
Geographical variation Monotypic.
Similar species Sympatric Ridgway's Pygmy Owl (172) is generally larger and long-tailed, with a streaked crown. Geographically separated Tamaulipas (177) and Colima Pygmy Owls (178) both have more pale bars on the tail.

◀ Central American Pygmy Owl is generally redder than Amazonian *Glaucidium hardyi* or Subtropical Pygmy Owls *G. parkeri*. Forecrown finely spotted buff to white, with this often extending to the nape. Tail has 2–4 broken whitish-buff bars. Costa Rica, April (*Alex Vargas*).

◀◀ Different morphs have not previously been described, but this seems to be a clear red morph. Costa Rica, April (*Alex Vargas*).

180. SICK'S PYGMY OWL
Glaucidium sicki

L 14–15cm, Wt *c.* 50g; W 85–91mm

Identification A tiny owl without ear-tufts. No data on any sexual size differences. Has dusky cinnamon-brown to warm-brown crown and upperparts, with tiny whitish flecks on the crown and a prominent 'occipital face' on the nape, but plain warm brown mantle and back; a few pale speckles are visible on the wing-coverts. Flight feathers are dark brown with whitish spots on each web, forming light bars across the open wing. The dark brown tail has three or four visible broken bars of large, rounded whitish spots. The throat has a rounded whitish area bordered above by a narrow rufous band, and dense reddish-brown mottling forms lateral patches with a few pale spots. The off-white underparts are streaked buffish-rufous on the flanks. Light greyish-brown facial disc has some indistinct reddish-brown concentric lines, with whitish eyebrows. The eyes are yellow, cere and bill yellowish-horn with slight greenish tint. Tarsi are feathered, the yellowish toes bristled, with horn-coloured claws with darker tips. *Juvenile* Downy chick is whitish. Mesoptile has an unspotted rufous crown, and some pale flecks on the forehead. *In flight* Shows dark rufous plumage and distinct occipital face with strong 'false eyes'.

Call Gives a hollow and slightly drawn-out *hew-hew-hew-hew*, the first note lasting 0.25 seconds and the second 0.2 seconds, with a 0.35-second pause between the notes. It calls just before dawn, less often after dusk.

Food and hunting Not well studied, but feeds mostly on insects. Possibly takes also small vertebrates.

Habitat Occupies primary evergreen rainforest and forest edges from near sea level up to 1100m.

Status and distribution Found from E. Brazil south to E. Paraguay and E. Peru, possibly extending to NE Argentina. It is locally not rare but is very little studied.

Geographical variation Monotypic. Earlier described as part of the 'Least Pygmy Owl *Glaucidium minutissimum*', but has now been separated and renamed. Much uncertainty surrounds the nomenclature of Sick's and Pernambuco Pygmy Owls, and ongoing research may lead to changes.

Similar species Overlapping Ferruginous Pygmy Owl (183) is larger with shaft-streaks on the crown rather than spots.

◄ The crown is warmer brown than in Amazonian *Glaucidium hardyi* or Central American Pygmy Owls *G. griseiceps*, with tiny white spots which are not black-bordered as in Pernambuco Pygmy Owl *G. minutissimum*. São Paolo, Brazil, January (*Guilherme Gallo-Ortiz*).

◄◄ Sick's Pygmy Owl is relatively short-tailed and dark rusty-brown overall. Tail has four visible whitish bars. São Paolo, Brazil, April (*Arthur Grosset*).

181. PERNAMBUCO PYGMY OWL
Glaucidium minutissimum

L 14–15cm, Wt 51g (1♂); W 87–90mm

Identification A tiny owl without ear-tufts. No data on sexual size differences. Has an umber-brown crown, paler than the back, fairly densely spotted with small whitish flecks which are finely edged with sepia or blackish; lacks the 'occipital face' and 'false eyes', but has a whitish band with very few or no dark-edged feathers forming a nuchal collar, this bordered below by a narrow area of fulvous or sienna-buff that intergrades with the darker, uniformly dusky-brown mantle. Some paler spots are evident on outer webs of scapulars. Primaries and secondaries are darker brown than the back, their outer webs with rows of irregular-shaped light cinnamon dots and inner webs with rows of large dull yellowish-white spots. The dusky to blackish-brown tail has five incomplete but visible bands of white spots. Below, a white area covers the centre of the underparts from throat to belly; there are a very few small, inconspicuous whitish spots on the sienna-brown sides of the upper breast, prominent fulvous streaks on sides of lower breast and flanks, and white undertail-coverts and leg feathering with fulvous or russet streaks. The pale whitish facial disc has visible rufous concentric lines, with whitish eyebrows above. The eyes are yellow and the bill greenish-yellow. Tarsi are feathered to the base of the orange-yellow bristled toes, which have dark horn-coloured claws. *Juvenile* Unknown.

Call Gives a series of some six yelping notes, *gwoi-gwoi-gwoi-gwoi-gwoi-gwoi*, lasting 1.4 seconds, the series repeated at intervals of 4.5–15 seconds.

Food and hunting Not well documented, but feeds mainly on insects, and possibly takes some small vertebrates.

Habitat Occupies primary lowland rainforests with tall trees, from sea level to 150m.

Status and distribution Occurs only in Pernambuco State, in NE Brazil. This owl is very rare and is listed by BirdLife International as Critically Endangered; the entire known population lives in a protected forest reserve.

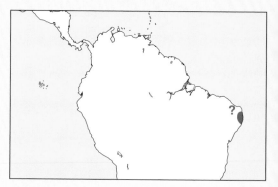

Geographical variation Monotypic. Previously the Tamaulipas (177), Colima (178), Central American (179), Sick's (180) and Pernambuco Pygmy Owls (181) were treated as subspecies of what was called the 'Least Pygmy Owl *Glaucidium minutissimum*', but they have now been separated mainly on the basis of vocal differences. This split led to the disappearance of the vernacular name, as both the present species and Sick's Pygmy Owl were given new names. The Pernambuco Pygmy Owl originally was renamed '*G. mooreorum*', before being given the present name of *G. minutissimum*. There is still some confusion over the precise location of the holotype discovery, and Brazilian ornithologists have recommended that the population in the Pernambuco region of Brazil be referred to as *Glaucidium mooreorum*, and that those in SE Brazil (Sick's Pygmy Owl) be called *Glaucidium minutissimum*. Thus, much uncertainty continues to surround the nomenclature, and until the critical papers are officially published the names given in König *et al.* (2008) have been accepted here.

Similar species Geographically separated Sick's Pygmy Owl (180) is darker, with a distinct occipital face. Also allopatric, Amazonian Pygmy Owl (182) has a more greyish head and a prominent occipital face.

182. AMAZONIAN PYGMY OWL
Glaucidium hardyi

L 14–15cm, Wt 52–63g; W 89–96mm

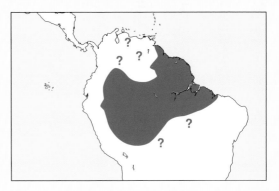

Identification A tiny owl without ear-tufts. No data on sexual size differences. Two colour morphs exist. The far commoner brown morph has greyish-brown crown marked with many very small off-white spots and some larger scaly markings, and on nape two large blackish spots surrounded by a pale area form a typical 'occipital face', with a narrow ochre-buff nuchal band below this. Slightly reddish-brown to earth-brown upperparts contrast with greyish head; there are a few, not very prominent whitish spots on wing-coverts. Dark earth-brown flight feathers have broken bars of whitish spots on each web, and the dark brown tail usually has five (only three visible) broken bars of large whitish spots. Below, plumage is off-white, with a large unmarked patch from throat to central breast; densely rufous-mottled sides of upper breast have a few whitish spots, and flanks and rest of underparts are boldly streaked reddish-brown. The pale greyish-brown facial disc has fine brownish flecks, and short whitish eyebrows above. The relatively small eyes are bright yellow; bill yellowish-horn. Tarsi are feathered and the golden-yellow toes bristled, with horn-coloured claws with darker tips; talons are rather small. Rare rufous morph is in general cinnamon-brown or rusty-brown, without whitish markings on head and tail, and with fairly indistinct dull yellowish shaft-streaks on cinnamon-rufous crown, fairly large whitish-buff areas on outer webs of scapulars, and some whitish-yellow spots on wing-coverts; has largely uniform reddish-brown sides of upper breast, and white belly with contrasting dark rufous streaks. The dark brown tail has about seven rows of dull reddish-

yellowish bars. *Juvenile* Resembles the adult, but has unspotted crown and less clear streaking below. *In flight* Shows the greyish head contrasting with mantle.

Call Gives a melodious trill, *bybybybyby*, consisting 10–30 fluted notes in a rapid staccato, some 10–13 notes per second; each phrase lasts 2.5–3 seconds, and phrase is repeated at variable intervals. Relatively loud and ringing trill is unlike the soft trills given by Pernambuco (181) and Sick's Pygmy Owls (180). Calls frequently during the day.

Food and hunting Not well documented, but eats chiefly insects, and perhaps also reptiles, small tree-dwelling mammals and birds.

Habitat Inhabits primary rainforest, where it lives in the epiphyte-rich forest canopy. Occurs in the lowlands and foothills, up to 850m.

Status and distribution Occurs in Amazon area from Venezuela, Guyana and N. Brazil south to E. Ecuador and NE Bolivia. Its status and exact distribution are poorly known, because this tiny owl is difficult to observe in the rainforest canopy.

Geographical variation Monotypic. Voice and DNA data show close relationship with Andean (185) and Yungas Pygmy Owls (186), rather than with the 'G. minutissimum complex'.

Similar species Sympatric Ferruginous Pygmy Owl (183) is larger and long-tailed, the tail either plain chestnut or with many bars varying from buffish-white to pale reddish-brown, and its crown is marked mainly with shaft-streaks, not spots.

◀ Amazonian Pygmy Owl is relatively long-winged and has only four bands in the tail. Guyana, December (*Jon Hornbuckle*).

183. FERRUGINOUS PYGMY OWL
Glaucidium brasilianum

L 17–20cm, **Wt** 46–107g; **W** 92–106mm, **WS** 38cm

Identification A very small owl without ear-tufts. Females are 15–30g heavier than males. Highly polymorphic, occurring in grey, brown and red morphs, with intermediates; in nominate race, brown morph is the commonest, red fairly common, and grey rather rare. Brown morph is warm earth-brown above, the crown with narrow pale buffish shaft-streaks, and occasionally with some whitish dots, particularly on crown sides and behind head. The nape has black 'false eyes' bordered whitish above and pale ochre below, but not forming a proper nuchal collar. The upperparts are either plain or irregularly spotted with dull yellow to whitish, the uppertail-coverts having a rufous tint. Pale yellowish outer edges of scapulars form a thin row across the shoulder, and whitish or whitish-buff spots are visible on wing-coverts. Flight feathers are barred incompletely light and dark, with pale rows of spots on both webs of feathers, and the dark brown tail has six or seven broken whitish bars, not reaching the feather shafts. The off-white underparts have dense brown mottling on sides of upper breast, with some paler flecks, and brown streaking below. Light ochre-brown facial disc with darker flecks is bordered above by whitish eyebrows. The eyes are yellow, and the bill and cere greenish-yellow. Tarsi are feathered brown and dull yellow, and the light yellow, bristled toes have orange-yellow soles, and dark horn-coloured claws with blackish tips; the talons are powerful. Red morph has

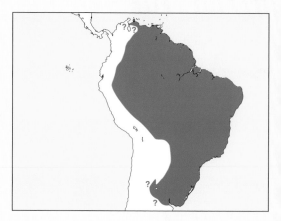

▼ Race *phaloenoides* of Ferruginous Pygmy Owl is similar to *medianum* in Venezuela and has relatively large yellow eyes and toes. Trinidad, April (*Robin Chittenden*).

rusty-brown upperparts variably buff-spotted, and pale buffish shaft-streaks on crown, sometimes appearing rather plain but with small yellowish spots and very fine light shaft-streaks; rusty-brown tail has narrow dark brown bars, but is sometimes unmarked reddish-brown; dull yellowish underparts have distinctly darker russet streaks. Grey morph is similar to brown morph in plumage pattern, but in general is greyish-brown, with whitish pale markings. *Juvenile* Downy chick is whitish. Mesoptile resembles adult, but with less clear markings and unspotted crown.

Call Gives a long series of bell-like ringing *whoip* or *poip* notes, equally spaced at three per second, and repeated 10–60 times; these phrases are given at intervals of several seconds. Usually jerks its tail when uttering the notes. Sings during daytime as well as at night, often attracting mobbing from small birds.

Food and hunting Feeds mainly on insects, small birds and other small vertebrates. This owl is able to catch prey larger than itself. It hunts normally from a perch, but will also seize a bird or insect among foliage in sudden dashing flight.

Habitat Inhabits mainly lowland humid primary and secondary forests, forest edges and clearings with thorn scrub, and pasturelands, but also parks and gardens near human habitations. Normally in lowlands below 1500m, avoiding mountain and cloud forests; in Venezuela ascends to 2250m.

Status and distribution Widely distributed in South America from Colombia and Venezuela south to Uruguay and C. Argentina. This owl is locally common, but little is known of its biology and ecology.

Geographical variation Seven subspecies are recognised: nominate *brasilianum* occurs from NE Brazil south to N. Uruguay; *G. b. medianum*, from N. Colombia and N. Venezuela east to N. Suriname and Cayenne, is relatively small; *G. b. phaloenoides*, from Margarita Island (off N. Venezuela) and Trinidad, is similar to *medianum*; *G. b. duidae*, from Mt Duida, in S. Venezuela, is very dark; *G. b. olivaceum*, from Auyán-Tepui, in SE Venezuela, is tinged greyish-olive; *G. b. ucayalae*, from S. Venezuela and Amazonian Colombia and Brazil south to E. Ecuador, E. Peru and Bolivia, has a more hollow song; *G. b. stranecki*, from S. Uruguay to C. & E. Argentina, is the largest subspecies. It is possible that *ucayalae* is specifically distinct from the Ferruginous Pygmy Owl, and it is also questioned whether *medianum*, *phaloenoides* and *duidae* really are races of this species. Clearly, this whole complex of forms is in need of full taxonomic revision.

Similar species Geographically overlapping Sick's (180), Pernambuco (181), Amazonian (182) and Subtropical Pygmy Owls (184) are all smaller and short-tailed, with a spotted crown. Partly sympatric Chaco Pygmy Owl (189) is smaller, with a slightly streaked and more spotted crown, mostly unspotted mantle and back, and generally without whitish on outer webs of scapulars, and it lives in arid habitats.

▲ Race *ucayalae* of Ferruginous Pygmy Owl is very white below like the nominate but has fewer white head streaks. Ecuador (*Murray Cooper*).

▲▶ Nominate Ferruginous Pygmy Owl's dark morph is very dark brown with a few spots on the head, scapulars, wings and chest. Misiones, Argentina, September (*Martjan Lammertink*).

▶ Nominate Ferruginous Pygmy Owl's red morph. Dorsal view showing the plain rufous tail. Pousada Xaraes, Brazil, September (*Ronald Messemaker*).

◄ Most races of Ferruginous Pygmy Owl (this is the nominate) have a very white breast and belly; only the darkest subspecies, *duidae*, is less white below. Brazil, March (*Kleber de Burgos*).

184. SUBTROPICAL PYGMY OWL
Glaucidium parkeri

L *c.* 14cm, Wt 59–64g; W 90–97mm

Identification A tiny owl without ear-tufts. No data on sexual size differences. The plumage is dark brown above, a little greyer on the crown, densely marked on sides of head and crown with thinly dusky-edged whitish dots; nape has a prominent 'occipital face' with an often concealed whitish collar below. Plain dark brown mantle has a dull olivaceous wash, which becomes more intense towards wing-coverts. Outer webs of scapulars and upperwing-coverts have bold whitish spots edged dusky at base. Flight feathers are marked with irregularly shaped white spots, and the blackish-brown tail has five incomplete white bands of similar spots. Below, the plumage is white, including a large white patch between throat and centre of upper breast, with rufous sides of neck and upper breast with small whitish flecks, and clearly olive-washed chestnut streaks on flanks and lower underparts. Pale greyish-brown facial disc is finely speckled white and brown. The eyes are yellow, and the bill and cere greenish-yellow. Tarsi are feathered, and the yellow toes bristled, with dark horn-coloured claws. *Juvenile* Undescribed.

Call Utters short phrases of usually four rather high-pitched notes, *hüw-hüw-hüw-hüw*, lasting about two seconds, the phrase given at intervals of several seconds. This call is unique among the New World pygmy owls.

Food and hunting Not documented, but probably feeds chiefly on insects.

Habitat Occupies humid mountain forest and cloud

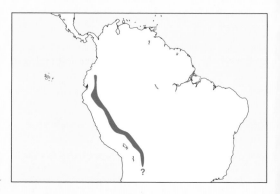

forest with many epiphytes, mainly between 1450m and 1975m.

Status and distribution Found on the eastern slopes of the Andes from SE Colombia to Peru and N. Bolivia. Status and distribution are poorly known but it could be endangered by habitat destruction.

Geographical variation Monotypic. This owl was described relatively recently (1995), and its relationships with other Neotropical pygmy owls remain to be studied.

Similar species Partly geographically overlapping Ferruginous Pygmy Owl (183) is much larger and long-tailed, with a mostly streaked crown. Sympatric Andean Pygmy Owl (185) usually lives at higher altitudes, has longer wings with pointed tips, a pale-flecked mantle and is mottled rather than streaked below.

185. ANDEAN PYGMY OWL
Glaucidium jardinii

L 15–16cm, Wt 56–75g; W 95–101mm

Identification A tiny owl without ear-tufts. Females are on average well over 10g heavier than males. Red and brown morphs are known, and intermediates occur. Brown morph has rather warm-coloured dark earth-brown crown and upperparts, crown having tiny whitish or dull yellowish spots, sometimes slightly drop-shaped, and sometimes also very thin shaft-streaks; 'occipital face' on nape, and below it a nuchal half-collar slightly paler than the mantle. The mantle and back, which are a little more rufous in tone than the crown, have some irregular pale spots. Flight feathers have broken pale rows of whitish-buffish spots, creating a pattern of light and dark barring, and the blackish tail has five or six bands of irregular whitish spots on both webs of each feather, but not reaching the shaft. Below, a narrow brown band separates white throat from a large white patch on foreneck and upper breast; dark brown sides of upper breast are mottled light and dark, some dusky mottling and some indistinct streaking are present on flanks, and the centre of the belly is plain whitish. Pale greyish-brown facial disc has darker concentric lines, and is bordered above by rather prominent whitish eyebrows. The eyes are yellow, and the bill and cere yellowish-horn. Tarsi are feathered brown and white, and the yellow, bristled toes have blackish-tipped dark horn-coloured claws. Red morph is brown with an orange-rufous or orange-buff tinge. *Juvenile* Downy chick is whitish. Mesoptile resembles the adult, but has unspotted crown and less clearly marked underparts. *In flight* Shows clearly pointed wings.

Call This owl's song is made up of two different phrases: the first consists of four or five short, stuttering trills, *puéehtututu-puéehtututu*, at 0.35-second intervals, and the second of five to ten staccato *tew-tew-tew-tew* notes in rapid succession, four per second at 0.15-second intervals. A new phrase begins only after several seconds.

Food and hunting The diet is of small birds, insects and other arthropods, as well as small mammals and other vertebrates. It normally catches its prey from a perch.

Habitat Found in semi-open cloud and mountain forests and humid woodlands from 900m to 4000m. In Colombia, generally at 2100–2800m.

Status and distribution Occurs from Venezuelan and Colombian Andes south through Ecuador to NC Peru.

This species is rather frequent locally, but is threatened by extensive logging.

Geographical variation Monotypic. This species and the Costa Rican Pygmy Owl (174) were formerly considered to be conspecific, but they differ considerably in vocalisations and have been separated as being specifically distinct. Further studies are, however, needed on the taxonomy, bioacoustics and biology of all pygmy owls inhabiting the Central and South American mountains.

Similar species Partly sympatric Subtropical Pygmy Owl (184) is clearly smaller with longer wings.

▶ Andean Pygmy Owl brown morph. Ecuador (*Tasso Leventis*).

186. YUNGAS PYGMY OWL
Glaucidium bolivianum

L c. 16cm, Wt 53–70g; W 94–103mm

Identification A tiny owl with no ear-tufts. Females are on average 10g heavier than males. Grey, brown and red morphs are known, brown being the most common; grey morph seems to be restricted to the S. Yungas. Above, the grey morph is dusky grey-brown with many pale spots, the crown fairly densely marked with roundish whitish dots and also larger light ochre scaly markings; on the nape two large black spots, whitish-edged above and bordered pale ochre below, form a narrow brownish-yellow nuchal collar below the prominent 'occipital face'; the mantle and back have irregular whitish and pale brownish-yellow spots, these often spear-shaped. Partly whitish outer webs of scapulars are not particularly noticeable. Flight feathers have whitish-buff spots on both webs, creating a pattern of light and dark barring, and the blackish tail has five or six broken whitish bars formed by oval spots, not reaching the shafts. The white throat is bordered below by a narrow dusky band, beneath which is a large plain whitish area extending through centre of the breast to the belly; rest of underparts are off-white, with dusky greyish-brown mottling at sides of upper breast and distinct dark streaking on the flanks and lower regions. The ill-developed, fairly flat, light greyish-brown facial disc has dark flecks, with white eyebrows above. The eyes are golden-yellow, and the bill and cere greenish-yellow. Tarsal feathers are mottled dusky and whitish, and the bristled dirty-yellow toes have dark horn-coloured claws with blackish tips. Brown morph is similar in pattern to grey morph, and in general warm

dark brown, the blackish tail with only five whitish bars. Red morph is rusty orange-brown in general, with light markings above similar to other morphs. *Juvenile* Downy chick unknown. Mesoptile has unspotted crown slightly greyer than upperparts, and is less distinctly streaked below. *In flight* Shows more rounded wings and longer tail than Andean Pygmy Owl (185).

Call The song consists of two different phrases: first two or three melodious *wüühurrr* whistles, and secondly a series of hollow staccato notes, *whüp-whüp-whüp-whüp*, this part at fairly slow tempo, less than two notes per second. This owl calls only during dark.

Food and hunting Preys mainly on insects, other arthropods and small birds. Forages primarily in the canopy and in dense foliage below it.

Habitat Occupies humid mountain and cloud forests with many epiphytes, between 900m and 3000m; sometimes found even higher, depending on the timberline.

Status and distribution Occurs from N. Peru to Bolivia and N. Argentina, mainly on E. Andean slopes. Its status is not known, but in Argentina it is endangered as a result of local forest destruction. This owl is less diurnal than its congeners, making studies more difficult.

Geographical variation Monotypic. As this species was first described fairly recently (1991), comparative studies including it and Peruvian, Subtropical and Andean Pygmy Owls would be of great interest.

Similar species Partly overlapping Subtropical Pygmy Owl (184) is smaller, with much shorter tail, mantle unspotted, and white underparts with clear dark rufous streaking. Largely allopatric Andean Pygmy Owl (185) has pointed wingtips and shorter tail, less densely marked crown with smaller spots, is more mottled and barred below, and no grey morph is known.

◀ Yungas Pygmy Owl taking off. Manu-Cusco, Peru, March (*Matthias Dehling*).

◄ Yungas Pygmy Owl is similar in size and plumage to Andean *Glaucidium jardinii*, with five broken bars on tail, and flanks more striped than barred. Here the white throat is puffed up. Salta, Argentina, August (*James Lowen*).

▼ Yungas Pygmy Owl showing the clear occipital 'face'. Salta, Argentina, August (*James Lowen*).

▼ ► The red-brown morph of Yungas Pygmy Owl. Manu-Cusco, Peru, March (*Matthias Dehling*).

187. PERUVIAN PYGMY OWL
Glaucidium peruanum

L 15–17cm, Wt 58–65g; W 98–104mm

Other name Pacific Pygmy Owl

Identification A tiny to very small owl without ear-tufts. Females are a little larger and 5g heavier than males. This is a highly polymorphic species, with grey, brown and red morphs. Grey morph is dark greyish-brown above, with very short and narrow pale shaft-streaks on forehead, and some fine speckling and whitish spots of various sizes on crown to nape and sides of head; 'occipital face' on nape is whitish-edged above and with a narrow ochre nuchal collar below; variably sized whitish spots, often roundish rather than triangular, are present on the mantle and back, and larger whitish areas on outer webs of scapulars. Flight feathers are dark greyish-brown with rows of whitish or dull yellowish spots on each web, forming incomplete light and dark barring, and the dark greyish-brown tail has six or seven broken whitish bars (but only four or five visible), not reaching shafts of rectrices. Whitish below, with a relatively large whitish throat patch, and densely dark greyish-brown mottling and a few whitish speckles on sides of upper breast, the rest of underparts having bold dark greyish-brown streaking. The facial disc, not well developed, is bordered above by whitish eyebrows. The relatively large eyes are yellow, and the bill and cere greenish-yellow. Tarsi have greyish-brown

feathering mottled off-white, and the yellow toes are bristled, with dark horn-coloured claws with blackish tips. Brown morph is similar in pattern to grey morph, but dark earth-brown in general, and with more finely light-spotted crown. Red morph has rusty-brown upperparts with dull yellowish or whitish flecks and dots, normally pale buffish shaft-streaks on crown, and usually rufous tail with some seven paler russet or orange-buff bars reaching the shafts; off-white to light yellowish underparts with orange-brown markings, but not so distinct as on grey and brown morphs. *Juvenile* Downy chick is whitish. Mesoptile resembles adult, but has plain, unspotted crown.

Call Emits a very rapid, upslurred staccato *toitoitoitoitoi toitoit*, six or seven notes per second, the phrase varying in length, and usually repeated at short intervals.

Food and hunting Feeds mainly on insects and other arthropods; small birds are also important prey, as are other small vertebrates, at least locally.

Habitat Inhabits riparian thickets, mesquite, semi-arid woodlands and bushlands with cacti and thorny shrubs, but occurs also in city parks and agricultural areas with trees. From lowlands to 2400m, locally even higher, to 3000m.

Status and distribution Found from W. Ecuador through Peru to N. Chile, but only on the western slope of the Andes. This owl is apparently not so rare locally.

Geographical variation Treated as monotypic, but populations of higher and lower altitudes may belong to different, as yet undescribed, subspecies. Individuals from lower elevations often have crown with some spots and many prominent elongated drop-shaped shaft-streaks. Until 1991 this owl was regarded as conspecific with Ferruginous Pygmy Owl (183), but DNA data have now indicated that it is a distinct species, although very little specific information is available on its biology. In Apurimac area of Peru there is a distinctive *Glaucidium* owl which could be a new species, or at least a new subspecies (of this owl?).

Similar species Allopatric Ferruginous Pygmy Owl (183) lives east of the Andes, and lacks an ochre nuchal collar below the occipital face. Also allopatric, Yungas Pygmy Owl (186) lives in E. Andes, has rather rounded wing-tips and smaller, often triangular pale spots on the back and mantle, and is less clearly streaked below.

▲ This Peruvian Pygmy Owl from Ecuador seems to have very different tail-bars to Peruvian birds – is this a different race perhaps? Machalilla, Ecuador, July (*Karen Hargreave*).

▲▶ Peruvian Pygmy Owl has bright yellow eyes and legs. Limon, Peru, June (*Paul Noakes*).

▶ Pair of Peruvian Pygmy Owls. Both are red morphs. Lambayeque, Peru, August (*Christian Artuso*).

◀ Typical brown morph of Peruvian Pygmy Owl. Lima, Peru, October (*Dubi Shapiro*).

188. AUSTRAL PYGMY OWL
Glaucidium nana

L 17–21cm, Wt 55–100g; W 95–108mm

Identification A very small owl without ear-tufts. Females are on average up to 20g heavier than males. Red and grey-brown morphs, as well as intermediates, are known. Grey-brown morph is dark greyish-brown above, marked with variably sized and variably shaped whitish spots and dots, and a prominent 'occipital face' on nape is whitish-bordered above and below. Has irregularly light-spotted mantle and back, and large white areas on outer webs of scapulars. Flight feathers are incompletely barred with rows of whitish or pale buffish dots, and the tail has eight to ten very thin rufous bars reaching the shafts of the rectrices, the light bars in general thinner than the dark ones. Below, the plumage is off-white, with whitish-speckled dark greyish-brown patches at sides of upper breast, a relatively small white patch on the throat and white continuing into a narrow zone in centre of belly, with another narrow white flank panel from about breast to thighs; area between the central and lateral white panels is densely marked with thin and relatively short shaft-streaks and light and dark mottling. The pale greyish-brown facial disc has fine dark flecks and streaks, with whitish eyebrows and lores. The eyes are yellow, and the bill and cere greenish-yellow. Feathered tarsi are whitish, mottled with greyish-brown, and the bristled toes are yellow, with dark horn-coloured claws with blackish tips. Red morph is similar in pattern, but has rufous-brown as base colour. *Juvenile* Downy chick is whitish. Mesoptile has deep yellow eyes, often with orange tint, an unspotted crown, and in general less distinct patterns than adult. *In flight* Shows the densely barred tail.

Call Gives a rapid, harsh staccato *kü-kü-kü-kü-kü...*, each note with upward inflection, 3.5–5 notes per second, and some 20–30 notes per sequence, the whole repeated at intervals of several seconds.

Food and hunting Insects make up half of the diet, small mammals 32%, and birds 14%. Dragonflies, scorpions, spiders, lizards and reptiles are also captured.

Habitat Prefers open woodland and forests with thorn shrubs and groups of trees, and is found locally also in parks and farmland areas near human habitations. From sea level to 2000m.

Status and distribution Occurs from Chile south to Tierra del Fuego. Said to be not rare locally. Reported as wintering in N. Chile and C., N. & E. Argentina.

Geographical variation Monotypic. Often regarded as a subspecies of Ferruginous Pygmy Owl (183), as it is similar in both plumage and voice; new DNA data support its status as a distinct species, but its biology, ecology and behaviour require further study.

Similar species Allopatric Ferruginous Pygmy Owl (183) of southernmost subspecies (*stranecki*) is almost equal in size, but has relatively larger eyes and less densely barred tail, and is more boldly streaked below. Also geographically separated, Peruvian Pygmy Owl (187) has a less densely barred tail; its red morph has wider pale bars on the tail.

▶ Austral Pygmy Owl's tail has 8–11 narrow dark-rufous or buffish bars. Ferruginous Pygmy Owl *Glaucidium brasilianum* has a much less densely barred tail. Los Glaciares, Argentina, December (*Stefan Hohnwald*).

◀ Two grey-brown immature Austral Pygmy Owls. Los Glaciares, Argentina, December (*Stefan Hohnwald*).

189. CHACO PYGMY OWL
Glaucidium tucumanum

L 16–18cm, Wt 52–60g; W 90–100mm

Other name Tucuman Pygmy Owl

Identification A very small owl without ear-tufts. Females are less than 5g heavier than males. Three morphs are known, grey being the most common. Dark greyish-brown above with a slight olive cast or fairly dark slate-grey; has very thin, short whitish shaft-streaks on forehead and forecrown, grading into almost rounded whitish spots on hindcrown, as well as on sides of head; a prominent 'occipital face' on nape is bordered whitish, without nuchal collar, and it has unspotted mantle and back. Irregularly shaped whitish dots are present on greater wing-coverts, and flight feathers are incompletely whitish-barred. The very dark grey-brown tail has five or six broken whitish bars. Off-white underparts have sides of upper chest dark greyish-brown with some whitish flecks; otherwise heavily but somewhat diffusely streaked dark greyish-brown below, sometimes with a slight violet tinge. Dusky facial disc has fine whitish or pale buff flecks and whitish eyebrows. The eyes are light to bright yellow, and the bill and cere greenish-yellow. Tarsi are feathered, the bare toes yellowish-green above and yellow below, with blackish claws. Brown morph is similar in pattern to grey morph, but is in general more earth-brown. Red morph is reddish-sandy to light reddish-brown, with plain mantle, paler-edged scapulars. *Juvenile* Downy chick is whitish. Mesoptile is like adult, but with unmarked crown, and yellowish-grey eyes.

Call Utters a series of upward-inflected staccato notes, *toik-toik-toik-toik*, two notes per second and 10–30 notes in sequence, repeated at intervals of several seconds.

Food and hunting Small birds seem to be the main prey, at least in the dry season, but it also eats small mammals, reptiles, lizards and insects. Hunts mainly from perches, but takes small birds among foliage in dashing flight.

Habitat Prefers semi-open, dry forests with giant cacti and thorny scrub; also gardens or parks near human settlements. From 500m to 1800m, but seems to avoid high mountain forests.

Status and distribution Found from Bolivia and Paraguay to CW Brazil and N. Argentina. This owl is locally not so rare, but is inadequately studied.

Geographical variation Two subspecies are listed: nominate *tucumanum* occurs from Paraguayan Chaco to N. Argentina; *G. t. pallens*, from Bolivian Chaco to SW Brazil, occurs in brown and red morphs. Race *stranecki* of Ferruginous Pygmy Owl (183) interbreeds with this owl, but all known hybrids so far have been infertile. Vocal and DNA evidence, as well as differences in ecology, have justified the treatment of Chaco Pygmy Owl as a separate species.

Similar species Partly sympatric Ferruginous Pygmy Owl (183) is larger and overlaps near southern limit of distribution, but has crown more streaked than spotted. Geographically separated Yungas Pygmy Owl (186) is very similar, with rounded or drop-shaped spots on crown and an ochre nuchal collar.

◀ Chaco Pygmy Owl is a small, earth-brown pygmy owl. In the south grey morphs tend to be predominant, but this bird is a brown morph. Crown is densely spotted, and brown tail has 5–6 incomplete white bars. Underpart-stripes are dark brown. Salta, Argentina, November (*G. Armistead*).

190. JUNGLE OWLET
Taenioglaux radiata

L c. 20cm, Wt 88–114g; W 120–136mm

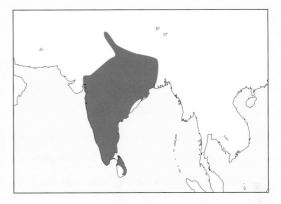

Identification A very small to small, round-headed owl without ear-tufts. No data on sexual size differences. Evidently two morphs exist. Normally has dark greyish-brown crown and upperparts, including wings and tail, densely patterned with narrow pale ochre or reddish-brown bars, often almost pure white bars on back, rump and uppertail-coverts. No 'occipital face'. White underparts, somewhat rufous-tinged to lower breast, with grey-brown barring on breast, belly sides and flanks. Has an inconspicuous facial disc with pure white short eyebrows above, and white moustachial streak, chin and breast patch. The eyes are bright lemon-yellow, the bill greenish-yellow to yellowish-grey and the cere bluish. Feathered and finely bristled toes are dirty greenish-yellow, with soles yellowish, and dark horn-brown claws. Grey morph is much greyer, particularly on lower back and tail. *Juvenile* Mesoptile resembles adult, but plumage is fluffier, with browner-barred tail and less distinct bars below.

Call Utters a loud, musical, wooden trill, *praorr-praorr-praorr-praorr*, 1.5–2.5 notes per second; single phrase has three to ten notes, and is repeated at intervals of several seconds.

Food and hunting Feeds mainly on locusts, grasshoppers, cicadas and other larger insects, but also takes molluscs, lizards, small birds and mammals.

Habitat Inhabits moist deciduous forests and secondary jungle with bamboo, locally to 2000m in Nepal.

Status and distribution Occurs in Himalayan foothills east to Bhutan and West Bengal, and south through India to Sri Lanka, and also in NE India (Arunachal and Assam), Bangladesh and Burma. Not rare locally, and fairly common in Burma.

Geographical variation Two subspecies are described:

▶ The greyish-brown morph of Jungle Owlet is typical. This male *malabarica*'s white throat is puffed up prior to calling. Goa, India, January (*Hira Punjabi*).

nominate *radiata* is found in most of the species' range; *T. r. malabarica*, from Malabar coast just north of Goa south to Kerala, in India, is much darker and more rufous than nominate. The Malabar subspecies is said to differ also in vocalisations, suggesting that it may possibly merit full species status, but a taxonomic study is required.

Similar species Largely allopatric Collared Pygmy Owl (168) is much smaller, with a rufous half-collar and an occipital face. Partly overlapping Asian Barred Owlet (193) is larger, with much broader barring and dark streaks on belly.

▲◄ Rufous morph *malabarica* Jungle Owlet is much darker than the nominate, and more rufous to chestnut on the crown, nape, lower back and wings. Kerala, India, February (*Amano Samarpan*).

▲ Jungle Owlet of race *malabarica* from behind, showing the minimal tail and reddish-brown wings and neck. Kerala, India, January (*John and Jemi Holmes*).

◄ Greyer nominate Jungle Owlet from behind, showing the finely barred upperparts and bright lemon-yellow eyes. Goa, India, November (*Amano Samarpan*).

► Portrait of grey-brown morph of Jungle Owlet. Nepal, January (*Neil Bowman*).

191. CHESTNUT-BACKED OWLET
Taenioglaux castanonota

L 17–19cm, Wt *c.* 100g; W 122–137mm

Identification A very small owl without ear-tufts. Females are somewhat larger and heavier than males. Has a dark brown head with narrow rufous-ochre bars, and contrasting bright chestnut mantle and back with a few blackish bars; no 'occipital face'. Flight and tail feathers have a barred pattern of ochre or whitish and dark rufous or chestnut. Dark brown and ochre barring on upper breast forms a pectoral band; lower breast, flanks and belly are white with blackish streaking. The facial disc is not well developed. The eyes are bright yellow, the bill yellowish or greenish-horn, and the cere dusky greenish. Tarsi are feathered, and the yellowish-olive toes sparsely bristled, with dark horn-coloured claws. ***Juvenile*** Mesoptile resembles adult in plumage but is fluffier, with barring and streaking more diffuse.

Call Utters a long, far-carrying, musical, purring vibrato *krrraw-krrraw-krrraw*, *c.* 2.5 notes per second, each lasting 0.15–0.25 seconds. One phrase has four to nine notes, repeated at intervals of several seconds. Calls also during the day.

Food and hunting Feeds chiefly on insects, but also takes small mammals, birds and reptiles.

Habitat Occupies dense, humid forests from lowlands up to 1950m. Sometimes occurs at edge of rubber plantations and in thickly planted gardens, as, for instance, in the vicinity of Colombo city.

Status and distribution Endemic to Sri Lanka. This owl is less common than it was earlier, its decline being due to habitat destruction.

Geographical variation Monotypic. Many authors have regarded this owl as a subspecies of Jungle Owl (190), but it has recently been suggested that it merits full species status; its biology, habits and vocalisations, however, are poorly studied.

Similar species Allopatric Jungle Owlet (190) is very similar in shape, size and coloration but has the mantle and back distinctly barred dark brown and white, and lacks blackish shaft-streaks below.

◀ Chestnut-backed Owlet is similar to Jungle Owlet *Taenioglaux radiata* but has a bright chestnut back, as the name suggests. Also more rufous on the wings, with a few dark cross-bars. Kitulgala, Sri Lanka, February (*Gary Thoburn*).

◀◀ Chestnut-backed Owlet has three clear, white vertical stripes below; lower chest and flanks have large dark brown shaft-streaks (not barred as in Jungle Owlet). Kitulgala, Sri Lanka (*Gary Thoburn*).

192. JAVAN OWLET
Taenioglaux castanoptera

L 23–25cm; **W** 144–150mm

Identification A small, round-headed owl without ear-tufts. No data on sexual size differences. The head, throat and sides of the neck are barred dark brown and pale brownish-yellow, the chin whitish, and white cheek feathers narrowly brown-tipped. Rufous-chestnut upperparts contrast the white outer webs of the scapulars, which form a distinct line across the shoulder. Flight feathers are chestnut-brown with broken pale ochre bars, and the dark brown tail has seven rather thin ochre bars narrowly blackish-bordered at their upper edge. Mostly white underparts have bright rufous-chestnut streaking, and the reddish-brown feathers of the breast sides are broadly margined with white on both webs; flank feathers have white inner webs and chestnut outer webs. The eyes are yellow, the bill greenish-yellow, more yellow at tip and on cutting edges, and the cere olive-green. Tarsi are feathered, and the yellowish-tinged olive-green toes are sparsely bristled, with greenish-yellow or yellow soles and dark horn-coloured claws with blackish tips.
Juvenile Downy chick is whitish. Fledglings resemble adults, but are much duller.
Call Utters a rapid trill, somewhat laughter-like when loudest, repeated at regular intervals. Calls at dawn and dusk.
Food and hunting Takes mainly insects, but also spiders, scorpions, myriapods (including centipedes), lizards, small mammals, birds, and rarely even small snakes.

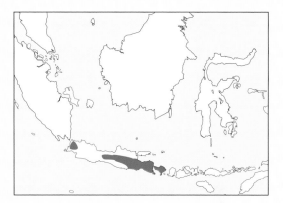

Habitat Occupies dense primary lowland rainforests and thick bamboo jungle, from sea level to 2000m; optimum elevation appears to be somewhere between 500m and 900m.
Status and distribution Endemic in Java and Bali, in Indonesia. Although generally rare, it is still rather common in optimal habitat. The biology and behaviour of this species are in great need of further research.
Geographical variation Monotypic.
Similar species No other owlet of this size lives in Java and Bali. Sympatric Javan Scops Owl (62) has fine wavy rufous cross-bars and black shaft-streaks on underparts, as well as visible ear-tufts (when erected).

◄ Javan Owlet is similar in size to Asian Barred Owlet *Taenioglaux cuculoides* but with rufous-chestnut upperparts. Breast-sides are barred brown and ochre, centre of breast and belly whitish, with chestnut-striped flanks. Carita, Java, May (*James Eaton*).

193. ASIAN BARRED OWLET
Taenioglaux cuculoides

L 22–25cm, Wt 150–240g; W 131–168mm

Identification A small, round-headed owl without ear-tufts and 'occipital face'. Females are larger and more than 35g heavier than males. Has dull brown or olive-brown crown, upperparts, sides of head, neck and wing-coverts all faintly russet-tinged and closely barred with fulvous-white or dull rufous-white. The flight feathers are similar to the upperparts, and the blackish tail has about six widely spaced whitish bars. Throat has a distinct white patch; the breast has dark brown and dull tawny-white barring, the upper belly is barred light brown and white, and lower underparts are rather more streaked than barred. Has an indistinct facial disc, white eyebrows extending to behind the eyes, and white moustachial streaks. The eyes are lemon-yellow, the bill yellowish-green and the cere greenish-horn. Tarsi are feathered. the bare toes greyish olive-yellow and sparsely bristled, with chrome-yellow soles and horn-brown claws. *Juvenile* Downy chick is white. Mesoptile is in general more rufous, with barred wings and tail, plain or weakly barred mantle, and has fine pale buff spots on head and nape.

Call Utters a series of musical *kwühk kwühk kwühk kwühk* notes, the rate increasing after a soft beginning from one note to four notes per second, and the phrase lasting from five to 20 seconds.

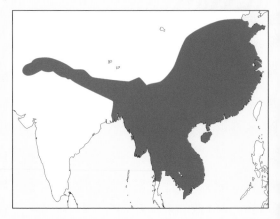

Food and hunting Feeds mainly on insects, such as beetles, grasshoppers and cicadas, but also takes lizards, small mammals and birds. Sometimes seizes birds in the air in the manner of a hawk.

Habitat Inhabits pine forests with rhododendron at higher elevations, as well as evergreen jungle at lower elevations, but occurs also in gardens and parks near human habitations. From mountain foothills up to higher levels; to 2700m in Himalayas of N. Pakistan.

Status and distribution Found from W. Himalayas east to Burma, E. Bangladesh and E. & SE China, south to Hainan and SE Asia (except S. peninsular Thailand and peninsular Malaysia). Reported as rather common.

Geographical variation Five subspecies are listed: nominate *cuculoides* occurs from NE Pakistan and Kashmir to E. Nepal and W. Sikkim; *T. c. rufescens*, from E. Sikkim and Bhutan through Bangladesh to N. Laos and N. Vietnam, is much richer rufous; *T. c. bruegeli*, from Tenasserim, in S Burma, Thailand, S. Laos, Cambodia and S. Vietnam, is clearly smaller and darker brown above; *T. c. whitelyi*, from Sichuan, Yunnan and SE China, is long-winged; *T. c. persimilis*, from Hainan Island, is even more rufous than *rufescens*. This owl is given species status according to the biological species concept, but its biology, behaviour and vocalisations are in need of further research. Study is required also of the taxonomy of the whole Jungle Owlet and Asian Barred Owlet complex.

Similar species Sympatric Collared Pygmy Owl (168) is much smaller, with a spotted crown and an occipital face. Slightly overlapping Jungle Owlet (190) is a little smaller, and is barred (not streaked) below.

◀ Asian Barred Owlet *whitelyi* is the largest race, and very dark like the nominate. Hong Kong, April (*Martin Hale*).

◀ ▶ Nominate Asian Barred Owlet is partly sympatric with slightly smaller Jungle Owlet, which is unstreaked below. Northern India, January (*Harri Taavetti*).

◀ This owl is race *rufescens* which is much richer rufous than the nominate. Arunachal Pradesh, India, February (*Neil Bowman*).

▶ Race *bruegeli* has very dark brown upperparts, a white belly and large white scapular patches. Kaeng Krachan, Thailand, January (*Neil Bowman*).

194. SJÖSTEDT'S OWLET
Taenioglaux sjostedti

L 25–28cm, Wt *c.* 140g; W 152–168mm

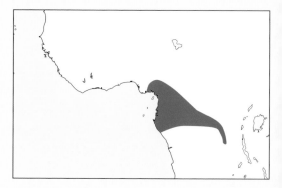

Identification A small owl without ear-tufts. Females are reported as being a little larger and heavier than males. Has dusky brown forehead, crown and nape, all finely and densely whitish-barred, contrasting deep chestnut mantle and back, some feathers on upper back with light edges. Dark brown greater upperwing-coverts are whitish-tipped and washed with chestnut, and outer webs of scapulars have fine pale cinnamon to whitish barring. Brown or blackish flight feathers are finely whitish-barred, and the blackish-brown tail has very narrow whitish bars. The throat is plain white, and the underparts are cinnamon-buff with narrow dark brown barring; light yellowish-rufous underwing-coverts and undertail-coverts. Indistinct facial disc is dark brown with thin whitish barring, and thin white eyebrows. The eyes are yellow, and the bill and cere pale yellow. Tarsi have cinnamon-rufous feathering to the base of the light yellow, sparsely bristled toes, which have horn-coloured claws with darker tips. *Juvenile* Downy chick undescribed. Mesoptile resembles adult, but is paler, with barring on upper breast and flanks darker, and has buff-edged scapulars.

Call Not well known, but male utters two to four *kroo-kroo-kroo-kroo* notes, the phrase lasting two seconds,

and repeated at intervals of one or two seconds. Sings mainly at dawn and dusk, but occasionally by day (as do all African members of the genus).

Food and hunting Feeds mainly on insects, such as grasshoppers and dung beetles, but also on crabs, spiders, small rodents, reptiles and birds. It plunders open nests of small birds with young. Otherwise, normally catches its prey from a low perch.

Habitat Inhabits humid primary lowland forests, avoiding forest edges. Reaches higher elevations locally, for instance on Mt Cameroon.

Status and distribution Distributed in Nigeria, Cameroon, Equatorial Guinea, Gabon, Congo and DR Congo. It is said to be rather frequent in Gabon, but is doubtless threatened by forest-logging as it prefers the interior of primary forests. Generally very little studied.

Geographical variation Monotypic.

Similar species Allopatric African Barred Owlet (196) is smaller, with whitish underparts dark-barred only on upper breast and boldly dark-spotted below. Also geographically separated, Chestnut Owlet (198) has a spotted crown, plain rufous-chestnut back, spotted whitish lower breast, and the tail densely barred brown and buff.

◀ Sjöstedt's Owlet is a very large owlet with a barred dusky brown and white head, neck and face. Breast is cinnamon with some fine brown barring. Belly almost free of barring, and cinnamon-white. Red-chested Pygmy Owl *Glaucidium tephronotum* is much smaller and has a plain crown, an occipital 'face', and is spotted below. Cameroon, March (*Nik Borrow*).

195. ETCHÉCOPAR'S OWLET
Taenioglaux etchecopari

L 20–21cm, **Wt** 83–119g; **W** 123–132mm

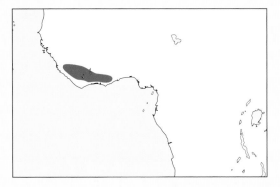

Identification A small owl without ear-tufts and 'occipital face'. Females are at least 10g heavier than males. The plumage is relatively dark brown above, spotted and slightly barred pale buff on the forehead, with fine, broken light bars on crown and nape, and the dark brown mantle and back marked profusely with indistinct chestnut bars; no clear nuchal collar. Narrow whitish areas on outer webs of the scapulars form an inconspicuous line across the shoulder, and indistinctly whitish-edged webs of innermost feathers of greater wing-coverts are barely noticeable. Primaries and secondaries are rather indistinctly barred lighter and darker brown, and the dark brown tail is slightly marked with narrow brownish-yellow bars. The brown foreneck and upper breast are barred with buffish-ochre, and the whitish underparts have relatively large, almost triangular dark brown spots, particularly on sides of breast and flanks. Inconspicuous brownish facial disc has paler concentric lines. The eyes are yellow, and the bill and cere greenish-yellow. Tarsi are feathered off-white to the base of the greenish-yellow toes, which have horn-coloured claws with darker tips. *Juvenile* Undescribed.

Call Not well known, but the male has a series of purring notes, reported as being different from those of Chestnut Owlet (198).

Food and hunting Not studied, but feeds mainly on large insects and small vertebrates. Hunts normally from a perch.

Habitat Inhabits primary and old secondary forests with high trees, but lives also in heavily logged forests.

Status and distribution Occurs only in Liberia and Ivory Coast, and possibly in Ghana as well. This owl's status is not well known, but it may be not so rare in Ivory Coast. Research is needed on all aspects.

Geographical variation Monotypic.

Similar species Partly overlapping Red-chested Pygmy Owl (167) has an occipital face and unmarked dark greyish head, as well as orange-rufous breast sides. Allopatric Chestnut Owlet (198) has bright chestnut back, is more spotted than barred, with a ferruginous head, and has a prominent nuchal collar of ochre and brown bars.

▶ Etchécopar's Owlet is patchily distributed in Liberia and Ivory Coast. This photo was taken in Ghana; it has a barred head and seems good for Etchécopar's, extending the known range of this owl. December (*Ian Merrill*).

196. AFRICAN BARRED OWLET
Taenioglaux capense

L 20–22cm, Wt 81–139g; W 131–150mm, WS 40–45cm

Identification A small owl without ear-tufts or 'occipital face'. Females are on average 10g heavier than males. Has greyish-brown to dark earth-brown head and nape fairly densely marked with fine whitish bars, and dark yellowish-brown mantle to uppertail-coverts narrowly buff-barred. Large dark-tipped whitish areas on outer webs of yellowish-brown scapulars form a distinct whitish row across the shoulder. Flight feathers are barred yellowish-brown and reddish-brown, and the greyish-brown tail has rather dense pale buff barring. Greyish-brown throat and upper breast have dense buffish-whitish barring, and the rest of the off-white underparts have a dull yellowish wash, and large dark brown dots at tips of many feathers; whitish-buff underwing-coverts with some brown spots. Pale brownish facial disc has white concentric lines, with not

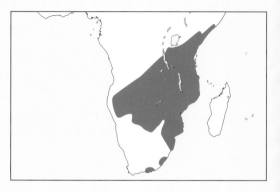

very prominent whitish eyebrows. The eyes are yellow, and the bill and cere greenish-grey with a yellowish tint. Tarsi are feathered whitish with a rufous wash, and

◀ This bird is the race *ngamiense* which is smaller and paler than the nominate. Chobe, Botswana (*Martin B. Withers*).

▶ The east African race *scheffleri* is sometimes considered specifically distinct due to morphological differences. Mikumi, Tanzania, April (*Adam Scott Kennedy*)

▶▶ Supposed *schefferi/ngamiense* intergrade. Has very prominent white scapular spots and yellowish-barred flanks and upper breast. Tanzania, November (*Nik Borrow*).

▶▶ Nominate *capense* is an isolated and rarely seen race. It is larger and darker than *ngamiense* with chocolate-brown upperparts; the wings are less prominently barred, the head is more finely spotted (not barred), and the tail has narrower barring. Eastern Cape, South Africa, November (*John Carlyon*).

bristled toes are brownish-yellow to yellowish-olive with horn-coloured claws with darker tips; talons are relatively small. *Juvenile* Downy chick is white. Mesoptile is like adult, but browner and less barred on back, and less spotted below. Juvenile has blackish tongue and gape. *In flight* Flies low with whirring wingbeats.

Call Utters a series of six to eight whistles, *cow-cow-cow…*, at rate of one per second, rising and falling in pitch and repeated after 15–20 seconds. The call of the nominate race is slower and clearer.

Food and hunting Takes small mammals, birds, reptiles, frogs, insects and other arthropods, such as caterpillars and scorpions. Hunts normally from a perch.

Habitat Inhabits riverine *Acacia* forests and open woods with large trees, and forest edges, but will also accept secondary growth. Normally lives below 1200m elevation. Nominate *capense* favours coastal scrub.

Status and distribution Found from S. Somalia through C. African woodland to Angola and Namibia, and in southeast to Mozambique and to Eastern Cape, in South Africa; range includes Mafia Island, off C. Tanzania. This species' status is not well known, and it could be threatened by forest destruction.

Geographical variation Three subspecies are listed: nominate *capense* occurs in the Eastern Cape, in South Africa; *T. c. ngamiense*, from C. Tanzania, Mafia Island and SE DR Congo to S. Angola, Namibia, N. Botswana, NE South Africa (Mpumalanga) and S. Mozambique, is paler and smaller than nominate; *T. c. scheffleri*, from extreme S. Somalia south through C. & E. Kenya to NE Tanzania, is also smaller and has a brown head, evenly spotted white facial disc with faint greyish barring, more prominent white eyebrows, darker brown upperparts with less barring, and creamy-white upper breast with rusty-brown bars. The subspecies *ngamiense* and *scheffleri* are sometimes considered specifically distinct but nominate *capense* is poorly known. Thus, the taxonomy of this owl, as well as its biology, ecology and behaviour, needs further research.

Similar species Sympatric Pearl-spotted Owl (166) is smaller and is streaked below, with a spotted crown and back. Geographically separated Sjöstedt's Owlet (194) is larger and long-tailed, with heavier talons, a chestnut mantle, dense barring below, and tail with narrow whitish bars. Overlapping scops owls have longer wings, small ear-tufts and a cryptic plumage below.

◄ African Barred Owlet pair of the race *ngamiense*. Paler and smaller than nominate, the female is a little larger with more spots on the breast. Mpumalanga, South Africa, May (*Niall Perrins*).

197. ALBERTINE OWLET
Taenioglaux albertina

L 21cm, Wt 73g (1♀); W 126–138mm

Identification A very small owl without 'occipital face' and ear-tufts. No data on sexual size differences, as only one female weighed. Above, the plumage is predominantly warm maroon or maroon-brown, with variously sized creamy spots on forehead, crown and nape, and yellowish-white spots on hindneck several of which are horizontally elongated, almost forming bars or fine scales, and intergrading into creamy bars on uppermost mantle; otherwise, mantle and back are plain maroon-brown. Small yellowish-white areas on outer webs of scapulars form a line across the shoulder. Flight feathers are barred lighter and darker brown and creamy-spotted, but outermost three primaries plain brown; dark brown tail has seven narrow whitish or yellowish bars. White chin contrasts with brown throat, and similarly brown upper breast has broad yellowish-white bars; whitish remaining underparts are marked with fairly large maroon spots, particularly on flanks. The brownish facial disc has some paler flecks, and narrow whitish eyebrows. The eyes are yellow, and the bill and cere yellowish. Tarsi are feathered, and bristled yellowish toes have horn-coloured claws with darker tips. *Juvenile* Unknown.

Call Undescribed.

Food and hunting Not studied, but some insects have been recorded as food.

Habitat Found in humid mountain forests with rich undergrowth, from 1000m to 1700m.

Status and distribution Endemic in Albertine Rift region of NE DR Congo and N. Rwanda. Thus far

known only from five specimens. Listed as Vulnerable by BirdLife International. The entire biology, behaviour and vocalisations of this owl require research.

Geographical variation Monotypic. Taxonomic relationships are unknown.

Similar species Partly sympatric Pearl-spotted Owl (166) has a prominent occipital face and streaked underparts. Also partly overlapping Red-chested Pygmy Owl (167) has a dark grey head with unspotted crown, plain dark brown mantle and back, and tail with large, rounded white spots. Allopatric African Barred Owlet (196) is a little larger, with pale-barred crown, dull yellow barring on mantle and back, and the tail densely barred brown and buff. Sympatric Chestnut Owlet (198) has a plain chestnut mantle and more prominent creamy outer webs of the scapulars.

198. CHESTNUT OWLET
Taenioglaux castanea

L 20–21cm, **Wt** *c.* 100g; **W** 128–139mm

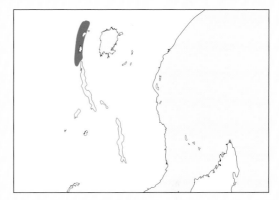

Identification A small owl without 'occipital face' or ear-tufts. Females are reported as being mostly larger and heavier than males. The plumage is deep reddish-brown above, the crown and mantle whitish-spotted and back plain chestnut, with whitish outer webs to scapulars. Flight feathers are barred brown and yellow, and the brown tail has dull yellow barring. The upper breast is densely barred brown and buff, and the remaining off-white underparts are heavily brown-spotted. The brownish facial disc has paler bars and flecks, and whitish eyebrows. The eyes are yellow, and the bill and cere greenish-yellow. Tarsi are feathered, and the bristled toes are dirty yellow to greenish-yellow, with horn-coloured claws with darker tips. *Juvenile* Undescribed.

Call Emits a melancholic series of whistled, rolling *kyurr-kyurr-kyurr...* notes, accelerating towards the end.

Food and hunting Feeds on small mammals, birds and other vertebrates, but also on insects and other arthropods. It hunts normally from a perch.

Habitat Found in humid lowlands, rainforests and other primary forests, from near sea level up to 1700m.

Status and distribution Occurs only in NE DR Congo and SW Uganda. Its status is not known.

Geographical variation Monotypic. The taxonomy requires confirmation, as this owl was earlier considered only a race of African Barred Owlet (196). The African owlets of the genus *Taenioglaux* may be found, after bioacoustical and molecular studies, to consist of more species than are currently recognised, or the results of such studies could indicate that we should include all these poorly known species as subspecies of *T. capense*. Etchécopar's Owlet (195) has been treated more firmly as a distinct species because of its well-separated geographical distribution.

Similar species Sympatric Red-chested Pygmy Owl (167) has a dark grey head contrasting with dark brown back, plain upperparts and a tail with large white spots. Slightly allopatric African Barred Owl (196) is larger and has a mainly pale-barred crown, with spotting (if any) limited to forehead, and a buff-barred mantle and back. Sympatric Albertine Owlet (197) has creamy-spotted forehead and crown, yellowish-white bars on nape and upper mantle, plain maroon-brown upperparts, and a brown tail with seven thin whitish bars.

◄ Chestnut Owlet is fairly similar to race *scheffleri* of African Barred Owlet *Taenioglaux capense* but it has a brighter chestnut-coloured back. Head is spotted and not barred. Tail is much longer than that of Albertine Owlet *Taenioglaux albertina*. Specimen from BMNH (Tring), collected in Bwamba, W. Uganda, September (*Nigel Redman*).

199. LONG-WHISKERED OWL
Xenoglaux loweryi

L 13–14cm, **Wt** 46–51g; **W** 100–105mm

Other name Long-whiskered Owlet
Identification A tiny owl without 'occipital face' or ear-tufts, but with long, fan-like whiskers around base of bill and on sides of facial disc. No known sexual size differences. Above, the plumage is warm brown from crown to uppertail-coverts, all densely vermiculated dark brown to blackish; large whitish spots form a collar on the lower nape, and distinct whitish subterminal spots are evident on outer webs of scapulars. Primaries are dull black, the edges of the outer webs with small light spots and the bases of the inner webs with irregular whitish areas; the dull brown tail is mottled lighter and darker. The underparts have a ground colour similar to that of the upperparts, but with many whitish vermiculations, these denser towards the belly. The brown facial disc is not prominent, but specialised long whiskers are present around the bill and even longer whiskers at the sides of the disc project clearly beyond its edge; has narrow white eyebrows. The eyes are orange-brown to amber-orange, with blackish-brown eyelids, the bill is yellow-tipped greenish-grey and the cere pinkish-grey.

The bare tarsi and toes are flesh-pink in colour, with dark-tipped horn-coloured claws. *Juvenile* Unknown.
Call Not well known. A deep, husky *woh*, dropping very slightly at end (sounding almost disyllabic), repeated at rate of one per three seconds, has been recorded. A series of three to five similar whistles followed by a series of faster, slightly higher-pitched notes could be the song.
Food and hunting Not studied, but diet may consist chiefly of insects.
Habitat Inhabits humid cloud forests with very dense undergrowth, from 1900m to 2350m.
Status and distribution Endemic in Río Mayo valley, in the E. Andes of N. Peru. This species has been recorded at just two remote localities; so far, five specimens have been collected, and one or more others have been mist-netted during the night. Reports of calling individuals and one sight record made at other sites in the same area. Very little studied, and listed as Endangered by BirdLife International.
Geographical variation Monotypic. Proposed relationship with *Glaucidium* owls seems unlikely.
Similar species The only owl of its kind. Within its tiny range, all pygmy owls are larger and long-tailed, with a distinct occipital face and streaked underparts.

◄ Long-whiskered Owl lives in humid cloud forests. Long, fine whiskers around the base of the bill and on the sides of the face project beyond the edge of the facial disc. Tarsi and toes fleshy pink. Abra Patricia, Peru, October (*Roger Ahlman*).

► This is the second smallest owl in the world; it has entirely finely-barred, warm brown plumage, with clear yellowish-white 'eyebrows'. Bill is greyish with a light yellow tip. Abra Patricia, Peru, November (*Dubi Shapiro*).

200. ELF OWL
Micrathene whitneyi

L 12–14cm, **Wt** 36–48g; **W** 99–115mm

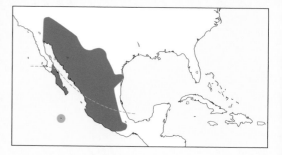

Identification A tiny owl without 'occipital face' and ear-tufts. By weight surely the smallest owl in the world. Females are on average some 4–5g heavier than males. The general coloration varies between more greyish and more brown, but no real morphs have been separated. The plumage is greyish-brown above, densely vermiculated lighter and darker, with some light yellow spots on the forehead, and narrow whitish nuchal collar on nape. Whitish spots are apparent on the wing-coverts, and more obvious whitish outer webs of scapulars form a prominent white row across the shoulder. Flight feathers are barred whitish and ochre-buff, and the tail, which has only ten rectrices, has three or four narrow pale bars. Whitish underparts are densely mottled and vermiculated with greyish-brown and cinnamon, from a distance appearing fairly plain grey-brown. The brownish facial disc is diffusely vermiculated, with narrow whitish eyebrows. The eyes are yellow, with blackish-edged eyelids, the bill greyish-horn with yellowish-horn tip, and the cere greyish-brown. Bristled tarsi and toes are greyish-brown, with dark horn-coloured claws. *Juvenile* Downy chick is whitish. Mesoptile resembles adult, but has unspotted greyish forehead and crown, indistinct light spots on upperparts, and is mottled grey and whitish below.

◄ A browner Elf Owl partly showing the dark brown upperparts; yellow eyes are prominent. Arizona, USA, May (*Paul Bannick*).

► Elf Owl in flight shows clean white spots on the flight feathers and the tail is spotted buffish-white. Arizona, USA (*Scott Linstead*).

Call Utters a rapid sequence of short, yelping whistles, *guwewiwiwiwiwiwiwirk*, each phrase containing up to 20 slightly accelerating notes, and phrases repeated at intervals of several seconds.

Food and hunting Diet consists almost entirely of insects and other arthropods, such as spiders; it sometimes takes small snakes and rodents. Hunts usually from a perch, but also hawks prey in flight or comes to lights or to camp fires to catch insects. Sometimes hovers over prey before swooping down.

Habitat Prefers open semi-deserts with giant cacti, but occurs also in semi-open bushlands or swampy grounds, from sea level up to 2000m.

Status and distribution Found from SW United States south to C. Mexico, including Baja California and Socorro Island. This owl is locally not rare but is almost extirpated in SE California. It winters in N. & C. Mexico. May suffer adversely from the use of pesticides.

Geographical variation Four subspecies are listed: nominate *whitneyi* occurs from SW USA to N. Mexico, wintering to C. Mexico; *M. w. idonea*, from S Texas to C. Mexico, is greyer above than nominate; *M. w. sanfordi*, from S. Baja California and W. mainland Mexico, is darker below than *idonea*; *M. w. graysoni*, from Socorro Island, off W. Mexico, is more olive-brown above and its tail has broad cinnamon-buff bars. This last subspecies, however, is suspected to be extinct. Further research is needed on the taxonomic relationships of this species.

Similar species All sympatric screech owls are larger and have ear-tufts, and long-tailed pygmy owls have a distinct occipital face, feathered tarsi and are streaked below.

▲ A greyish Elf Owl showing hardly any brown on the facial disc or below (*S & D & K Maslowski*).

▶ Elf Owl from the side, showing the dark greyish-brown upperparts and yellow eyes. Arizona, USA, May (*Paul Bannick*).

201. FOREST SPOTTED OWL
Heteroglaux blewitti

L 20–23cm, **Wt** 241g (1♂); **W** 145–154mm

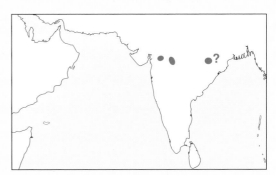

Other name Forest Owlet

Identification A small owl without ear-tufts. No data on sexual size differences. Has plain brownish-grey crown with a few miniature whitish speckles, the nape almost lacking 'occipital face' and with greatly reduced hind-collar; the head is very sparsely spotted and on many individuals appears unspotted, as are the mantle and back; a few whitish spots on scapulars and wing-coverts. Primaries are barred white and blackish-brown, and the dark brown tail has whitish bars and terminal band more than 5mm broad. Plumage is white below, with distinct dark throat-band (hidden in normal posture) and nearly uniformly grey-brown pectoral band on upper breast, pectoral band sometimes partially broken in centre, contrasting with large whitish throat patch and plain whitish centre of lower breast and belly; broad dark brown bars on flanks and sides of lower breast. The mostly white facial disc (almost

invisible in the field) has fine light brown to dark brown barring, with thin and straight white eyebrows; pale brown ear-coverts lack white rear border. The eyes are yellow, with black eyelids, the bill yellowish, and the cere black-tinged yellow. Tarsi are feathered pure white, and upper surface of the long, dirty-yellowish toes have soft whitish feathering, toes having relatively very heavy blackish claws. *Juvenile* Undescribed. *In flight* Direct, agile, strong and non-undulating flight, in which it appears larger, shorter-winged and longer-tailed than Spotted Little Owl (208).

Call Gives a series of short, melodious and querulous *kowóoh* whistles, each whistle lasting 0.14–0.4 seconds, series uttered at intervals of 4–15 seconds. Calls also during day.

Food and hunting Not well documented, but, because its skull and legs are more massive, it possibly takes larger prey than does Spotted Little Owl. Lizards, small mammals, larger insects and other invertebrates probably form the main diet.

Habitat Inhabits humid riparian jungle, mango groves and open deciduous forests in plains and hills, up to 500m above sea level.

Status and distribution Endemic in NC Indian peninsula. This owl was for a long time believed to be extinct, but was rediscovered in November 1997 in W. Kandesh and at Khaknar Forest. Later, 25 records (in 2000) from an intermediate area demonstrate that the owl still persists, but is obviously extremely rare. Listed as Critically Endangered by BirdLife International. In Melghat Tiger Reserve, in Maharashtra, fertile hybrids

◄ Forest Spotted Owl in an 'anxiety' position, with wings dropped, showing well the white throat, brown nuchal band and large white chin. Maharashtra, India, March (*Jayesh Joshi*).

have been found between Forest Spotted Owl and Spotted Little Owl, which co-exist in the forest.

Geographical variation Monotypic. This owl requires research on its biology and systematics, as its tail-flicking behaviour would indicate closer relationship with *Glaucidium* pygmy owls, rather than with *Athene* little owls, notwithstanding proven hybridisation with *Athene brama*.

Similar species Allopatric Little Owl (203) has streaked belly and Grey-bellied Little Owl (206) has grey and brown belly and no horizontal or longitudinal stripes on underparts. Allopatric Northern Little Owl (207) has clearer occipital face, spotted (rather than streaked) underside and even more feathered white legs. Sympatric Spotted Little Owl (208) is smaller, but has longer wings, much more spotted head, especially on crown, white behind auriculars, curved eyebrows, heavily barred and spotted centre of lower breast and belly, scattered subterminal spots on back, and lacks plain band across breast.

◄ Forest Spotted Owl in flight, showing its rounded wings and very short tail; note it has shorter wings and a longer tail than Spotted Little Owl *Athene brama*. Maharashtra, India, December (*Jayesh Joshi*).

▲ ► Immature Forest Spotted Owl has a less developed facial disc, but strong reddish-brown horizontal markings on the belly. Maharashtra, India, March (*Jayesh Joshi*).

◄ Forest Spotted Owl has a strong dark band on the chest, and a clear white collar below the yellow bill. Maharashtra, India, March (*Jayesh Joshi*).

► Forest Spotted Owl demonstrating leg- and wing-stretching. It has feathered tarsi and more massive feet than Spotted Little Owl *A. brama*. Maharashtra, India, December (*Ian Merrill*).

202. BURROWING OWL
Athene cunicularia

L 19–25cm, **Wt** 120–250g; **W** 142–200mm

Identification A small owl without ear-tufts and lacking an 'occipital face'. Females are clearly larger and nearly 30g heavier than males. Brown forehead and crown are marked with whitish streaks and dots, and brown upperparts are irregularly marked with whitish to pale ochre dot and relatively large, fairly rounded spots. Flight feathers are barred light and dark, and the brown tail has three or four pale bars. Below, the plumage is whitish to dull yellow, densely barred dusky brown. The pale brownish facial disc has a distinct whitish throat-band below and prominent white eyebrows above. The eyes are bright yellow, the bill greyish-olive and the cere greyish-brown. Conspicuously long tarsi are sparsely feathered, and the bristled toes olive-grey with dark horn-coloured claws with blackish tips. *Juvenile* Downy chick is light grey-brown to whitish. Mesoptile resembles adult, but has unspotted crown and more diffusely marked underparts, brownish-buff sides of breast, and whitish facial disc with dusky zones on outer edge of eyes.

◄ The ground-dwelling Burrowing Owl will sometimes perch in a tree. This subspecies, *grallaria*, is relatively dark above with a rufous wash and small scapular spots. Dark below, with a distinctly spotted chest-band. Lower breast and belly have a few rusty bars and a buffish wash. Itirapina, Brazil, September (*José Carlos Motta-Junior*).

Call Gives a hollow, plaintive *coo-coo oo*, repeated at intervals of several seconds.

Food and hunting Feeds mainly on beetles and other insects, spiders, scorpions, small mammals, amphibians, reptiles and occasionally small birds. Normally hunts from a fence post or pole, or by walking or hopping on the ground. It will also hawk insects in the air.

Habitat Occurs in open country, savanna, desert, grasslands, pastureland and agricultural land, and also airports, golf courses and large city gardens. From sea level up to 4500m.

Status and distribution Found from W. North America south to Tierra del Fuego, in extreme S. Argentina. In many areas it is rather frequent but it is suffering from the effects of pesticide use and transformation of prairie landscapes into agricultural land.

Geographical variation Races vary mainly in biometrics, in depth of coloration and in strength and extent of markings; it is reported that North American subspecies are larger than Caribbean and tropical ones, but datasets for *floridana* and *hypugaea* show little size difference. The following 15 extant subspecies are listed: nominate *cunicularia* occurs from N. Chile and S. Brazil south to Tierra del Fuego; *A. c. grallaria*, from dry interior of Brazil to Paraná, is relatively dark above and has rusty bars and buffish wash below; *A. c. hypugaea*, from British Colombia east to C. Manitoba, in Canada, south to Mexico and W. Panama, has a broad chest-band and is whitish below; *A. c. floridana*, found patchily in E. USA, widely in Florida and Bahama Islands, has a narrow chest-band and is rather white below; *A. c. troglodytes*, from Hispaniola, including islands of Beata and Gonâve, is smaller and darker than *floridana*; *A. c. rostrata*, from Clarion Island, off west coast of Mexico, is poorly known; *A. c. brachyptera*, from N. & C. Venezuela, including Margarita Island, is very short-winged; *A. c. tolimae*, from W. Colombia, is very dark like *pichinchae*, but smaller; *A. c. carrikeri*, from E. Colombia, is extremely pale; *A. c. minor*, from savanna of upper Rio Branco, in Brazil, to French Guiana, Suriname and Guyana, is poorly known; *A. c. pichinchae*, from Andes of W. Ecuador, is dark grey-brown above and with rather dense dusky barring below; *A. c. punensis*, from coastal areas from SW Ecuador to NW Peru, is small and pale, with less extensive barring below; *A. c. nanodes*, from Pacific coast of Peru to northernmost Chile, is a small desert form; *A. c. juninensis*, from the Andes of Peru south to W. Bolivia and NW Argentina, has pale buff upperparts; *A. c. boliviana*, from arid habitats of Bolivia and N. Argentina, intergrades with nominate in Tucumán. Two subspecies from the Lesser Antilles, *amaura* (from Nevis and Antigua) and *guadeloupensis* (from Guadeloupe and Marie-Galante), are extinct. In captivity, hybrids between Little Owl (203) and Burrowing Owl have proved to be infertile, perhaps supporting that Burrowing Owl should have been left in its own genus *Speotyto*, in which it was placed until recently. New DNA evidence, however, indicates that *Speotyto* should be subsumed in *Athene*.

Similar species All sympatric screech owls have small ear-tufts. Long-tailed pygmy owls are smaller, with a prominent occipital face. Partly overlapping Cuban Bare-legged Owl (105) is much smaller, with totally bare legs.

▶ Race *nanodes* is small with large white spots on the back. This is said to be a 'desert' form, though it is not very pale. Lambayeque, Peru, August (*Christian Artuso*).

▼ Race *punensis* is paler than *nanodes*. Manabi, Ecuador, March (*Roger Ahlman*).

▲ Race *hypugaea* has a broad dark brown chest-band and almost white belly. Montana, USA (*Donald M. Jones*).

▲ ▶ Race *boliviana* is warm brown above with a yellowish tint on the brown-barred belly. Formosa, Argentina, October (*James Lowen*).

▶ A pair of Burrowing Owls of race *floridana* – the female is larger with fully feathered legs. This subspecies is a darker warm-brown than *hypugaea*, with a narrow, more sparsely spotted chest-band and a less white belly. Florida, USA, October (*Lesley van Loo*).

203. LITTLE OWL
Athene noctua

L 21–23cm, **Wt** 105–260g; **W** 146–181mm, **WS** 53–59cm

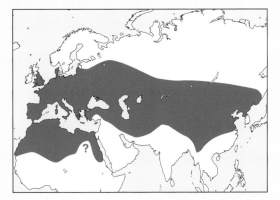

Identification A small, flat-headed owl without ear-tufts. Females are on average only 20–30g heavier than males in Europe, but in Asia they are up to 50g heavier. More rufous and more greyish morphs exist. The plumage is generally dark brown above, with whitish streaks and spots on forehead and crown, an indistinct 'occipital face' on nape, and heavy whitish spotting on upperparts. Flight feathers are barred whitish and dark brown, and the dark brown tail has a few whitish and pale ochre bars. Plain whitish throat is separated by a narrow brown collar from the diffusely light- and dark-spotted neck and upper breast; whitish underparts are boldly streaked dark brown, with the belly relatively unmarked. The rather flat facial disc is greyish-brown with light mottling, a relatively distinct ruff, and prominent whitish eyebrows. The eyes are sulphur-yellow to pale yellow, the bill greyish-green to yellowish-grey, and the cere olive-grey. Whitish-feathered tarsi are relatively long; the bristled toes are light grey-brown, with dark horn-coloured claws with blackish tips. *Juvenile* Downy chick is white, slightly grey-mottled on the upperside. Mesoptile resembles adult, but has fluffier plumage with more diffuse markings, less distinct spotting, and yellowish-grey eyes. *In flight* It has an undulating flight in which rapid wingbeats alternate with glides.

Call A fluted, slightly nasal *goooek hoót*, with upward inflection, is repeated at intervals of several seconds.

Food and hunting Feeds chiefly on insects, especially beetles and grasshoppers, but also on other arthropods, small reptiles and frogs, earthworms and small mammals and birds. Hunts normally from a perch, by swooping down on to prey, but can also run after prey on the ground.

Habitat Inhabits open country, steppes, rocky terrain, pastureland, and gardens with fruit trees, often near farmhouses and human settlements, even in towns. Occurs from sea level to 4600m.

Status and distribution Widely distributed from Denmark south to N. Africa and east to C. Asia. Introduced in New Zealand and England, from where

► Little Owl of the race *bactriana* is clearly white-spotted above and white below with sandy-brown streaks. Iran (*Ali Sadr*).

it spread to S. Scotland and Wales. This owl is locally quite common but it suffers from effects of pesticide use, road traffic and severe winters. In C. Europe the estimated population is 25,000 pairs.

Geographical variation Seven subspecies are listed, separated mainly by colour and size: nominate *noctua* occurs from C. Europe east to NW Russia; *A. n. vidalii*, from W. Europe (Spain to Belgium) and also introduced in UK and New Zealand, is darker than nominate; *A. n. glaux*, from N. Africa (Morocco and Mauritania to Egypt), is more cinnamon-brown; *A. n. orientalis*, from Siberia to NW China, is not well known; *A. n. ludlowi*, from Tibet to N. Bhutan, is larger than nominate and chocolate-brown above; *A. n. indigena*, from Crete and Greece to Ukraine, east to Caspian Sea and south to Asia Minor, is paler and more russet-grey; *A. n. bactriana*, from E. Iraq to SE Azerbaijan, Afghanistan, east through C. Asia to Tien Shan and Lake Balkash, in Kazakhstan, is sandy-brown above and white-spotted. Taxonomy requires research, as some subspecies may deserve full species status (note that *A. lilith* (209), *A. spilogastra* (205) and *A. plumipes* (207) have already been upgraded to the rank of full species).

Similar species Partly overlapping Lilith Owl (204) is smaller and has smaller, lemon-yellow eyes, and much paler or whitish crown and nape than any desert races of Little Owl; it is locally sympatric with races *indigena*, *glaux* and *bactriana*. In China, sympatric Grey-bellied Little Owl (206) is similar in length but much lighter in weight, and, as the name suggests, has a grey belly. Partly overlapping Northern Little Owl (207) is similar in size, but has bright yellow eyes and heavily feathered legs, with toes more densely covered with plumes (rather than bristles). Largely allopatric Spotted Little Owl (208) has bright yellow eyes and is diffusely barred below.

▼ The very pale Little Owl race *saharae*; this race is often placed within the darker race *glaux,* covering the whole of North Africa. Morocco, April (*Juan Matute*).

▼▼ Race *ludlowi* of Little Owl is larger than the nominate; it has a much wider distribution in southern and central China than was previously known, and overlaps with Grey-bellied Little Owl *Athene poikilis,* but it has a much shorter tail. Sichuan, China, August (*Christian Artuso*).

▼ Race *indigena* from Turkey is much paler than the nominate Little Owl and interbreeds freely with Lillith Owl *Athene lilith*. Muradiye, Turkey, May (*Daniele Occhiato*).

▼ This Little Owl from the Netherlands is of the very dark race *vidalii*, which is even darker than the nominate. It is dark fuscous-brown above with white spots, and umber-brown below with white streaks and spots. June (*Lesley van Loo*).

204. LILITH OWL
Athene lilith

L 19–20cm; W 152–164mm

Other name Lilith Owlet
Identification A small owl without ear-tufts. No data on sexual size differences. Has a pale yellowish-red to whitish-buff crown indistinctly streaked, blurred mottling with some indistinct streaks on sides of head, and distinct 'occipital face' on nape. Mantle, back and upperwing-coverts are pale sandy brown, many outer webs of feathers whitish; relatively large whitish spots on outer webs of scapulars do not form a very clear row. Slightly darker primaries and secondaries have whitish bars, and the sandy-brown tail has about six pale bars. White throat is bordered above by some fine darker lines; neck and upper breast are mottled with ochre-buff and yellowish-white, below which the creamy-white to whitish-buff underparts have not very prominent light brownish-buff streaks. The rather indistinct facial disc is pale buff with some brownish mottling and speckles, a very inconspicuous ruff and white eyebrows. The eyes are yellow, with prominent blackish rims of eyelids, the bill greenish-yellow with a yellow tip. Whitish-buff tarsi (shorter than those of

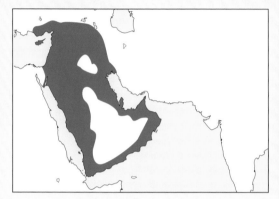

Little Owl) are feathered to the base of the yellowish-grey, sparsely bristled toes, which have horn-coloured claws with darker tips. *Juvenile* Undescribed.
Call Utters a drawn-out and slightly hoarse *gwüüh* or *gwüah*, without Little Owl's upward inflection and nasal character.
Food and hunting Feeds mainly on larger insects and other arthropods, such as spiders and scorpions, but also takes reptiles, frogs, and small rodents and birds.
Habitat Found in semi-deserts, rocky deserts, river valleys, ruins and buildings near human settlements. From sea level up to mountains with sparse vegetation.
Status and distribution Occurs from Cyprus and S. Turkey to Sinai and Arabian Peninsula. Appears to be locally not rare, but studies on its biology, vocalisations and ecology are urgently needed. In captivity, Little and Lilith Owls have successfully produced fertile young, which, in turn, have produced fertile offspring.
Geographical variation Monotypic. The production of fertile hybrids with Little Owl would be expected if Lilith Owl were only a subspecies of the latter, but recent DNA data from S. Turkey indicates that *lilith* is specifically distinct from *Athene noctua*; in addition, the vocalisations of S. Turkey *lilith* differ from those of the Little Owl.
Similar species Partly sympatric Little Owl (203) is a little larger and much darker, with a more distinct ruff, and longer legs. Geographically separated Ethiopian Little Owl (205) is slightly smaller and browner, and has creamy or whitish-buff underparts with darker mottling on the upper breast and some pale brown streaks.

◄ Adult Lilith Owl; note the long, fully feathered legs. Sanliurfa, Turkey, May (*Daniele Occhiato*).

▶ All very pale *Athene* owls from Israel to southern Turkey are considered to be Lilith Owls; other forms represent integrading populations between *A. lilith* and Little Owls of races *A. noctua indigena, saharae* and *glaux*. Nizzana, Israel, March (*David Jirovsky*).

▼ Lilith Owl has big yellow eyes, and is less streaked below than Little Owl *A. noctua*. Tel Arad, Israel, October (*Amir Ben Dov*).

▼▶ Note how well this pale Lilith Owl blends with its surroundings. Mirbat, Oman, March (*Hanne and Jens Eriksen*).

205. ETHIOPIAN LITTLE OWL
Athene spilogastra

L 18–19cm; W 129–147mm

Identification A very small owl without ear-tufts. No data on sexual size differences. The sandy-brown forehead and crown have distinct darker brownish streaks, the sides of the head are mottled brownish and creamy, and there is an indistinct 'occipital face' on the nape. Light brownish-buff mantle and back are mottled with a number of yellowish-white dots. Large dull yellowish areas on outer webs of scapulars form a comparatively indistinct light row across the shoulder, and pale brownish-buff feathers of the upperwing-coverts have creamy-buffish distal parts, forming two visible whitish rows on the closed wing. Flight feathers are a little darker than wing-coverts, and barred yellowish-white, and the brown tail has about five distinct yellowish-white bars. The throat is whitish, and the upper breast densely mottled brownish-buff and creamy, shading into dull yellowish; some indistinct brownish streaks

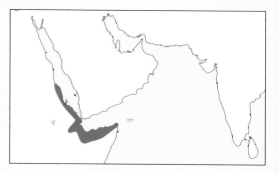

are present on sides of breast and flanks, but largely unmarked below. Fairly indistinct facial disc is sandy-coloured with fine brown mottling, the ruff not conspicuous, but white eyebrows prominent and reaching far beyond the eyes. The eyes are yellow with blackish rims of eyelids, the bill greenish-yellow, and the cere dirty greenish-grey. Brownish-buff tarsi are sparsely feathered, distally more bristled than feathered, and the light brownish-grey toes have blackish-tipped dark horn-coloured claws. *Juvenile* Undescribed.

Call No detailed information available, but said to differ from that of Little Owl (203).

Food and hunting Feeds mainly on insects, centipedes, spiders and scorpions, but obviously also takes some reptiles and small mammals. May plunder nests of ground-breeding birds.

Habitat Prefers open areas with termite mounds and rocky areas in semi-deserts.

Status and distribution Occurs on western coast of Red Sea from E. Sudan and south to N. Somalia. Status is not known, but this owl could be endangered by the effects of pesticides.

Geographical variation Monotypic. Research is required in order to confirm the correctness of separating this owl from the Little Owl complex.

Similar species Allopatric Little Owl (203) is larger and stouter, with a more distinct ruff around the facial disc. Also geographically separated, Lilith Owl (204) is a little larger and much paler, with distinctly shorter toes.

◀ Race *somaliensis* from Somalia is like a 'dwarf' form, with thinly bristled legs. It is darker than nominate *spilogastra*, with fewer white markings on the breast. Note strong white 'eyebrows' above the white face and large, yellow eyes. N. Somalia, September (*Nik Borrow*).

206. GREY-BELLIED LITTLE OWL
Athene poikilis

L 22–24cm, **Wt** 87–115g; **W** 166–169mm

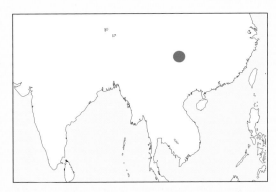

Identification A very small owl without ear-tufts. Females are nearly 30g heavier than males, but males have longer wings; the sexes are very similar in plumage, but the female has the brown colour more intense and it has less white-grey under the dark brown throat. The top of the relatively small head is covered with grey-white spots, and the forehead and cheeks are white, with blackish-brown ear-coverts. The dark brown upperparts have greyish-white spots, and the wing-coverts have white and grey dots on a dark brown background; the underwing is whitish. The 3rd and 4th primaries are the longest. The dark brown tail is about two-thirds the length of the wing and has greyish-white bars less than 4mm wide. Under the neck there is a grey band and a white collar, which are not solid; the chin is dark brown, and the chest, abdomen and flanks are brown mixed with grey, but without longitudinal or horizontal stripes. The colour of the black-rimmed eyes is not known; the bill is pale yellow, the surrounding whiskers black. The feathered legs are white, dashed with brown and grey, and the toes also are feathered, with black-brown claws; legs and claws are thinner and weaker than those of Little Owl (203). *Juvenile* Undescribed.

Call Not known.

Food and hunting Not studied.

Habitat Lives in mountain forests between 2200m and 3100m. Little Owl is said to occur in W. Sichuan at 3500m or higher; Spotted Little Owl (208) is normally found from sea level up to 1500m.

Status and distribution This recently described owl occurs in Yahong and Baosin districts of W. Sichuan Province, in China. Two holotypes collected in 1964 (1♂) and 1982 (1♀), are kept in the Suzhou Agronomic Institute (male) and in the China Academy of Sciences, Kunming Institute of Animal Research (female). It is not known if any recent records exist.

Geographical variation Monotypic. The holotype description compares this new owl carefully with the Chinese subspecies of Little Owl, and with Northern Little Owl (207; at that time treated as a race of Little Owl), as well as with Spotted Little Owl (208) and Forest Spotted Owl (201); the clearest differences from all are in measurements and the complete lack of horizontal or longitudinal markings below. It was decided to include this owl as a new species because the holotype description is very professional, and it is possible that it has been overlooked (since 1988) only because of language and political barriers. In any case, the entire group of *Athene* owls is in need of further taxonomic study.

Similar species Allopatric Forest Spotted Owl (201) has no 'occipital face' but is much heavier, and has broad dark brown bars on the flanks and sides of lower breast. Sympatric Little Owl (203) is similar in size, but has white longitudinal spots above a white nuchal band and is boldly streaked dark brown below. Geographically separated Northern Little Owl (207) has slightly longer wings, a clear occipital face, and white underside with large brown spots. Allopatric Spotted Little Owl (208), living farther south and at lower elevations, has white double-spots above a white nuchal band, and grey to brownish scaly barring on creamy-white underparts.

207. NORTHERN LITTLE OWL
Athene plumipes

L *c.* 22cm; W 158–179mm

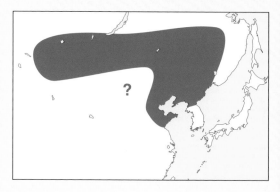

Identification A small owl without ear-tufts, although strong and long white eyebrows can be raised to give appearance of small ear. Females are larger than males, with wings on average 10mm longer. Above, the plumage is reddish-brown with large white spots, especially on scapulars. The head is darker brown than the back and with numerous small white dots; the neck has an 'occipital face' composed of yellow-white longitudinal spots above a slightly less clear yellowish-white nuchal band, with two white nuchal collars and a brown collar between them. The wings have whitish-yellow spots in six or seven lines, and the tail has three or four yellowish spot-lines. Below, it is very white with broad brownish spots, the flanks have more reddish-brown than the breast, and the belly and undertail are fluffy white. The facial disc is not well developed, but has strong white comma-shaped patches around the large eyes, the white area with a very thin black zone next to the eyes, and long white eyebrows. The eyes are bright yellow, and the bill is horn-yellowish and surrounded by red-brown bristles. The legs are fully white-feathered, and the toes also are more densely covered with plumes rather than bristles, with dark claws. *Juvenile* Likely the same as Little Owl (203).

Call Not described.

Food and hunting This owl's diet has not been studied, but its northerly distribution suggests that small mammals may be important items.

Habitat Not described.

◄ Northern Little Owl has thickly feathered legs; even the toes are densely covered with plumes. This is a clear adaptation to the cold climate of Mongolia and northeastern China. Hebei, China, October (*Ian Fisher*).

► Northern Little Owl has two white nuchal collars, with brown between. It is not known how much this owl may overlap with Grey-bellied *Athene poikilis* in central China, *A. noctua orientalis* in northwest China or *A. n. ludlowi* in central and southern China. Hebei, China, October (*Ian Fisher*).

►► NE Mongolia is the northernmost part of this bird's distribution; here it does not overlap with any other *Athene* owls. Galshir, Mongolia, June (*Mathias Putze*).

Status and distribution Occurs from the Russian Altai to Inner and Outer Mongolia and to W., C. & N. China. Its status is not known.

Geographical variation Monotypic. Only recently separated from Little Owl, when it was found in molecular studies to have a distinct genetic lineage. It is not known how this new species, first described in 1870, is related to Grey-bellied Little Owl (206), described only in 1988, the molecular phylogeny of which has not yet been studied.

Similar species It is not known if this owl overlaps with somewhat larger Little Owl (203) of race *orientalis* in extreme NW China and adjacent Siberia; Little Owl, however, has lemon-yellow eyes, darker brown back and more horizontal streaks below. All other *Athene* owls have a more southerly distribution.

▶ On Northern Little Owl's wing, the 6th and 7th primaries are the longest. In Grey-bellied *Athene poikilis* the 3rd and 4th primaries are longest. Photo shows also the occipital 'face'; this is not mentioned at all in the description of *A. poikilis*. Hebei, China, October (*Ian Fisher*).

208. SPOTTED LITTLE OWL
Athene brama

L 19–21cm, **Wt** 110–115g; **W** 134–171mm

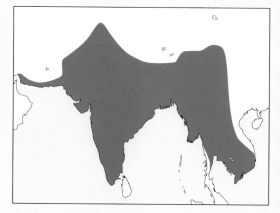

Other name Spotted Owlet

Identification A very small to small owl without ear-tufts. No data on sexual size differences. The earth-brown crown is heavily white-spotted, and the sides of head and the upperparts are grey-brown with irregularly shaped and scattered white spots (instead of streaks); an indistinct 'occipital face' is present on the nape. Broad white edges on the scapulars, but not forming a distinct row across the shoulder. The wings are banded and spotted with whitish, and the short tail has narrow white bars and a conspicuous white tip. Below, it has whitish-buff or cream-coloured chin, throat and sides of neck, and an indistinct pectoral band of brown and pale mottling is usually narrow and broken; creamy-white underparts have short grey to brownish bars, giving a scaly appearance. The darkish facial disc has brown concentric lines, and is bordered by dark brown cheeks and a rather prominent whitish ruff, and distinct curved white eyebrows meeting above the bill; the ear-coverts are white. The eyes are pale yellow to golden-yellow, the bill greenish-horn and the cere dusky greenish. Feathered tarsi and bristled toes are yellowish-green, with yellowish soles, and relatively small dark horn-coloured claws. *Juvenile* Downy chick

is pure white. Mesoptile has softer plumage than adult, less white and virtually unspotted crown and mantle, and below is uniformly dusky with indistinct darker streaks. *In flight* Deeply undulating flight consists of a few rapid flaps, followed by a drop on closed wings, then rising again by rapid fluttering.

Call Territorial call is a sequence of plaintive double whistles, *plew-plew*, lasting two to three seconds, these 'couplets' uttered at intervals of 15–25 seconds. It

▶ Spotted Little Owl differs from Little Owl *Athene noctua* in nape pattern and barred (not streaked) underparts. The large eyes are yellow, and the bill is greenish-yellow. Karnataka, India, January (*Ram Mallya*).

◀ The southeast Asian race *pulchra* of Spotted Little Owl is smaller and darker than the nominate. Thailand, March (*HY Cheng*).

also gives a medley of harsh screeches, chatters and chuckles.

Food and hunting Feeds mainly on insects, but also takes earthworms, lizards, small rodents and small birds. Hunts from a perch, and also seizes insects, such as winged termites, in flight.

Habitat Inhabits open forests and semi-open areas, including farmland and semi-deserts. Occurs also near villages and in towns, but avoids dense forests. From lowlands up to 1500m.

Status and distribution Found from S. Iran eastwards through India to SE Asia, excluding islands and the Malay Peninsula. This species is rather common, but its biology and vocalisations need more study.

Geographical variation Five subspecies are listed: nominate *brama* is found in S. India (but not in Sri Lanka); *A. b. indica*, from N. & C. India, partly intergrades with nominate; *A. b. albida*, from Iran and S. Pakistan, is slightly paler; *A. b. ultra*, from NE Assam to Brahmaputra and Luhit rivers, is somewhat larger; *A. b.*

pulchra, from Burma to SW Vietnam, is small and dark. Subspecies *ultra* has a different voice and a longer tail, details sometimes regarded as reasons for according it full species rank. In the UK, captive Spotted Little Owls have been mated with female Little Owls, and four or five such pairs all successfully hatched and reared chicks; all first-cross hybrids appeared to be sterile, confirming that these two owls represent separate species, despite their fairly close similarity in appearance.

Similar species Sympatric Forest Spotted Owl (201) is heavier-bodied, with more uniform upperparts, a sparsely spotted head with thin white eyebrows, an obsolete hindcollar, much broader, whitish bars on the tail, and heavily white-feathered, stout tarsi. Little Owl (203) overlaps in range in Baluchistan, but is a little larger, with longitudinally streaked crown and belly. Allopatric Grey-bellied Little Owl (206) has neither an occipital face on the nape nor any barring on the grey-brown belly.

▼ Adult Spotted Little Owl of race *indica* at the nest. Gujarat, India, January (*Martin Gottschling*).

▼ Spotted Little Owl has a broad white nuchal collar and a white throat patch. Karnataka, India, August (*Subharghya Das*).

▼ Spotted Little Owl of the southern Indian nominate race; this subspecies is darker than *indica*. Uttar Pradesh, India, February (*Hugh Harrop*).

A pair of Spotted Little Owl of the race *indica*. There is no visible difference between the sexes; *indica* is paler than the nominate *brama*. Haryana, India, February (*Amano Samarpan*).

These immature Spotted Little Owls are brownish, but close to the adult plumage. Rajasthan, India, January (*Harri Taavetti*).

209. TENGMALM'S OWL
Aegolius funereus

L 22-27cm, **Wt** 90-215g; **W** 154-192mm, **WS** 55-62cm

Other name Boreal Owl

Identification A small, round-headed owl without ear-tufts. During the breeding season females are on average 65g heavier than males. Sexes alike, but often coloration of facial disc gives a hint as to sex: female more uniformly ochre-whitish, male tinged greyish towards ruff. The plumage is greyish-brown to dark earth-brown above, with small white dots on forehead and crown, and larger white dots on back and wing-coverts; hindneck is mottled and spotted with dark and whitish markings. Rounded whitish spots are present on the primaries and secondaries, and the dark brown tail has four or five rows of whitish dots. Whitish under-parts have greyish-brown mottling and streaking. The whitish facial disc is round and surrounded by a dark ruff marked with tiny white spots, with a small dark zone between the eyes and the base of the bill. (If the owl is alarmed, the facial disc becomes elongated, with pointed corners at upper edge suggesting small ear-tufts, but this is due to lateral compressing of facial

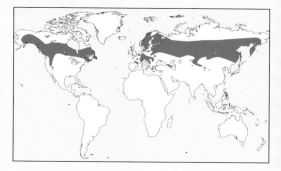

disc and is not analogous with true ear-tufts of other owl species.) The eyes are pale to bright yellow, with blackish-edged eyelids, and the bill and cere yellowish-horn. Densely feathered tarsi and toes, with dark horn-coloured to blackish-brown claws with extremely sharp tips. *Juvenile* Downy chick is white, but reddish skin shows through the thin down. Mesoptile is fairly uniformly dark chocolate-brown, with whitish eyebrows

► Tengmalm's Owl is very large-headed. Nominate *funereus* has dark brown stripes and white blotching below. Finland, April (*Harri Taavetti*).

◄ The Finnish name is 'pearl owl' due to the pearly-white spotted upperparts. Finland, April (*Harri Taavetti*).

and a streak on each side of chin. Juvenile initially has pale yellowish-grey eyes, these becoming more yellow before fledging. *In flight* Noiseless flight, unhurried, straight and with soft wingbeats.

Call Utters a rapid succession of *poo* hoots, normally five to seven notes in one phrase lasting two to three seconds, with inter-phrase interval of normally three to four seconds; perhaps reminiscent of the 'winnowing' or 'drumming' sound made by a displaying Common Snipe *Gallinago gallinago*. Singing bouts frequently last for 20 minutes, and can extend to two or three hours.

Food and hunting Feeds mainly on voles and shrews, but during 'poor' vole years it also takes birds. Swoops on to prey from a perch.

Habitat Prefers boreal and subalpine coniferous forests with mature trees, but occurs locally also in deciduous forests. Most important is the availability of suitable nesting holes. Occurs from plains up to 3000m, often at higher elevations than the Tawny Owl, which is not so resistant to cold and snow.

Status and distribution Has a Holarctic distribution in the coniferous belt, and locally in the mountains of Europe, Caucasus, Tien Shan, Himalayas and WC China; in North America (where this species is known as the Boreal Owl) it occurs south as far as New Mexico. It winters irregularly south to Japan and N. USA. This owl is generally fairly common, but some southern populations are threatened by deforestation; it suffers also from the effects of rodenticide use. In C. Europe the average estimated population is 9,800 pairs; in Finland there are 15,000 pairs during 'good' vole years and only 1,000 during the poor years. In the Old World it has suffered somewhat from the range expansion of Ural Owls, but it is not known if Boreal Owl has declined in North America because of the increase in Barred Owls (158) there.

Geographical variation Five subspecies are listed: nominate *funereus* occurs from Scandinavia to the Pyrenees, and east to Greece and Russia north of Caspian Sea; *A. f. magnus*, from NE Siberia to Kamchatka (and accidental in Alaska), is paler and larger; *A. f. pallens*, from W. & S. Siberia and Tien Shan east to Sakhalin and NE China, is paler and more spotted than nominate; *A. f. caucasicus*, from Caucasus east to NW Himalayas, NW India and mountains of W. China, is smaller and darker; *A. f. richardsoni*, from Alaska to New Mexico, is larger, darker and more boldly patterned than the nominate.

Similar species In Eurasia sympatric Little Owl (203) is similar in size, with a fairly flat facial disc marked with light and dark spotting and blotching, and bristled toes. Northern Saw-whet Owl (210) overlaps in North America, but is much smaller and has an indistinct ruff and blackish bill.

▲ Race *caucasicus* is smaller
and darker than the nominate
unereus. Gansu, China, June
(*John and Jemi Holmes*).

◀ Race *richardsoni* is warmer
brown above with smaller
white spots than the nominate.
Northern Minnesota, USA
(*Thomas Mangelsen*).

Fledgling Tengmalm's Owl is
uniform chocolate-brown, with
a darker face and indistinct
white eyebrows, lores and
bill-base. Germany, April
(*Friedhelm Adam*).

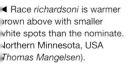

210. NORTHERN SAW-WHET OWL
Aegolius acadicus

L 17–19cm, Wt 54–124g; W 125–144mm

Identification A very small, round-headed owl without ear-tufts. During the breeding period females are on average more than 50g heavier than males, but in autumn difference is only 16g. Has a warm rusty-brown or grey-brown head with fairly dense white shaft-streaks, particularly on forehead, and coffee-brown mantle and rest of upperparts with white spots. Similarly, flight feathers are white-spotted, and relatively short tail has three rows of white spots on both webs of rectrices. The underparts are whitish with broad reddish-buff streaks. The brownish facial disc has a blackish spot between base of bill and eye, and a whitish area around the eyes in the form of radial white streaks extending towards the disc edge, which lacks a dark ruff but has a narrow rim of light and dark spots. The eyes are orange-yellow, with blackish edge of eyelids, and the bill and cere are blackish. Slightly feathered toes have dark horn-coloured claws with blackish tips. *Juvenile* Downy chick is whitish. Coffee-brown mesoptile has facial disc indistinctly white-rimmed, and white eyebrows, forehead and lores forming a light 'X' on dark brown face. Juvenile has yellowish eyes. *In flight* Noiseless and soft flight.
Call As the name suggests, aggressive call is a rasping

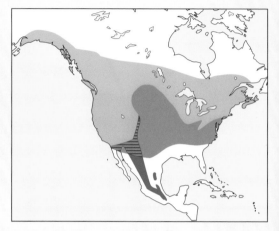

shrill *screerrave*, like the sound of a saw being sharpened. Territorial song is a long sequence of more mellow *tew-tew-tew...*, some two notes per second, and repeated after a short break.
Food and hunting Feeds mainly on small rodents and occasionally small birds, frogs and insects. Queen Charlotte subspecies often takes crustaceans and other intertidal arthropods.
Habitat Lives mostly in coniferous forests, but often moves in autumn into deciduous woods. Occurs also in tamarack swamps, cedar groves and humid forest edges. Occurs from lowlands up to 2800m, and mainly above 1500m. In Mexico mostly between 1350m and 2500m.
Status and distribution Found from Alaska and S. Canada south to SW Mexico; winters irregularly south to S. USA. This species is locally frequent and not thought to be endangered in any way.
Geographical variation Two subspecies are listed: nominate *acadicus* occurs from S. Alaska to the Mexican highlands; *A. a. brooksi*, confined to Queen Charlotte Island, off British Columbia, W. Canada, is very dark and much less spotted above, and rich orange-buff below. The latter is a distinctive form, often named 'Queen Charlotte Owl', but it is not given species status as no DNA differences were found between it and the nominate race.
Similar species Sympatric Tengmalm's (Boreal) Owl (209) is larger, with a yellowish bill and spotted crown. Unspotted Saw-whet Owl (211) may overlap slightly in Mexico, but is more uniform above and below.

▲ This west coast Northern Saw-whet Owl is much darker than the east coast one opposite but mainland birds from British Columbia have not yet been assigned to a different subspecies. British Columbia, Canada (*Tim Fitzharris*).

◄ Northern Saw-whet Owl is smaller than Tengmalm's Owl *Aegolius funereus*; it is white below with reddish streaks and a streaky-looking reddish face without a dark ruff. Note the white spots on the scapulars. New York, USA, March (*Matt Bango*).

► An almost-independent juvenile Northern Saw-whet Owl. Manitoba, Canada, August (*Christian Artuso*).

211. UNSPOTTED SAW-WHET OWL
Aegolius ridgwayi

L 18–20cm, Wt *c.* 80g; W 133–146mm

Identification A very small, round-headed owl without ear-tufts. No data on sexual size differences. Has earth-brown head and upperparts, the head and mantle sometimes slightly darker than the rest, occasionally with fine whitish shaft-streaks on crown. The brown wings have narrow white edges on alula and primaries, and white spots on inner secondaries; uniformly brown tail has a few white spots on inner webs of rectrices, but these are normally invisible. Below, the dull yellowish-brown breast forms an indistinct but broad pectoral band, contrasting the plain yellowish to pale ochre-buff rest of underparts. The brownish facial disc, lighter around the eyes, has a whitish rim, with whitish eyebrows, chin and lores contrasting well with dark face. The eyes are brown to honey-yellow, and the bill and cere dark horn-coloured. Feathered tarsi and fleshy toes with some buffish bristles, and dark brownish-horn claws. *Juvenile* Downy plumage unde-scribed. Mesoptile resembles adult, but with softer, more downy plumage, and breast sometimes with pale and faint streaking.

Call Gives a sequence of melancholy notes, like those of Northern Saw-whet Owl (210) but mellower and lower-pitched; easily distinguished on sonograms, but not so easily in the field.

Food and hunting Virtually nothing known.

Habitat Mountain forests and cloud forests from 1600m to 3000m, occasionally down to 1400m in Guatemala.

Status and distribution Ranges from S. Mexico to W. Panama. This owl is very poorly known. It is listed as Near-threatened by BirdLife International, but it could be partly overlooked as it is very little studied.

Geographical variation Two doubtful races described, *tacanensis* (from Chiapas, S. Mexico) and *rostratus* (from Guatemala), but it has recently been suggested that both of these are hybrids between Unspotted Saw-whet and Northern Saw-whet Owls; *A. ridgwayi* is here tenta-tively listed as monotypic, implying that further inves-tigation is needed to clarify this matter. The taxonomy also requires more research, because this species was formerly treated as a subspecies of Northern Saw-whet Owl.

Similar species Allopatric Tengmalm's (Boreal) Owl (209) is larger, with rounded white spots on the crown and thickly feathered toes. Partly sympatric Northern Saw-whet Owl (210) has lemon or orange-yellow eyes, a whitish-streaked head and crown, and is streaked below.

◀ Unspotted Saw-whet Owl is similar to a young Northern Saw-whet Owl *Aegolius acadicus,* but has a whitish-rimmed facial disc. It is unspotted above, and has brown underwing-coverts. Bill is black, and the eyes brown. Guatemala, December (*Knut Eisermann*).

212. BUFF-FRONTED OWL
Aegolius harrisii

L 18–23cm, **Wt** 104–155g; **W** 142–167mm

Identification A very small to small owl without ear-tufts, but can appear to have small tufts (particularly when alarmed and 'folding' the facial disc). Females are larger and heavier than males. The crown is plain blackish-brown, and the upperparts dark coffee-brown with a few rounded white spots and some dull yellowish ones; nuchal collar is narrow, buffish-ochre, contrasting with dark back. Several large yellowish-brownish spots are visible on outer webs of scapulars. The wings are marked with rows of whitish rounded spots, and the blackish tail has a white tip and two visible rows of rounded white spots on each web of the feathers. The underparts are plain yellowish-tawny to ochre-buff. The almost circular creamy-coloured facial disc has a narrow blackish ruff with dull yellowish border, and a blackish-brown area from eyes to edges of disc borders the ochre-buffish forehead; the chin has a dark brown or blackish bib nearly merging into the thin blackish ruff. The eyes are yellow, the bill yellowish to pale bluish-green. Tarsi feathered to the base of the pale yellow and totally bare toes, which have dark brown claws. *Juvenile* Downy plumage undescribed. Young fledglings are similar to adults, but with unspotted back and less brightly coloured plumage.

Call Utters very rapid whistled trills with a quivering character, *gürrrrrürrrrrrrürrr…*, some 15 notes per second and lasting seven to ten seconds.

Food and hunting Not studied.

Habitat Inhabits mountain forests and cloud forests, dry forests and stunted alpine forests. Occurs also at lower altitudes, but in the Andes lives between 1700m and 3900m.

Status and distribution Found from Venezuela to Ecuador and patchily to N. Argentina; also a disjunct population from SE Brazil to Uruguay, which lives from sea level up to 1000m. Appears to be uncommon to rare, but may be partly overlooked owing to its secretive habits.

Geographical variation Three subspecies are listed: nominate *harrisii* lives in the Andes from Venezuela and Colombia south to E. Bolivia and perhaps including Paraguay; *A. h. dabbenei*, from W. Bolivia to NW Argentina, is darker above, with pale yellow eyes; *A. h. iheringi*, from SE Brazil to Uruguay, is dark above

and the scapulars are variably buff-edged; face and underparts deeper orange. The isolated subspecies *iheringi* could be specifically distinct, but molecular and biological research is required in order to determine this. A possible fourth subspecies, not yet officially described, from Cerro Neblina in S. Venezuela, is only known from a couple of specimens collected in 1985.

Similar species No other *Aegolius* owls occur on the South American continent.

▶ Buff-fronted Owl is the only *Aegolius* in South America. This is the nominate *harrisii*. It has a bluish bill, and distinct blackish-marked face and ruff. Ecuador, August (*János Oláh*).

213. RUFOUS OWL
Ninox rufa

L 40–57cm, Wt 700–1300g; W 260–383mm, WS 100–120cm

Identification A fairly large, reddish-brown owl with no ear-tufts and a rather flat crown. Males are always larger and up to 300g heavier than darker females. The forehead, crown, nape, upperparts and upperwings are dark rufous, finely barred light brown. Long tail has broader dark bars above and is broadly barred yellowish-white below. Throat, breast and belly are rich reddish-brown, finely creamy-barred, and underwings are light brown. The blackish-brown facial disc is indistinct. The eyes are golden-yellow, and the bill grey with short black bristles at base. Reddish-brown legs are feathered, feathering extending to about proximal half of the dirty-yellowish toes, which have blackish-tipped dark horn-coloured claws. *Juvenile* Downy chick is whitish. Mesoptile has whitish head and underparts, face with blackish-brown mask around the eyes. Young adults are more broadly barred than older birds.

Call Utters a deep double hoot, *wooh-hoo*, lasting about one second, or six or seven single hoots at one-second intervals.

Food and hunting Takes a large variety of prey, from beetles and crayfish to birds and flying foxes. Prey pounced on from perches or seized in aerial chases.

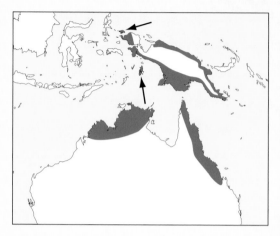

Habitat Occupies lowland rainforests, and gallery forests along creeks and well-wooded savanna; less commonly seen in upland rainforests or in swampy woodlands. In New Guinea lives at up to 2000m above sea level.

Status and distribution Occurs in N. & NE Australia and New Guinea, having a patchy distribution in suitable habitats. Rare to uncommon.

Geographical variation Four subspecies are described, but a revision would be sensible. Nominate *rufa* is found in tropical N. Australia; *N. r. humeralis*, from New Guinea, including Waigeo Island and Aru Islands, is smaller and browner; *N. r. queenslandica*, from E. Queensland, is large and dark, with paler underparts; *N. r. meesi*, from NE Queensland (very few records), is small.

Similar species Unlikely to be confused with any other owl. Allopatric Powerful Owl (214) is larger and geographically separated Morepork (218) is smaller; both are very different in plumage, as described in the accounts for these species. Sympatric Barking Owl (215) is more similar in size, but with white spots on the wings, and is dark brown above and whitish with dark brown streaks below.

▶ Nominate Rufous Owl has a clear white area between the yellow eyes, and a rufous-brown barred head and underparts. Darwin, Australia, October (*Rohan Clarke*).

◀ Race *queenslandica* is darker above with dark cheeks and colder brown barring below. Queensland, Australia (*Martin B. Withers*).

214. POWERFUL OWL
Ninox strenua

L 45–65cm, **Wt** 1050–1700g; **W** 381–427mm, **WS** 115–135cm

Identification Australia's largest owl, with no ear-tufts; males are considerably larger and up to 250g heavier than females. The plumage is grey-brown to dark brown above, the crown and nape with fine pale spotting, the upperparts and wings with irregular creamy-white barring, and the tail with about six narrow yellowish-white bars. The underparts are dull white with broad brownish chevrons in irregular bars. Dark brown facial disc has prominent eyebrows. The large eyes are golden, and the powerful bill is bluish-horn with bristly feathers at base. Tarsi are feathered to the base of the dull yellowish and bristled toes, which have massive blackish-tipped dusky horn-coloured claws. *Juvenile* Downy chick is whitish. Mesoptile has back and wings paler, more heavily white-barred, white face with dark eye patches, whitish crown with fine dark speckles, and white underparts with some fine dark streaks and faint barring on flanks.

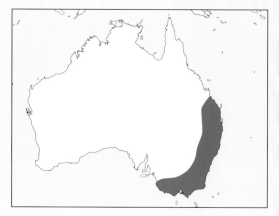

Call Gives an impressive deep, far-carrying double hoot, *whoo-hooo* or *ooo-hoo*, each note lasting just over half a second, with a brief pause between.

Food and hunting Hunts slow-moving arboreal mammals and large birds, occasionally insects. Hunts from perches or on ground. Often catches birds at their night-time roosts.

Habitat Favours wet and hilly, heavily timbered forests with dense gullies adjacent to more open forests. From coast up to 1500m.

Status and distribution Occurs only in SE Australia from SE Queensland (south from Dawson River), south through E. New South Wales to S. & E. Victoria. Regarded as uncommon, but it requires a huge territory. Loss of old-growth forest in SE Australia could make this owl vulnerable if it is not able to adapt to using pine plantations and coastal scrub.

Geographical variation Monotypic.

Similar species No other owl of this size is found in Australia. No other sympatric, large *Ninox* has bold grey-brown V-bars on whitish underparts.

◄ Powerful Owl is a fierce predator with powerful talons. Queensland, Australia, August (*Stuart Elsom*).

▲ Powerful Owl is easily the largest Australian owl. Upperparts show irregular creamy-white barring. (*David Hosking*).

▲▶ Two young Powerful Owls. These are very different from the adults, having a grey face and downy white underparts. Wings and tail are more or less like the adult. Victoria, Australia, October (*Rohan Clarke*).

▶ This Powerful Owl has captured a Little Raven *Corvus mellori*. Victoria, Australia, January (*Tadao Shimba*).

215. BARKING OWL
Ninox connivens

L 35-45cm, **Wt** 425-510g; **W** 244-325mm, **WS** 85-100cm

Identification A medium-sized brown owl without ear-tufts. Very little colour difference between the sexes, but male is some 100g heavier than female. The plumage is brown to grey-brown above, back and wings with large white spots, scapulars with several large spots forming a whitish row across the shoulder, wing-coverts with variable smaller spots. Relatively long tail is inconspicuously brown-barred above, broadly barred white and brown below. Brown throat is white-streaked, and erectile throat feathers can give bearded appearance. Underparts are whitish with variable amount of dark grey to rusty streaks. The large eyes are yellow and the bill black. Powerful legs are feathered, the toes dull yellow and the claws dusky horn-coloured with blackish tips. *Juvenile* Downy chick is whitish. Mesoptile is similar to adults, but eyebrows more white and body more fluffy. *In flight* Reluctant to fly by day, but has bouncing flight followed by goshawk-like glides.
Call Gives a noisy dog-like barking *wuf wuf* or *wook-*

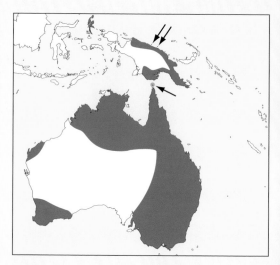

wook, sharper and quicker than call of Morepork (218); the bark is preceded by a short, low growl, audible only at close distance. Female and male often duet.
Food and hunting Takes rabbits, gliders, small possums, bats, rodents, and a wide variety of diurnal birds. Like other *Ninox* owls, it eats many insects. Most prey is taken on the ground or from perches.
Habitat Inhabits open savanna and lowland country with large trees, preferably close to water. Sometimes nests close to farm buildings and even in towns. In New Guinea it occurs up to 1000m above sea level.
Status and distribution Found in Australia and New Guinea, where its distribution is very patchy. Absent from most of W. Australia and the arid interior. Locally endangered by the use of pesticides.
Geographical variation Five subspecies are listed: nominate *connivens* is found in SW, SE & E. Australia; *N. c. rufostrigata*, from N. Moluccas (Bacan, Halmahera, Morotai, Obi), is clearly browner above; *N. c. assimilis*, from C. & E. New Guinea (including Manam and Karkar islands), is very small; *N. c. occidentalis*, from N. & W. Western Australia east to NW Queensland, is much browner and smaller; *N. c. peninsularis*, restricted to Cape York Peninsula, in N. Queensland, is slightly darker and smaller than the nominate race, with chestnut-brown streaks below.

◄ Pair of Barking Owls; the male is always heavier than the female. Australia (*Tom and Pam Gardner*).

Similar species Geographically overlapping Rufous Owl (213) is larger, and densely barred below. Although sympatric, the present species is unlikely to be confused with the much larger Powerful Owl (214) or the somewhat smaller Southern Boobook (217), which has paler yellowish eyes.

▶ Race *occidentalis* has a bluish-grey head and differs from the nominate in being much browner and less greyish above. Underside is much browner, and streaked. Northern Territory, Australia, January (*Adrian Boyle*).

▼ This male *assimilis* Barking Owl is puffing up its reddish-brown, finely streaked throat before calling. Sabai Island, Australia, July (*Rohan Clarke*).

▼▶ Race *assimilis* is the smallest subspecies. It is very brown with white 'eyebrows' and black rimmed yellow eyes. It has a strongly streaked breast. Saibai Island, Australia, July (*Rohan Clarke*).

216. SUMBA BOOBOOK
Ninox rudolfi

L 30–40cm, Wt 222g (1); W 227–243mm

Identification A medium-sized owl without ear-tufts. No sexed weight measurements are available, but females are said to be slightly larger and heavier than males. Dark brown crown is densely white-spotted, and the nape, mantle and back are dark brown with a rufous wash, spotted and mottled with white. Wing-coverts are barred dark and whitish, the dark brown primaries and secondaries have rows of whitish-buff spots, and the tail has about five pale bars. Plain white throat forms a well-defined patch, contrasting the reddish-brown, heavily barred underparts (single feathers have rufous and whitish bars). Eyebrows are whitish, but not prominent, and ear openings dark brown. The eyes are brown and the bill yellowish-brown. Tarsi heavily feathered, with pale dirty-yellow toes and blackish claws. *Juvenile* Undescribed.

Call Utters a long cough-like series of *cluck-cluck-cluck-cluck…* notes, 0.23 seconds each, at rate of two per second.

Food and hunting Unstudied, but possibly takes many insects.

Habitat Inhabits monsoon forest and rainforest, coastal swamps and farmlands. Lives in lowlands up to 930m.

Status and distribution Confined to Sumba, in the

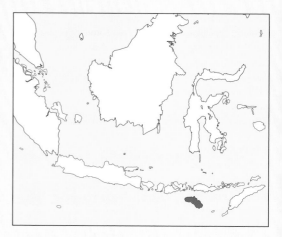

Lesser Sundas, where it is generally rare or uncommon. Small numbers were recorded at five localities during 1989 and 1992 surveys. Listed as Near-threatened.

Geographical variation Monotypic.

Similar species Geographically overlapping Little Sumba Hawk Owl (235) is smaller, with yellow eyes and very prominent white eyebrows, and its underparts have dark V-shaped vermiculations.

◄ Sumba Boobook is larger than Little Sumba Hawk Owl *Ninox sumbaensis* and has dark brown eyes. Both species live only on Sumba Island in Indonesia. June (*James Eaton*).

217. SOUTHERN BOOBOOK
Ninox boobook

L 25–36cm, **Wt** 146–360g; **W** 188–261mm

Identification A small to medium-sized brown owl without ear-tufts. Females are some 65g heavier, slightly larger and darker than males. There is clinal variation in size, smaller in warmer areas; very variable also in coloration, with paler desert forms and darker forest forms. Has pale to dark brown crown and upper-parts, with irregular light-coloured or white spots on wing-coverts, and an indistinct whitish row across the shoulder. Primaries and secondaries are barred reddish-brown and dark-brown, and the dark rufous tail has paler bars. The throat is plain whitish, and the underparts are whitish with broad reddish-brown streaks and cross-bars. The light facial disc is not very prominent, but has nearly blackish large patches behind the eyes. The eyes are pale greenish-yellow and the bill bluish-grey. Tarsi feathered, the brownish-grey toes rather bare, with bristles on upperside, and the claws horn-coloured with darker tips. *Juvenile* First down is whitish or pale buff. Mesoptile has nearly black

'spectacles' around hazel-brown eyes. *In flight* Looks dark and deceptively large.

Call The onomatopoeic name alludes to the distinctive song, *boo-book* or *mo-poke*, a brief double hoot with the second note lower-pitched than the first; each note lasts *c.* 0.25 seconds and there is a half-second gap between the two. Call repeated some 20 times in one or two minutes.

Food and hunting Takes mostly insects and other arthropods, more so than any other Australian owl; when breeding it also eats small birds and mice. Hunts from a perch, but takes flying insects by hawking flight through and above trees or in light of streetlamps.

Habitat Can be seen almost anywhere where there are trees, but generally absent from dense rainforest; found also in town suburbs, orchards, parks and streets with abundant trees. Generally in lowlands, but on Timor ascends to 2300m.

Status and distribution Occurs throughout most of Australia, and in E. Lesser Sundas and S. New Guinea. Generally common. It is the most common of all Australian owls, but can be locally endangered by the effects of pesticide use.

Geographical variation Nine subspecies are recognised: nominate *boobook* occurs from S. Queensland south to South Australia and Victoria (introduced on Norfolk Island, but has since disappeared); *N. b. rotiensis*, from Roti Island, in C. Lesser Sundas, is

◀ Race *fusca* is a somewhat darker and smaller race of Southern Boobook. It is cold grey-brown without any trace of rufous or warm-brown. Timor, August (*Filip Verbelen*).

smaller than the nominate and has a totally different voice; *N. b. fusca*, from Timor, Romang and Leti, in C. Lesser Sundas, is dark grey-brown above and heavily spotted below; *N. b. plesseni*, from Alor, in C Lesser Sundas, is generally similar to *fusca* and was for a long time known only from holotype (1929), but this book includes the first photos taken in the wild on Alor; *N. b. moae*, from Moa, in EC Lesser Sundas, is darker than *ocellata*; *N. b. cinnamomina*, from Babar Islands, in E Lesser Sundas, is very distinctive, deep cinnamon dorsally and with heavy cinnamon streaking below; *N. b. remigialis*, from the Kai Islands, in extreme E Lesser Sundas, is similar to *moae*, but has less pronounced barring on the flight feathers; *N. b. pusilla*, from the lowlands of S. New Guinea, resembles *ocellata*, but is distinctly smaller; *N. b. ocellata*, from tropical N. Australia and W. & S. Australia, and Sawu Island (between Sumba and Timor), is paler and sandy brown, with buff-streaked underparts. DNA-sequencing indicates that *rotiensis* is highly distinctive, and should be given species status as 'Roti Boobook *Ninox rotiensis*'; full and proper bioacoustical and molecular analyses are awaited, however, before any subspecies is elevated to species rank. The main reason for the delay in proposing new species is that the taxonomy of the whole 'boobook group' requires further molecular data and comparative studies on voice, behaviour and biology; note that the 'official' separation of Southern Boobook *N. boobook* from Morepork *N. novaeseelandiae* is still awaited.

Similar species Brilliantly yellow-eyed Barking Owl (215) is sympatric, but is much larger and greyer, with grey to rufous streaks on a whitish breast. Geographically separated Red Boobook (219) is dark chestnut-brown above, with hardly any pale markings on the crown or nape; in addition, it has a rufous throat with some dark streaks, ocellated breast and belly, and the dusky-brown feathers of the mantle narrowly buffish-edged.

▼ The distinctive race *cinnamomina* lives in the Topa and Babar islands. It is browner on the crown and deep cinnamon dorsally, with deep cinnamon streaking below. Babar Island, Indonesia, August (*Filip Verbelen*).

► Race *plesseni* drooping its wings downwards when agitated. This behaviour does not seem to occur in *ocellata* or *cinnamomina,* This race was formerly known only from the holotype and this is the first time the bird has been photographed. Pantar Island (Alor Archipelago), Indonesia, June (*Filip Verbelen*).

▼ Race *plesseni*, the 'Alor Boobook', is smaller than the nominate *boobook*. Its general coloration is similar to *fusca*, but the entire upperparts are marked with white and pale brown spots. Breast has longitudinal stripes, tending to become ocellations on the lower underparts. Alor Island, Indonesia, June (*Filip Verbelen*).

▼ ► Race *plesseni* is another that may warrant specific status, but molecular confirmation is required. Pantar Island (Alor Archipelago), Indonesia, October (*Rob Hutchinson*).

▲ Southern Boobook *ocellata* is a small, highly variable owl, but it is generally paler than other Australian races. This bird has very pale, white-fringed yellowish eyes. Western Australia, August (*Adrian Boyle*).

► This Southern Boobook of race *ocellata* has a clear dark mask and is very reddish-brown. Western Australia, September (*Adrian Boyle*).

▲ Birds from Roti Island, *rotiensis*, could be a separate species, based on their remarkably different territorial call. The iris is yellow. Roti Island, August (*Filip Verbelen*).

▶ Young Southern Boobooks have nearly black 'spectacles' and hazel-brown eyes (closed here!). Cere is blue, and the bill black. New South Wales, Australia, November (*Rohan Clarke*).

218. MOREPORK
Ninox novaeseelandiae

L 26–29cm, Wt 150–216g; W 183–222mm

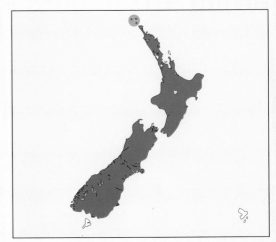

Identification A small to medium-sized owl without ear-tufts. Females are some 15g heavier than males. Both sexes are variable in plumage, and many colour morphs exist. The head and upperparts are generally dark brown, with ochre-buff mottling and streaks on head, neck and mantle, the wing-coverts and scapulars spotted cinnamon and whitish-buff. Blackish-brown flight feathers have narrow yellowish-brown bars, and the dark brown tail has about five to eight narrow buffish bars. The brownish-white to pale buff throat to upper breast is marked with dark brown flecks and streaks; below this the underparts have ocellated feathers, each whitish but with a dark shaft-streak and terminal bar. Dark brown facial disc has narrow whitish eyebrows and a dull yellow rim. The eyes are bright golden-yellow and the bill dark with pale tip. Tarsi are feathered yellowish-brown to reddish-buff, with bare yellow to brownish-yellow toes and dusky-brown to blackish claws. *Juvenile* Downy chick is whitish to whitish-grey, and mesoptile dark smoky-brown. When leaving the nest, young has spectacle-like black patches behind the eyes. *In flight* Shows clearly the relatively rounded wings and long tail when flying.

Call The Maori name for this owl ('ruru') comes from the haunting, melancholic song, *ru-ru* or rather *quor-quo*, which is repeated at intervals of several seconds.
Food and hunting Feeds largely on insects, but also takes small birds, lizards, rats, bats and mice. Prey are caught from a perch or hawked in flight. Aerial prey are seized with the talons before being transferred to the bill.
Habitat Inhabits forests and plantations up to the timberline, but occurs also in city parks and near farmhouses. Also on many offshore islands.
Status and distribution Confined to New Zealand, where it is the only surviving native owl. It is not uncommon, and has adapted, at least to some extent, even to exotic pine forests. Formerly occurred also on Lord Howe Island and on Norfolk Island, both off E. Australia, but now extinct on Lord Howe, possibly as a result of introduction of much larger American Barn Owl (2) and Tasmanian Masked Owl (20). Only one female was known to survive on Norfolk Island in 1986, and this female had at least four hybrid chicks with an introduced nominate-race male in 1989–90; this artificial *undulata* × *novaeseelandiae* population is now well into double figures.
Geographical variation Three subspecies are recognised: nominate *novaeseelandiae* in New Zealand; *N. n. albaria*, from Lord Howe Island, off E. Australia (now extinct), was pale brown above, with pale brown markings below; *N. n. undulata*, from Norfolk Island, even farther off E Australia (now surviving only as an inter-

racial hybrid population), is a little darker than *albaria* and has a few spots on the neck. This species was previously regarded as conspecific with Southern Boobook (217), and the taxonomy of the whole '*novaeseelandiae* complex' requires further analysis of molecular data, and comparative studies of voice, behaviour and biology, in order to verify the separation of the two at species level.

Similar species In New Zealand Morepork is the only medium-sized owl without ear-tufts. Geographically separated Southern Boobook (217) has pale yellowish-green eyes. Introduced and overlapping Little Owl (203) is smaller in size and much less brown in general coloration, with relatively short tail and less rounded head.

▲ Morepork is the only surviving native owl in New Zealand; the other endemics are all extinct. Little Barrier Island, New Zealand, February (*Rohan Clarke*).

▲▶ The Norfolk Island race *undulata,* photographed in 1986. This was the last surviving 'pure-bred' bird; it was a female and is slightly paler than the nominate. Norfolk Island (*John Hicks*).

▶ Race *undulata* is partly 'man-made'; the last female was crossed with an introduced nominate male. This is one of the subsequent hybrid population; it is very similar to the original *undulata*. Norfolk Island, January (*Rohan Clarke*).

◀ Morepork tilting its head inquisitively. Tiritiri Matangi, New Zealand, March (*Adam Riley*).

219. RED BOOBOOK
Ninox lurida

L 28–30cm, Wt 207–221g; W 244mm (1)

Identification A small to medium-sized dark owl without ear-tufts. No data are available on possible sexual size and colour differences. Crown and upper-parts are rather uniformly very dark chestnut-brown without whitish spots, the scapulars with narrow cinnamon-buffish feather edges. Flight feathers and tail are similar to the back or a little darker and more blackish, and generally unbarred. Underparts are gold-brown, ocellated with whitish and finely mottled with yellowish-brown. The facial disc is mottled rufous and dusky brown, with narrow whitish eyebrows (but no black mask around the eyes). The eyes are greenish-yellow, the eyelids with dark edges, and the bill and cere are greyish. Fawn-feathered tarsi, with bare greyish-brown toes and blackish claws. *Juvenile* First down is whitish. Mesoptile is similar to Southern Boobook (217), but without prominent dusky mask.

Call Song is said to be less clear and harsh than that of Southern Boobook. Further study is needed. It appears to be less vocal than other boobooks.

Food and hunting Little known, but diet is probably mainly of insects. Prey are caught from a perch or hawked in flight.

Habitat Found in mountain rainforests.

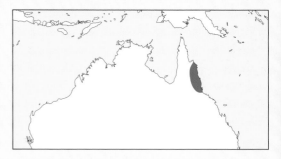

Status and distribution Occupies a fairly limited range in NE Queensland, in NE Australia. It could possibly be at risk from deforestation, but no information on its numbers is available.

Geographical variation Monotypic. This owl was formerly treated as a subspecies of Southern Boobook (217), but it is very distinctive when compared with all other boobooks, which are paler, with whitish spots above and a dark mask around eyes. Further studies are needed on the taxonomy, ecology and behaviour of this group.

Similar species Geographically overlapping Barking Owl (215) is much larger and has large, yellow eyes.

◀ Red Boobook has a limited range in northeast Queensland, and is very different from other boobooks. It is smaller and much darker than Southern Boobook *Ninox boobook*. It is uniform above, with pale-webbed scapulars. Throat patch is cinnamon. Jullaten, Queensland, September (*Rohan Clarke*).

220. TASMANIAN BOOBOOK
Ninox leucopsis

L 28–30cm; **W** 198–222mm

Identification A small to medium-sized owl without ear-tufts. There are no data on sexual colour and size differences. The crown and nape are reddish-brown to chestnut, fairly densely spotted with small whitish dots, and the chestnut upperparts, including the wing-coverts, are spotted whitish-buff to pale yellow. The brown primaries and secondaries have ochre bars, and the brown tail is densely marked with fine pale bars. The throat is white, contrasting the strongly ocellated rufous-orange and white underparts. Ill-defined facial disc has rufous areas around eyes with whitish radial streaking, and is bordered above by thin white eyebrows. The eyes are golden-yellow, the eyelids with dark rims, and the bill and cere are greyish. Tarsi are cinnamon-feathered to the base of the grey to greyish-brown toes, which have dark horn-coloured claws with blackish tips. *Juvenile* First down is whitish or pale buff, similar to Southern Boobook (217), but mesoptile has less distinct mask around eyes.

Call Emits double-noted hoots, similar to those of Morepork (218).

Food and hunting Feeds on insects and other invertebrates, but also small birds and rodents.

Habitat Found in semi-open landscapes with bushes and trees, farmlands with trees, and swampy woodlands, but also lives within human settlements.

Status and distribution Endemic and common in Tasmania and islands in Bass Strait.

Geographical variation Monotypic. Formerly regarded as a race of Southern Boobook, but DNA studies seem to show closer ties with Morepork of New Zealand. Further molecular studies are required in order to verify its relationships and species status.

Similar species Slightly larger Southern Boobook (217), also allopatric, has greenish-yellow eyes, the head not whitish-spotted, and the underparts without orange-rufous ocellations. Geographically separated Morepork (218) has crown and nape with cinnamon-buff streaking, darker plumage and less densely barred tail.

▶ Tasmanian Boobook is slightly smaller than Southern Boobook *Ninox boobook*. It has golden-yellow eyes, and the head is densely marked with small whitish dots (which Southern Boobook lacks). Back and mantle whitish-spotted, and underparts strongly white ocellated. Tarsi are feathered to the toes – which here are holding a mouse. Woodbridge, Tasmania, Australia (*Dave Watts*).

221. BROWN HAWK OWL
Ninox scutulata

L 27–33cm, **Wt** 170–230g; **W** 145–245mm

Identification A small to medium-sized owl without ear-tufts. Sexes are said to be similar in plumage, but females are slightly larger than males. This is a very hawk-like owl, with dark brown upperparts. Has irregular white patches on outer webs of scapulars, and long, pointed wings; the rather long, brown tail is paler-barred. The throat and foreneck are whitish-buff to pale fulvous, boldly rufous streaked, and the underparts whitish with large drop-like reddish-brown streaks. The relatively small, rounded head has a whitish forehead, but is characterised by the virtual absence of

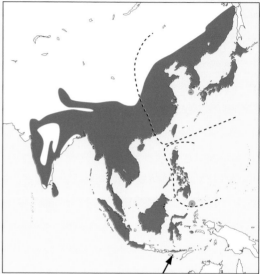

a facial disc; a narrow dark area is present around the eyes. The eyes are bright yellow, the bill bluish-black and the cere dull green or greenish-brown. Tarsi feathered, with yellow or yellowish-green toes and dark horn-coloured claws. **Juvenile** Downy chick is whitish. Mesoptile is more fluffy and less striated below than adult. **In flight** Swift flight resembles that of a hunting falcon, or a nightjar when hawking insects at night.

Call Three different types of song have been described, indicating the possibility that three distinct species, rather than subspecies, could be involved. The song of two northern subspecies (*japonica* and *totogo*) consists of two or three *whoop* notes, and seven other subspecies have a common song, *whoowúp-whoowúp*. Philippine subspecies (*randi*) has a similar song to the first two, but *whoop* notes in couplets are lower-pitched.

Food and hunting Feeds mainly on insects, but also takes bats, lizards, frogs, crabs, terrestrial rodents, flying squirrels and small birds. Hunts from a perch, generally on a tree stump or fence post. Often hawks insects in flight.

Habitat Inhabits forests and woodlands both at low and at high elevations, up to 1700m, and associates

◄ Brown Hawk Owl is extremely polymorphic and is likely to be split into at least three different species. This north Indian bird of the race *lugubris* is in the 'tall-thin' position. Rajasthan, India, January (*Harri Taavetti*).

with human settlements and well-wooded parks and gardens in urban areas. SE Asian races seem to avoid human habitations and live only in lowland mangroves and rainforests.

Status and distribution Has a wide distribution, ranging from Indian Subcontinent to E. Siberia and Japan and south to the Sundas. In northern parts of the range, from Indian Subcontinent east to Japan, this species is fairly common, but southern races suffer from destruction of primary lowland rainforests in Indonesia, Malaysia and the Philippines.

Geographical variation This is an extremely polytypic species, consisting of at least ten subspecies divided into three groups as follows:

BROWN HAWK OWL *N. scutulata*: nominate *scutulata*, occurring in S. Malay Peninsula, Riau Archipelago, Lingga Archipelago, Sumatra and Bangka Island, is a fairly dark race having short, rounded wings; *N. s. lugubris*, from N. & C. India to W. Assam, is somewhat paler and greyer; *N. s. burmanica*, from NE India to S China (S Yunnan) and south to SE Asia, is darker above and below than *lugubris*; *N. s. hirsuta*, from Bombay, India, south to Sri Lanka, is even darker than *burmanica*; *N. s. javanensis*, from Java, is a small, round-winged, very dark race; *N. s. borneensis*, from Borneo, is smaller and short-winged; *N. s. palawanensis*, from Palawan, in SW Philippines is also rather short-winged.

NORTHERN BOOBOOK *N. (s.) japonica*: *N. (s). japonica*, from S. Korea and Japan, is large and pale; *N. (s). florensis*, from SE Siberia (Ussuriland), N & C China to N. Korea, is even larger and paler than *japonica*; *N. (s). totogo*, from Ryukyu Islands, Taiwan and Lanyu Island, may be a synonym of *japonica*, and has similarly pointed wings.

CHOCOLATE BOOBOOK *N. (s.) randi*: *N. (s). randi*, from Philippines, and the Sulu and Talaud Archipelagos, is similar to nominate *scutulata*, but more rufous, and with darker streaking on the underparts; it could be a separate species, but molecular-genetic data are required.

BirdLife International has not adopted the above-mentioned splits and neither has the present author. Biology and movements of many subspecies are not well known either; northern races are migratory, moving to the Greater and Lesser Sundas, Philippines, Sulawesi and the Moluccas, but southern races are sedentary and occur alongside migrants in some places. Vagrant *japonica* found once on Ashmore Reef, in the Timor Sea off NW Australia.

Similar species Hume's Hawk Owl (222) is similar in size, but has very dark plumage all over; it is confined to Andaman and Nicobar Islands, as also is Andaman Hawk Owl (223), which is smaller and warmer brown, with bright rufous tint on wings and tail. Sympatric Philippine Hawk Owl (226) is pale cinnamon-brown above, with relatively large white spots on outer webs of scapulars, and wing-coverts spotted whitish. Geographically overlapping Ochre-bellied Hawk Owl (227), from Sulawesi, has an ochre-rufous breast and belly with faint darker spots. Allopatric Solomon Hawk Owl (228), confined to the Solomon Islands, has whitish-barred or plain dark brown mantle, and whitish underparts either with fine shaft-streaks and dark-mottled pectoral band or broadly barred with rufous scaly bars. Geographically separated Jungle Hawk Owl (229) is rather plain brownish-chestnut below. Sympatric Speckled Hawk Owl (230), from Sulawesi, is smaller, with crown, nape and mantle densely speckled white; has a large white patch on throat and another between two whitish-speckled brown pectoral bands. Allopatric Russet Hawk Owl (231) is bright rufous-brown above, with whitish spots, and has orange eyes.

► Race *randi* occurs in the Philippines and on Siau Island, Indonesia. It has a dark head and is whitish, but dark streaked below. This subspecies has been proposed to be specifically distinct, as Chocolate Boobook; however, it is very difficult to separate in the field from, for example, nominate *scutulata*. Luzon, The Philippines, October (*Alain Pascua*).

▲ Race *burmanica* of Brown Hawk Owl differs from *lugubris* in being darker above and below; the head is as dark as the back, or even darker. Beidaihe, China, May (*Ran Schols*).

▲▶ Race *japonica* is often considered a separate species, Northern Boobook. Japan, July (*Kazuyasu Kisaichi*).

▶ This is a Brown Hawk Owl from Siau Island. Its taxonomic status is unresolved; these birds have so far been considered to belong to race *randi* (and therefore perhaps fall within a separate species, Chocolate Boobook), or they may be a separate race on their own. March (*Filip Verbelen*).

222. HUME'S HAWK OWL
Ninox obscura

L 29–30cm; **W** 197–220mm

Identification A small to medium-sized owl without ear-tufts. No data are available on sexual colour and size differences. The plumage is coffee-brown above and below, becoming a little lighter and more rufous on the abdomen; a few small whitish spots or bars are present on the flanks and below (but often seen only if the bird raises the overlying dark feathers), and the undertail-coverts are faintly barred with white. Flight feathers and tail are deep brown, the tail feathers having about four narrow pale greyish cross-bands and a whitish tip. Feathers of the lores, forehead and chin are bristly, and whitish or white at the base and black at the tips. The eyes are yellow and the cere dull green. The large bill is dark blue. The yellow feet have black claws. *Juvenile* Undescribed.

Call Said to have a song like that of Brown Hawk Owl (221), but this requires verification.

Food and hunting Not studied, but probably mainly eats large insects.

Habitat Forests and woodlands at low elevations. Whether it shares habitat with Andaman Hawk Owl (223), which lives mainly in lowland forests, is not known.

Status and distribution Endemic to the Andaman Islands, in the Indian Ocean. It could be vulnerable because of its limited distribution, but its status is unknown. Vocalisations and ecology require study.

Geographical variation Monotypic.

Similar species Allopatric Brown Hawk Owl (221) is very similar in size, but generally much paler, with clear whitish spots on forehead and more distinct streaking on whitish underparts. Sympatric Andaman Hawk Owl (223) is smaller, and much paler in coloration.

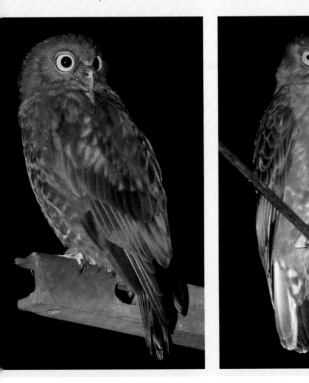

◀◀ Hume's Hawk Owl has only recently (2005) been separated from Brown Hawk Owl *Ninox scutulata*. It is only a little larger than Andaman Hawk Owl *N. affinis* but it is much darker with a uniform, almost bluish-black coloration and has a large, dark blue bill and only a little white between the eyes. Chidiya Tapu, Andaman Islands, January (*Niranjan Sant*).

◀ Hume's Hawk Owl shows some white on the undertail but is otherwise dark below. Wonder, Andaman Islands (*Rob Hutchinson*).

223. ANDAMAN HAWK OWL
Ninox affinis

L 25–28cm; **W** 167–170mm

Identification A small to medium-sized brown owl without ear-tufts. There are no data on sexual size or colour differences. It has distinct bright rufous streaking on brownish-white underparts, streaks appearing as long stripes from the neck to the belly. Crown and upperparts are rather plain brown, with indistinct fine ochre vermiculations, the scapulars with cinnamon-buffish outer webs. Flight and tail feathers are barred brown and dull yellow, secondaries with a rufous tint. The facial disc is greyish. The eyes are yellow, the bill blackish. Tarsi feathered, with yellowish toes and blackish-horn claws. *Juvenile* Fluffy mesoptile is less distinctly streaked below than the adults.

Call The song is a hollow, guttural *crauwu* lasting 0.45 seconds, and repeated at intervals of several seconds.

Food and hunting Known to hawk moths and beetles in the air, but otherwise food unstudied.

Habitat Found mainly in lowland forests, and accepts secondary forests.

Status and distribution Endemic to the Andaman Islands, in the Indian Ocean. Could be endangered owing to its restricted range.

Geographical variation Monotypic. Taxonomy requires study.

Similar species Allopatric Brown Hawk Owl (221) is larger and has clearer streaking on whitish underparts. Sympatric Hume's Hawk Owl (222) is only a little larger, but much darker and more uniformly coloured.

◄ Andaman Hawk Owl has big yellow eyes, white in the 'eyebrows' and throat, and a greyish head. It is smaller and much darker than Brown Hawk Owl *Ninox scutulata* but paler than Hume's *N. obscura*. Chidiya Tapu, Andaman Islands, January (*Niranjan Sant*).

▼ Andaman Hawk Owl has a much lighter bill than Hume's, paler on the culmen and the tip. Tarsi are feathered to the base of the yellowish toes, which are bare and sparsely bristled. Chidiya Tapu, Andaman Islands, January (*Niranjan Sant*).

224. MINDORO HAWK OWL
Ninox mindorensis

L 20cm, **Wt** 100–118g; **W** 157–175mm

Identification A very small owl without ear-tufts. The crown and nape are reddish-brown with fine yellowish and dusky barring, and the warm brown mantle has some darker bars. Scapulars have whitish spots on outer webs, and there are pale buffish spots on warm brown wing-coverts. Flight feathers and tail are brown, the remiges with rows of paler spots and the tail with thin buff bars. The wings are relatively long and pointed. Below, the plumage is orange-rufous, upper breast darker, and finely barred dark brown from neck to belly. The reddish-brown facial disc lacks a distinct rim, and has inconspicuous whitish eyebrows. The eyes are yellow and the bill light greenish. Incompletely feathered tarsi are orange-buff, leaving lower third bare; the bristled toes are yellowish-grey. *Juvenile* Undescribed.

Call Utters a series of high-pitched *cheehrr cheehrr* growls at intervals of one to two seconds; resembles the screech of Barn Owl (1) rather than the hoots of *Ninox* owls.

Food and hunting Not described.

Habitat Lives in forests and wooded areas.

Status and distribution Endemic to Mindoro, in the Philippines, but status uncertain.

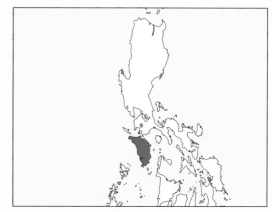

Geographical variation Monotypic. Formerly considered a subspecies of Philippine Hawk Owl (226), which does not occur on Mindoro, but separated on grounds of totally different vocalisation.

Similar species Philippine Hawk Owl (226) is somewhat smaller, and paler below, with indistinct brownish streaks rather than bars; it also differs vocally.

► Mindoro Hawk Owl has bright yellow eyes, an orange-rufous wash on densely barred and spotted underparts, and scapulars with a few large white spots. Lower belly is not marked. Siburan, Mindoro, February (*Markus Lagerqvist*).

▼ Mindoro Hawk Owl is often considered conspecific with Philippine Hawk Owl *N. philippensis*, but it differs vocally. Siburan, Mindoro, February (*Rob Hutchinson*).

225. MADAGASCAR HAWK OWL
Ninox superciliaris

L 23–30cm, Wt 236g (1); W 180–193mm

Other name White-browed Hawk Owl

Identification A small to medium-sized, round-headed owl without ear-tufts. No sexed measurements are available to determine whether any size differences exist between female and male. There are lighter and darker morphs. The crown and upperparts are either uniformly brown or with some whitish dots. Upperwing is sparsely white-spotted, the flight feathers barred pale and dusky, the brown tail with narrow pale bars. The wings are rather long and pointed. Chin and throat are brownish-white, and the underparts light yellowish-brown with bold brown bars, the barring becoming indistinct centrally, particularly on belly; the undertail-coverts and underwing-coverts are pure white. The greyish-tan facial disc is bordered by prominent white

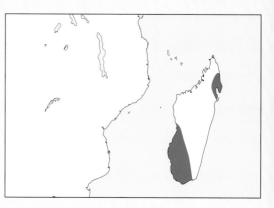

eyebrows. The eyes are dark brown, the bill whitish-horn and the cere pale yellowish. Tarsi are feathered brownish-yellow, but yellowish-white toes are bare, with the claws horn-coloured. *Juvenile* Undescribed.

Call Utters a muffled, rather howling *ho-o-o-hoo*, followed by 15–20 discordant *kiang-kiang* calls, rising in pitch and volume. Often extremely vocal at night.

Food and hunting Feeds mainly on insects, but probably also on small vertebrates. Hunts from a perch, keeping a lookout for its prey.

Habitat Found in well-wooded savannas and gallery and evergreen forests, but also in forest clearings and open terrain with thorny scrub, from sea level to 800m. Occurs also in the vicinity of human habitations. It is less dependent on forests than are many other owls.

Status and distribution Endemic in NE & SW Madagascar, where it is locally fairly common. It suffers from human persecution, as most local people regard all owls as birds of ill omen.

Geographical variation Monotypic. Taxonomy requires study. Cytochrome-b sequences place this owl in the *Athene* clade; as only one single sequence is available, however, such reclassification requires confirmation from further samples.

Similar species Only slightly sympatric Madagascar Red Owl (12) has a heart-shaped facial disc, orange-ochre plumage, and is not barred below. Partly overlapping Madagascar Scops Owl (41) is smaller and has yellow eyes, a row of whitish dots across the shoulder, and small but erectile ear-tufts. Also partly sympatric, Torotoroka Scops Owl (42) is smaller, with yellow eyes and erectile ear-tufts, and has no bars on underparts. All differ vocally.

◀ Madagascar Hawk Owl has very prominent white 'eyebrows'. Madagascar, September (*Ian Merrill*).

◀ Madagascar Hawk Owl has a pure white belly, with the white continuing towards the dark upper breast. It has dark brown eyes and a greenish-yellow bill with a pale brown chin and very white throat. Madagascar, October (*Mike Danzenbaker*).

▶ Madagascar Hawk Owls allopreening, to strengthen the pair bond. Madagascar, October (*Mike Danzenbaker*).

226. PHILIPPINE HAWK OWL
Ninox philippensis

L 15–20cm, Wt 120–142g; W 158–194mm

Identification A very small owl without ear-tufts. Males are about 10g heavier than females. Sexes said to be similar in plumage, but with extremely variable coloration. Above, the plumage is light yellowish-brown, the less marked head rather dark brown with ear coverts more dusky; scapulars have large white spots on outer webs and some inner ones ochre-barred, and the dark brown wing-coverts are white-spotted. The brown primaries also are ochre-barred, and the sepia-brown tail has six or seven narrow light brownish-yellow bars. Whitish chin and throat are marked with a few blackish streaks, and the whitish underparts with tawny wash are broadly marked with ochre-brown streaks. The pale brownish-yellow facial disc (feathers whitish at base) is bordered by narrow whitish eyebrows. The eyes are yellow and the bill greenish with yellow tip. Tarsi feathered almost to the base of densely bristled, dirty-yellow toes, with claws dark horn-coloured with blackish tips. *Juvenile* Natal down is white. Fledglings

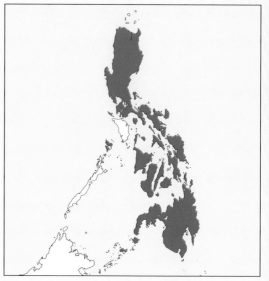

have almost uniformly dark brown upperparts, with a few buffish-white bars on outer scapulars, wings darker than the back, ochre-spotted and white on outer webs; whitish underparts give broadly streaked appearance owing to fawn feather centres.

Call Utters a long series of short, single *whoo* notes given at two-second intervals, in combinations of up to four notes, and entire calling series lasting two to three minutes.

Food and hunting Feeds mainly on insects and small birds. Frogs and colubrid snakes are also taken.

Habitat Occupies remnant patches of primary rainforest and gallery forest, but also secondary forests and even plantations. Locally up to 1800m.

Status and distribution Endemic to the Philippines, where it is locally common.

Geographical variation Six subspecies are listed: nominate *philippensis* occurs on Luzon, Polillo, Marinduque, Buad, Catanduanes, Samar and Leyte; *N. p. proxima*, from Masbate and Ticao, is darker brown

◄ Philippine Hawk Owl has six described subspecies, two or even three of which could be separate species. The nominate *philippensis* is the smallest race, pale cinnamon-brown and rufous-washed. It has large yellow eyes and clear white spots on the scapulars and wing-coverts. Whitish below, with ill-defined brown stripes. Luzon, Philippines, March (*Bram Demeulemeester*).

above and has darker streaks below; *N. p. centralis*, from Semirara, Carabao, Boracay, Panay, Guimaras, Negros, Bohol and Siquijor, is larger and duller brown above than *proxima*; *N. p. reyi*, from Sulu Archipelago, is large, without streaks below, and with rufous barring on head and back; *N. p. spilonota* from Tablas, Sibuyan, Cebu and Camiguin Sur, is barred below, with buff belly, and is also large, and very rufous above; *N. p. spilocephala*, from Dinagat, Siargao, Mindanao and Basilan, is paler and less spotted above (not crown), with chest rufous-washed. The two last-mentioned forms are sometimes regarded as separate species, and it seems likely that there are clear differences even within the *spilonota* subspecies on different islands (e.g. Sibuyan, Cebu and Camiguin, as shown in recent photos from these islands). Studies of the taxonomy and biology of the entire group are required. It will be interesting also to see if a newly discovered (but as yet undescribed) *Ninox* owl from Talaud Island, south of Mindanao, is closely related to these owls or whether it will merit species status.

Similar species Clearly larger Brown Hawk Owl (221) is sympatric but has much longer wings and no obvious white spots on the wing-coverts and is more coarsely streaked below. Philippine Hawk Owl does not occur in Mindoro where it is replaced by Mindoro Hawk Owl (224) which is densely vermiculated and barred on crown and mantle, and below; differs also vocally. Same-sized, sympatric scops owls have erectile ear-tufts.

► Race *reyi* from the Sulu Archipelago is similar to *spilocephala* from Mindanao but with prominent rufous barring on the head and back. It has brown-yellow eyes and streaks below, absent on the belly. This owl was photographed on Tawi Tawi Island; there are strong grounds for this being another full species (as Tawi Tawi Boobook *Ninox reyi*), though further study is required. Tawi Tawi, January (*Bram Demeulemeester*).

▼ Race *spilonota* from Camiguin Island has very dark eyes and no white 'eyebrows'. It has clear white scapulars but the white area on the throat is very narrow. Below it is very rufous-brown, with darker barring. If *Ninox spilonota* is proven to be a valid species, it could have different subspecies on Cebu, Sibuyan and Camiguin. Camiguin, January (*Bram Demeulemeester*).

▼ Race *spilonota* is also a possible split. In this pair, the male has a much larger white area on the throat; the female seems much darker rufous-brown below, with no white at all – like the nominate. Cebu, Philippines, March (*Bram Demeulemeester*).

▼ Birds of the race *spilonota* from Sibuyan Island may be distinct from other *spilonota*, and may represent a separate subspecies (or even a separate species altogether). It has yellow eyes and is not coarsely barred below, as on Camiguin *spilonota*. Sibuyan Island, January (*Bram Demeulemeester*).

▼ Race *spilocephala* from Mindanao is likely to be separated as a new species in the future. It has greenish-yellow eyes and a rufous-washed breast. Mindanao, January (*Rob Hutchinson*).

227. OCHRE-BELLIED HAWK OWL
Ninox ochracea

L 25–29cm; W 180–196mm

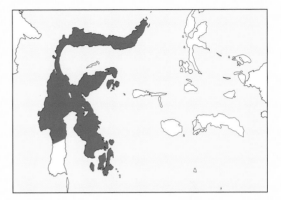

Other name Ochre-bellied Boobook

Identification A small to medium-sized owl without ear-tufts and with a relatively long tail. Female is usually a little smaller than the male. The head and upperparts are dark chestnut, tinged with brown, the crown duskier. Has whitish spots on outer webs of scapulars and some white dots on the wing-coverts. Primaries and secondaries have whitish spots on outer webs, with unmarked inner webs of primaries, and the darker brown tail has narrow whitish-buff bars. Tawny underparts have some indistinct lighter barring on upper breast, the lower breast to belly shading gradually into ochre-tawny with some diffuse darker dots. The brown facial disc becomes paler towards the eyes, has thin indistinct whitish eyebrows above, and is bordered below by a whitish chin and prominent white throat, which shows when the bird is calling. The eyes are yellow, and the bill and cere bluish-horn. Tarsi feathered to the base of the yellowish-grey toes, which have dark horn-coloured claws. *Juvenile* Undescribed.

Call Gives a series of hoarse, guttural *krurr-krurr*, each double note lasting 1.8 seconds.

Food and hunting Feeds mainly on insects. Hunts from perches.

Habitat Occupies dense, humid primary forests and mature secondary forests, also riverine forests, from lowlands up to 1000m.

Status and distribution Occurs on Sulawesi, Buton and Peleng Islands, in Indonesia. It is rather rare, and could be threatened by habitat destruction.

Geographical variation Monotypic.

Similar species Sympatric Brown Hawk Owl (221) is almost unmarked dark brown above, and whitish with prominent broad rufous streaks below. Geographically overlapping Speckled Hawk Owl (230) has brown eyes, and has dark brown upperparts with many small white spots, including on secondaries, wing-coverts and crown; its whitish underparts have a brownish pectoral collar below a large white throat patch. Sympatric Cinnabar Hawk Owl (233) is rich chestnut overall. Slightly allopatric Togian Hawk Owl (234) has a pale-spotted or pale-barred crown and is densely mottled dark and whitish below.

▶ Ochre-bellied Hawk Owl is sympatric with Cinnabar Hawk Owl but the latter is rich chestnut overall and has large white scapular spots. Tangkoko, Sulawesi, September (*Rob Hutchinson*).

228. SOLOMON HAWK OWL
Ninox jacquinoti

L 23–31cm, Wt 174g (1); W 157–228mm

Other name Solomon Islands Boobook
Identification A small to medium-sized owl without
ear-tufts. There are no data on sexual size and colour
differences. The plumage is dark rufous to dark brown
above, with numerous whitish flecks and pale edges on
feathers of crown, neck and mantle creating light bars
or vermiculations; the wing-coverts are whitish-spotted.
Dark brown flight feathers have narrow rows of small
whitish spots, and the dark brown tail has five to seven
narrow pale bars. The throat is whitish, and the upper
breast brown with indistinct lighter barring; below, the
underparts are whitish with narrow light brownish-buff
shaft-streaks. The brownish facial disc has some paler

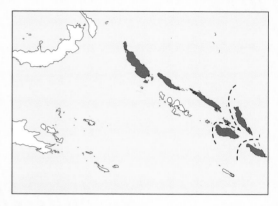

◄ Solomon Hawk Owl
of the nominate race
jacquinoti; rufous-brown to
blackish above, with thin
white 'eyebrows', rufous
belly and dark yellow eyes.
Santa Isabel, Solomons,
October (*Guy Dutson*).

► Solomon Hawk Owl of
the race *eichhorni* at a nest
hole. Smaller than nominate
and more coarsely barred
above. Arawa, Bougainville,
July (*Markus Lagerqvist*).

concentric lines, a whitish zone around the eyes and at base of bill, and narrow and very short white eyebrows. The eyes are yellow and the bill dark olive. Tarsi are feathered to the base of the yellowish toes, which have dark horn-coloured claws with blackish tips. *Juvenile* Undescribed.

Call Emits a series of *kwu-kwu* double notes at 0.6-second intervals. Duetting is common, and this owl is vocal throughout the year.

Food and hunting Feeds mostly on insects and other arthropods, and perhaps also some small vertebrates.

Habitat Prefers primary and tall secondary forests, but occurs also in forest patches and nearby gardens.

Status and distribution Widespread in Solomon Islands, but not so common on Makira (San Cristobal) and Malaita Islands. Deforestation and use of pesticides could affect the populations.

Geographical variation At least seven subspecies are listed: nominate *jacquinoti* occurs on islands of Santa Isabel and San Jorge; *N. j. eichhorni*, from Bougainville, Buka and Choiseul, is smaller and coarsely barred above; *N. j. mono*, from Mono Island, has less white on wings; *N. j. floridae*, from Florida Islands, is larger than nominate and has creamy face; *N. j. granti*, from Guadalcanal, is barred below and sometimes has brown eyes; *N. j. malaitae*, from Malaita Island, is the smallest race and relatively unmarked; *N. j. roseoaxillaris*, from Makira (San Cristobal), is also very small. A study of the taxonomy and biology of this group of subspecies is urgently needed, as it has recently been proposed that they constitute four separate species: 'West Solomons Boobook' *N. jacquinoti* (polytypic, with four subspecies), 'Guadalcanal Boobook' *N. granti*, 'Malaita Boobook' *N. malaitae* and 'Makira Boobook' *N. roseoaxillaris*.

Similar species Allopatric Jungle Hawk Owl (229) has dark brown upperparts and plain rufous-chestnut underparts, and the race living nearest to the Solomons (*N. theomacha goldii*, on Goodenough, Fergusson and Normandy Islands) is paler below and distinctly streaked brownish-buff. Much larger Fearful Owl (240) is sympatric, but is clearly paler ochre-buffish above, densely mottled, with fairly large dark brown spots and bars; it has a short tail, powerful talons and bill, and prominent white eyebrows.

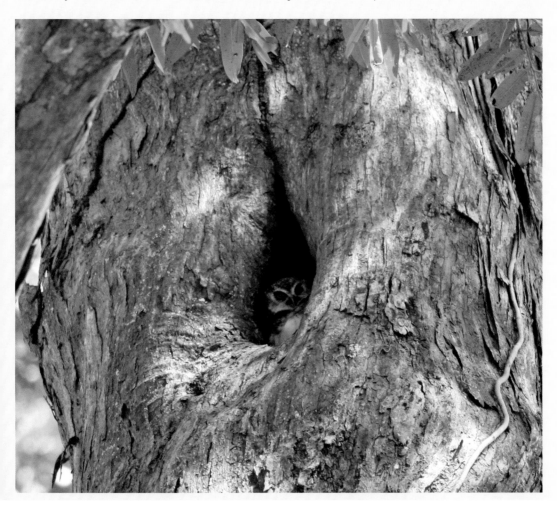

229. JUNGLE HAWK OWL
Ninox theomacha

L 20–28cm; **W** 175–227mm

Other name Papuan Boobook

Identification A small to medium-size owl without ear-tufts. Sexes are similar, females slightly larger than males. The crown and upperparts, including upperwings and uppertail, are uniformly dark brown, with a few white spots on secondaries. The underparts and undertail are uniformly chestnut, the underwing with some light or whitish bars. The blackish-brown facial disc has some white on forehead. The eyes are bright yellow or golden-yellow, the bill dark yellowish and the cere greenish-grey. Deep reddish-brown tarsi feathered to the base of the dull yellowish or brownish toes, which have black claws. *Juvenile* Natal down is grey, and mesoptile fluffy dull brown.

Call Utters a series of slightly descending double notes, *kru-kru*, repeated many times at intervals of a few seconds. Singing often continues all night.

Food and hunting Not studied, but has been observed hawking insects in the air, sometimes near streetlights.

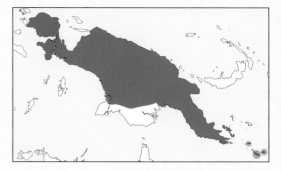

Habitat Frequents lowland rainforests and forest edges, but occurs also at up to 2500m in mountain forests.

Status and distribution Endemic to New Guinea and Papuan Islands. Locally not uncommon.

Geographical variation Four subspecies are listed: nominate *theomacha* occurs on mainland New Guinea; *N. t. hoedtii*, from West Papuan islands (Waigeo and Misool), is duller than nominate; *N. t. goldii*, from D'Entrecasteaux Archipelago (Goodenough, Fergusson and Normandy Islands), and *N. t. rosseliana*, from Louisiade Archipelago (Tagula and Rossel Islands), have white mottling on the lower breast and abdomen.

Similar species Much larger Rufous Owl (213) is partly sympatric, but has whitish to pale buffish underside with many rufous-brown bars. More robust Barking Owl (215), also partly overlapping in range, has creamy-buff to whitish underparts boldly streaked brown to chestnut. Southern Boobook (217) occurs only in S. New Guinea, and has heavily marked white underparts, breast mottled with red-brown and belly streaked. Russet Hawk Owl (231), confined to New Britain, is finely whitish-speckled above and has a white-barred rufous-brown pectoral band. Bismarck Hawk Owl (238), found only in New Hanover and New Ireland, has pale-spotted upperparts and rufous-barred whitish underparts. Sympatric Papuan Hawk Owl (239) is clearly larger, with a very long, prominently barred tail, brown upperparts and wings with dark brown to black bands, and buffish underside with bold brown streaking.

◄ A pair of Jungle Hawk Owls. There is little difference between the sexes. Papuan Hawk Owl *Uroglaux dimorpha* is larger and buffish below, with bold brown streaking. Papua New Guinea, August (*Stuart Elsom*).

230. SPECKLED HAWK OWL
Ninox punctulata

L 20–27cm, **Wt** 151g (1♂); **W** 157–177mm

Other name Speckled Boobook

Identification A small owl without ear-tufts. Sexes are reported to be similar in colour, but female is slightly larger, although only one sexed measurement available. The plumage is dull reddish-brown above, heavily spotted with white on crown, neck and mantle. Wing-coverts and secondaries also are white-spotted, the dark brown flight feathers have rows of whitish spots, and the brown tail has narrow pale bars. Rather variable underparts are often marked by a narrow white-spotted brown band across the upper breast with a whitish oval area below, and then a second similarly spotted brown band, the rest of the underparts usually being whitish with pale-barred flanks. The dark brown facial disc is bordered above by whitish eyebrows extending from base of bill to over the ear-coverts, and below by a large whitish throat patch or half-collar. The eyes are yellow and the bill greenish-yellow. Tarsi feathered tarsi, with yellowish-grey toes and dark horn-coloured claws. *Juvenile* Mesoptile is dark brown above and below, with eyebrows mixed dark brown and white; wing and tail feathers as adult.

Call Gives a loud and clear *toi-toi-toit* song, the first two notes short and the last one longer and lower-pitched. Very vocal at night throughout year.

Food and hunting Unknown, but probably eats mainly insects.

Habitat Prefers primary forests near streams, but is found also in open woodlands and cultivated lands, even near human settlements. Occurs from lowlands to 1100m, rarely to 2300m.

Status and distribution Endemic to Sulawesi and the islands of Kabaena, Muna and Buton, in Indonesia. It is common, but is probably threatened by forest transformation and the use of pesticides.

Geographical variation Monotypic. Taxonomy, vocalisations and biology require study.

Similar species Geographically overlapping Brown Hawk Owl (221) is larger, and boldly streaked below. Yellow-eyed Ochre-bellied Hawk Owl (227) is sympatric,

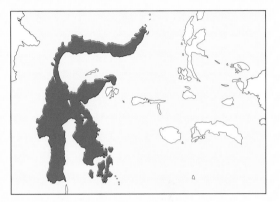

but is dark chestnut tinged with brown above, and without white spots; breast is cinnamon-tawny and belly pale brownish-yellow. Also overlapping, Cinnabar Hawk Owl (233) is bright chestnut all over. Marginally allopatric Togian Hawk Owl (234) has no prominent white throat patch, and is whitish and densely brown-mottled below.

▶ Speckled Hawk Owl comes to buildings and is vocal all year round. It has yellow eyes and a prominent white throat and neck-sides with small white dots on the dull reddish-brown upperparts, including the wing-coverts and secondaries. 'Eyebrows' also white, but narrow on blackish-brown facial disc. Dumoga Bone, Sulawesi, October (*Bram Demeulemeester*).

231. RUSSET HAWK OWL
Ninox odiosa

L 20–23cm, **Wt** 209g (1♀); **W** 170–187mm

Other name New Britain Boobook

Identification A small owl without ear-tufts. Sexes are similar in colour, but females are slightly larger than males. The forehead, crown, hindneck, side of neck and a wide band across upper breast are dark brown, spangled with small buffy-white spots; slightly lighter mantle is without markings, but the dark chocolate-brown wing-coverts have sparse white spots of varying size. Flight feathers have some white bars, and brown tail also has a few white bars. the lower breast and abdomen are heavily mottled whitish and light brown. The brown facial disc with short white eyebrows is bordered below by a large area of white on the throat and extending to the side of the neck. The eyes are bright yellow to orange and the bill pale greenish with a yellowish tip. Feathered tarsi have yellowish-brown toes, and dusky horn-coloured claws with blackish tips. *Juvenile* Undescribed.

Call Emits a long continuous song, *whoo-whoo-whoo-whoo...*, sometimes lasting for up to three minutes; the song starts low, then rises in pitch and becomes faster and louder.

Food and hunting Feeds on insects and small vertebrates, such as bats.

Habitat Frequents lowland and hill forests, but occurs

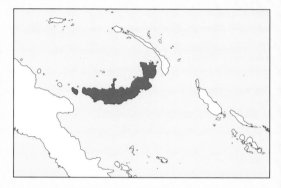

also in cultivated areas and even in towns. From low elevations to at least 800m.

Status and distribution Endemic to New Britain, in the Bismarck Archipelago. It is widely distributed and not too rare.

Geographical variation Monotypic. Taxonomy and biology require a detailed study.

Similar species Bismarck Hawk Owl (238) also occurs in New Britain but is larger and without a broad band on the upper breast; it has an unspotted crown and hindneck, and several broad reddish-brown bars from breast to belly.

◀ Russet Hawk Owl is similar to Speckled Hawk Owl *Ninox punctulata* but brighter, with a long tail. It has a white-barred rufous-brown chest band; the belly is white with heavy barring on the flanks and strong shaft-streaks below. Eyes are orange-yellow. Pokili, New Britain, May (*Nik Borrow*).

232. MOLUCCAN HAWK OWL
Ninox squamipila

L 25–39cm, Wt 140–210g; W 190–241mm

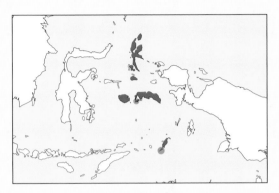

Other name Moluccan Boobook

Identification A small to medium-sized owl without ear-tufts. Sexes are similar in colour, but females often slightly larger than males. The plumage is dark reddish-brown above, darker on head, the wing-coverts with buffish spots, and outer webs of scapulars with short, bar-like whitish markings. Rufous-brown flight feathers have some four rows of light russet spots, and the reddish-brown tail is barred pale rusty. The upper breast is densely dusky-barred rufous, and the rest of the underparts are rather pale but with dense dark rufous barring. The facial disc is mainly reddish-brown, with thin whitish eyebrows. The cere is yellow and the bill pale grey; eye colour may be age-dependent (brown in younger individuals and yellow in older ones). Rufous tarsi are feathered to the base of the yellowish-brown toes, which have horn-coloured claws with darker tips. *Juvenile* Undescribed.

Call Emits a frog-like double-noted *kwaor-kwaor kwaor-kwaor*.

Food and hunting Feeds mainly on insects, such as grasshoppers. Recorded as hunting chiefly in mid-canopy.

Habitat Inhabits tropical rainforests at sea level in the coastal zone, and groves, thickets and mountain forests up to 1750m.

Status and distribution Endemic to the Moluccas and Tanimbar Islands, in Indonesia. It is locally rather common, especially on Buru.

Geographical variation Four subspecies are recognised: brown-eyed nominate *squamipila* is confined to Seram; yellow-eyed *N. s. hypogramma* occurs on Halmahera, Ternate and Bacan; *N. s. hantu*, from Buru, is very dark, small and with yellow eyes, and below is rufous-brownish and indistinctly barred; *N. s. forbesi*, from Tanimbar Islands, is paler rusty-brown and also has yellow eyes. There are grounds to split this species (e.g. into 'Halmahera Boobook' *N. hypogramma* and 'Tanimbar Boobook *N. forbesi*'), because vocalisations are reported to be very different throughout these islands. Further, the Bacan hawk owls are very different from those on Halmahera, and may represent at least one undescribed taxon. A possible new, as yet undescribed, *Ninox* owl is said to be present on Gebe Island, between Halmahera and West Papua (Vogelkop Peninsula).

Similar species Partly sympatric Barking Owl (215) is far less rufous and dark brownish-grey above, and the white underparts are not barred but streaked with dark brownish-grey. Allopatric Ochre-bellied Hawk Owl (227) lacks dense barring below. Non-overlapping Speckled Hawk Owl (230) has prominent white throat patch and white-speckled crown and mantle.

◄ Nominate Moluccan Hawk Owl is dark reddish-brown above, and darker on the head. The white-barred scapulars are well-shown here. Seram, January (*James Eaton*).

▲ Race *forbesi* of Moluccan Hawk Owl from the Tanimbar Islands has a much paler crown, back and chest than other races. This taxon may also merit specific status. Tanimbar (*James Eaton*).

▶ Moluccan Hawk Owl race *hypogramma* from Halmahera may be a separate species due to vocal and colour differences. Halmahera, October (*Rob Hutchinson*).

233. CINNABAR HAWK OWL
Ninox ios

L 22cm (1), **Wt** 78g (1); **W** 172mm (1)

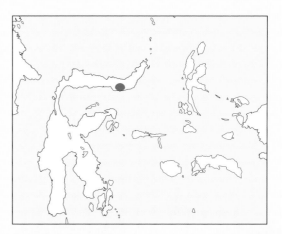

Other name Cinnabar Boobook

Identification A small owl without ear-tufts. No sexed measurements are available, but female is thought to be larger than male. The crown, nape, upperparts, secondaries and wing-coverts are deep reddish-brown, only scapulars having some triangular whitish dots, and primaries indistinctly barred chestnut and brown; the forehead is slightly paler than crown, shading into coloration of the latter. Has narrow, pointed wings, and the chestnut tail, which has some not very prominent brown barring, is relatively long. Bright rufous underparts have indistinct narrow, pale yellowish shaft-streaks. The head lacks facial patterning and the eyebrows are inconspicuous. The eyes are yellow, with pink rims of eyelids, the bill ivory-coloured and the cere dirty yellowish. Unusually short and slender tarsi have fawn feathering to near the base of the whitish-yellow toes, which have weak horn-coloured claws. *Juvenile* Undescribed.

Call Not well known, but thought to be disyllabic dry hoots, given in series with rising and falling pitch.

Food and hunting Unknown.

Habitat Holotype was found in mountain forest at 1120m. Probably occurs primarily at higher elevations than those where Ochre-bellied Hawk Owl (227) is found.

Status and distribution Has a very restricted distribution in N. Sulawesi, from where one specimen was collected in 1985; since then there have been only a few sight records. One mist-netted individual was larger than the holotype and therefore thought to be a female. Distribution and conservation status of this owl are not well known, and even less is known of its biology and vocalisations. The use of nets to catch bats for the local meat market may have an adverse impact on a small population of owls if some are accidentally caught as a result.

Geographical variation Monotypic.

Similar species This is the only *Ninox* with almost uniformly rufous-chestnut appearance. It is also smaller than any other hawk owls on Sulawesi.

▶ Cinnabar Hawk Owl is almost uniform rich chestnut, and the smallest hawk owl on Sulawesi; it normally occurs at higher elevations than Ochre-bellied Hawk Owl *Ninox ochracea*. Gunung Ambang, Sulawesi, March (*Filip Verbelen*).

234. TOGIAN HAWK OWL
Ninox burhani

L 25cm (1), **Wt** 98–100g; **W** 183–184mm

Other name Togian Boobook

Identification A small to medium-sized owl without ear-tufts. There are no data on sexual size and colour differences. Has dark brown crown, nape and upperparts with fine whitish or pale buffish spots and bars. Umber-coloured flight feathers have whitish and tawny-olive dots, and primaries are marked with triangular whitish spots; the rather long tail is dark greyish-brown and barred. Whitish underparts are mottled and brown-streaked. The brown face has rather prominent light tawny-olive eyebrows. The eyes are yellow to orange-yellow, and the bill creamy to grey with pale greenish culmen. Rather slender tarsi are not feathered, and the toes are bristled and have black claws. *Juvenile* Unknown.

Call Reported to give a croaking call of two to four notes, *ko-koko-ok*, but single-note croaks also heard. These calls differ clearly from any known call of congeners in the region.

Food and hunting Unstudied.

Habitat Found in tropical lowland forests, but also in mixed gardens near human habitations.

Status and distribution Endemic to the Togian Archipelago (in Gulf of Tomini), off C. Sulawesi. This owl is poorly known, but obviously occurs in moderate numbers in its very limited range. The islands are not

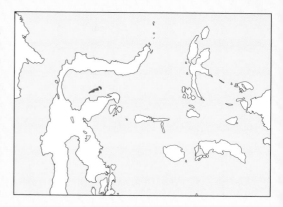

large, however, and forest clearance is increasing.

Geographical variation Monotypic.

Similar species Allopatric and larger Brown Hawk Owl (221) is prominently streaked below. Also not overlapping in range, Ochre-bellied Hawk Owl (227) has dark brownish-chestnut upperparts and dusky, unmarked crown, and is ochre-tawny below. Marginally allopatric Speckled Hawk Owl (230) has a prominent white throat and a large white patch on the upper breast separated by a white-spotted brown collar. Non-overlapping Cinnabar Hawk Owl (233) has no pale eyebrows and is overall bright chestnut.

◄ A pair of Togian Hawk Owls on Batudaka Island in Indonesia. Each has a little rufous on the chest, one more than the other. Batudaka Island, October (*Bram Demeulemeester*).

235. LITTLE SUMBA HAWK OWL
Ninox sumbaensis

L 23cm (1), **Wt** 90g (1); **W** 176mm (1); **WS** 57cm (1)

Other name Little Sumba Boobook

Identification A small owl without ear-tufts. No sexed measurements are available. The greyish-brown crown and nape have fine whitish barring and mottling, and the brownish-grey mantle and back have narrow blackish-brown vermiculations. Large whitish areas on scapulars, with a dark streak and bar on each feather, form a whitish row across the shoulder. Primaries and secondaries are barred rufous-grey and dark brown, and the greyish-brown tail has *c.* 16 dark brown bars. Rufous throat has dark vermiculations, and the buffish-white underparts fine, chevron-like dark vermiculations. Pale greyish-brown facial disc is rather indistinct, lacking a prominent rim, but with distinct white eyebrows above. The eyes are yellow. The bill is greenish-yellow with a yellow cere. Tarsi are feathered to the tops of the greyish-yellow, bristled toes, which have yellowish claws with greyish-black tips. *Juvenile* One fledged young observed was paler reddish, with no V-shaped vermiculations on its underparts.

Call Utters a single-note whistle, repeated at intervals of two to three seconds.

Food and hunting Unknown.

Habitat Inhabits primary and secondary forests, and avoids open areas outside the forest.

Status and distribution Endemic to Sumba Island, in the Lesser Sundas. This owl is becoming less common as a result of deforestation. Studies of its biology and vocalisations are needed.

Geographical variation Monotypic.

Similar species Much larger Sumba Boobook (216) is fully sympatric, but has brown eyes, a much darker bill and is densely white-spotted above and barred and streaked brownish-rufous below; vocally, the two species are very different. There are no other similar owls on Sumba.

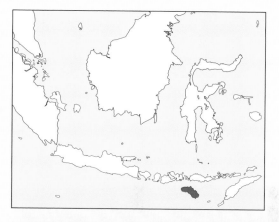

▶ Little Sumba Hawk Owl is yellowish-brown with prominent white 'eyebrows'. It is much smaller than the sympatric Sumba Boobook *Ninox rudolfi*. Sumba, July (*James Eaton*).

236. CHRISTMAS HAWK OWL
Ninox natalis

L 26–29cm, Wt 130–190g; W 188–200mm

Other name Christmas Boobook

Identification A small to medium-sized owl without ear-tufts. Sexes are similar in plumage, but the female is slightly larger than the male. The crown and nape are light tawny-brown with some pale buffish dots, and the brownish-yellow mantle has a few buffish-white dots and darker brown bars. Reddish-tawny wing-coverts have some darker and paler spots, and the scapulars are mottled buff and brownish-yellow. The flight feathers are ochre-tawny, with five to seven dark brown bars on primaries and four or five on secondaries, and the dark brown tail, which is relatively long, has ten light rufous-buffish bars. The rufous-ochre underparts are evenly barred reddish-brown. The brownish-chestnut facial disc has indistinct whitish eyebrows; the chin feathers are tawny with white shaft-streaks. The eyes are rich yellow, with prominent black eyelids, the bill is black and the cere light grey or lemon-yellow. Tarsi are feathered rufous-ochre with reddish-brown barring to the base of the straw-yellow toes, which have dark claws. *Juvenile* Has the head, neck and back spotted, the underparts

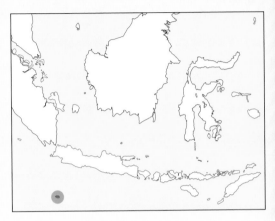

more widely barred, and the tail buffy chestnut.

Call Utters a double-noted *glu-goog glu-goog*, repeated at intervals of several seconds.

Food and hunting Large insects are the main prey, but skinks, geckos and small birds are also taken. Seen hawking moths around streetlights, but in forest it forages from perches.

Habitat Occupies primary and secondary rainforests, but hunts also in man-made clearings and around settlements. It is not known if mature trees are necessary for its continued survival.

Status and distribution Endemic to Christmas Island, 375km south of Java. This owl was formerly quite common, but decrease in extent of remaining rainforest may reduce the number of owls accordingly. It is listed as Vulnerable by BirdLife International. Latest survey, in 2004, found 1,000 individuals in primary forests and a small number also in second growth, so the species is obviously not yet in immediate danger.

Geographical variation Monotypic.

Similar species No other *Ninox* is known to occur on Christmas Island.

◀ Christmas Hawk Owl is a little smaller than Moluccan Hawk Owl with indistinct ruff and large yellow eyes. Head is tawny-brown. Buff-white below with rufous-brown bars. Kiritimati (Christmas Island), February (*David Hollands*).

▶ A pair of Christmas Hawk Owls. Male (above) looks a little smaller and a darker rufous than the female. Both have whitish supercilia, rich rufous-brown upperparts with white spots on the wing-coverts, and fine white bars on scapulars. Kiritimati (Christmas Island), January (*Eric Sohn Joo Tan*).

237. MANUS HAWK OWL
Ninox meeki

L 25–31cm; W 230–240mm

Other name Manus Boobook

Identification A small to medium-sized owl without ear-tufts. Females are reported to be slightly smaller than males, although no known sexed measurements are available. Male has uniformly rufous crown, but female has some barring or fine streaks. The reddish-brown nape has clear pale buffish barring, and the ochre-tinged rufous upperparts have some light bars; scapulars and wing-coverts are also pale-barred. Flight feathers and tail are barred dark brown and whitish. The throat is light tawny, and the whitish-buff neck and upper breast are densely marked with broad reddish-brown streaks, below which the rest of the underparts are whitish-buff with long rusty-brown streaks, streaks denser on lower breast than on belly. The rather uniformly brown facial disc becomes slightly paler around the eyes. The eyes are yellow, and the bill is slate-blue with paler tip and white rictal bristles. Feathered tarsi have creamy-yellow toes, the lower tarsi and toes bristled, with horn-coloured claws with darker tips. *Juvenile* Has an almost white throat, plain brown breast, heavier white bars on wing-coverts, and white-barred rump; wider and paler tail-bars, and narrower streaks on underside.

Call Utters a series of some ten gruff notes, slowly accelerating.

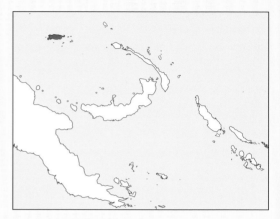

Food and hunting Unknown, but probably mainly eats insects.

Habitat Inhabits forests, but also occurs in degraded areas, riparian cultivations and open areas with some trees.

Status and distribution Endemic to Manus Island, in N. Bismarck Archipelago, New Guinea. Population size and biology are unknown but, as the island still has most of its forest cover, the owl is not immediately endangered, despite its very limited range.

Geographical variation Monotypic.

Similar species Slightly geographically separated Jungle Hawk Owl (229) is dark chocolate-brown above and plain, bright rufous-chestnut below. Also narrowly allopatric, Russet Hawk Owl (231) has white-speckled crown and is blotched or broadly barred below. Underparts of allopatric Bismarck Hawk Owl (238) are not streaked, but distinctly barred.

◀ Manus Hawk Owl is very similar to Bismarck Hawk Owl *Ninox variegata*, but it has prominent rufous-brown streaks instead of bars on the white underparts. There are sexual differences in crown colour, with the male uniform rufous-brown and the female barred. Manus, Papua New Guinea, July (*Jon Hornbuckle*).

238. BISMARCK HAWK OWL
Ninox variegata

L 23–30cm; **W** 192–224mm

Other name New Ireland Boobook

Identification A small to medium-sized owl without ear-tufts. No data exist on sexual size differences. Sexes reported to be similar in plumage. Dark brown upperside is faintly spotted with light and dark rufous, but mantle and back often more or less unmarked; scapulars and wing-coverts have small white bars and spots, but no clear row across the shoulder, and dark brown alula is unspotted. The brown primaries and secondaries have five rows of whitish spots, and the dark brown tail has paler bars. Whitish underparts have clear scaly barring, the dark morph with dark rufous-brown bars and the light morph with orange-rufous ones; neck is more densely marked, appearing darker than the breast and belly. The brown facial disc and head are more greyish-brown than mantle, with greyish-brown ear-coverts. The eyes are yellow, and the bill yellowish-horn with paler tip. Tarsi feathered to the base of the dull yellow and bristled toes, which have dusky horn-coloured claws. *Juvenile* Unknown.

Call Emits a double-noted, frog-like, croaking *kra-kra kra-kra*.

Food and hunting Little known, but probably eats insects.

Habitat Inhabits lowland forests up to 1000m.

Status and distribution Endemic to islands of New Hanover, New Ireland and New Britain, in Bismarck

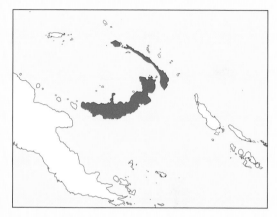

Archipelago, New Guinea. It is fairly common.

Geographical variation Two subspecies: nominate *variegata* occurs on New Ireland and New Britain; *N. v. superior*, on New Hanover, is paler brown and larger.

Similar species In New Britain could be confused with Russet Hawk Owl (231), which is also chocolate-brown, but has distinct buffish-white spots and speckles on forehead, crown, hindneck and neck-sides, broad dark brown band spangled with buff and white on upper breast, and fewer tail-bars. Narrowly allopatric Manus Hawk Owl (237) has prominently streaked rufous underparts.

▼ Bismark Hawk Owl of the nominate race; dark rufous-brown, spotted above and barred below. Crown and hindneck unspotted, and no broad band on upper breast. Russet Hawk Owl *Ninox odiosa* is smaller with distinct buffish-white spots and speckles on the forehead, crown, hindneck and neck-sides. Specimen from BMNH (Tring), collected in New Ireland, Bismarck Archipelago (*Nigel Redman*).

239. PAPUAN HAWK OWL
Uroglaux dimorpha

L 30–34cm; W 200–225mm

Other name New Guinea Hawk Owl

Identification A small to medium-sized owl without ear-tufts. No data for determination of any size differences between female and male. This species is very similar to *Ninox* owls, but head smaller and tail longer. The dark brown crown and hindneck are streaked brown to buffy brown; the upperparts, including wings and tail, are barred with even transverse bands of blackish-brown and light brown or rusty brown; and the underparts are buffy white to pale brownish-yellow with bold black streaking from throat to abdomen. The whitish facial disc is finely streaked with black. Eyes are yellow and the bill pale grey-blue with dark tip. Tarsi feathered to the base of the yellowish toes. Claws black. **Juvenile** Mesoptile is paler than the adults. Fledglings have whitish head and underparts.

Call Emits a drawn-out whistle, *poweeeeho*, repeated at intervals of several seconds.

Food and hunting Feeds mainly on insects, but also takes rodents and birds up to the size of doves.

Habitat Frequents rainforests and gallery forests in savannas, from sea level to 1500m.

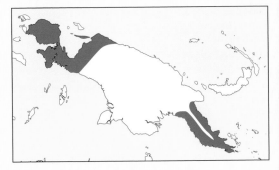

Status and distribution Endemic to New Guinea in two disjunct areas: SE Papua New Guinea, and Indonesian West Papua, including Yapen Island. This species' status is little known, but it could be threatened by forest destruction.

Geographical variation Monotypic.

Similar species All *Ninox* owls have a clearly shorter tail and lack heavy barring above. Sympatric Rufous Owl (213) is much larger, and has profuse but fine barring above and below.

▼ Papuan Hawk Owl is a slim and small-headed hawk owl. Underparts are streaked on a whitish to pale-rufous ground colour, darkest on the chest. Long black bristles around the dark-tipped, greyish bill. Relatively large eyes are yellow. The right-hand image shows the bird spreading its wings and tail in alarm. West Papua, July (*Edwin Collaerts*).

240. FEARFUL OWL
Nesasio solomonensis

L *c.* 38cm; W *c.* 300mm

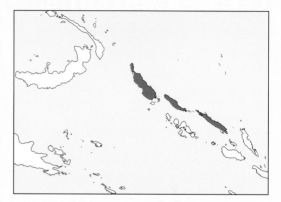

Identification A medium-sized owl without ear-tufts. No measurements are available to determine possible sexual size differences. The yellowish-brown or ochre-buff crown, nape and upperparts, with a rufous tinge, are densely streaked and barred dark brown; yellowish-tawny scapulars have brown shaft-streaks and a few brown cross-bars. Wings and tail are prominently barred light and dark brown. Deep ochre underparts are marked with narrow dark brown to blackish shaft-streaks. The rufous face has a prominent ruff and a mask-like dusky area around the eyes, and white eyebrows and lores form a prominent white X-shaped mark on facial disc. The eyes are dark yellowish, and the very powerful bill blackish-blue. Tarsi are feathered to the base of the ashy-grey toes, which have horn-coloured claws; the toes and claws are very strong. *Juvenile* Undescribed.

Call Said to utter a single drawn-out note sounding ghostly and mournful to human ears.

Food and hunting Hardly known. According to local inhabitants, it eats phalangers *Phalanger orientalis*, possums and small birds.

Habitat Occupies primary and secondary lowland forests.

Status and distribution Occurs on Bougainville and in Solomon Islands (Choiseul and Santa Isabel). Appears to be rare, but perhaps locally not uncommon. This species is listed as Vulnerable by BirdLife International, being threatened by forest destruction.

Geographical variation Monotypic.

Similar species Much smaller, hawk-like Solomon Hawk Owl (228) is fully sympatric, but has no white 'X' on face, is far less striped or barred below, and is rufous above with numerous spots and bars. The now extinct Laughing Owl *Sceloglaux albifacies* of New Zealand is said to have been similar to Fearful Owl in colour, voice and size. Both are S. Pacific species but there is no proved relationship between them and any similarity is likely to be due to convergence.

◄ Fearful Owl resembles Short-eared Owl *Asio flammeus* in general coloration, but has no ear-tufts, and its eyes are much larger and darker. White 'eyebrows' are very prominent. Tawny below, with dark streaking. Toes bristled and bill blackish. Santa Isabel, Solomon Islands, October (*Guy Dutson*).

241. JAMAICAN OWL
Pseudoscops grammicus

L 27–34cm; W 197–229mm

Identification A small to medium-sized owl with rather long, mottled brown and blackish ear-tufts. No data on sexual size or colour differences are available. The crown and upperparts are fairly uniformly warm tawny-brown, forehead and crown with dark mottling, and mantle, scapulars and wing-coverts with blackish arrow-shaped markings and fine vermiculations; there are no white spots on scapulars. Flight feathers and warm reddish-brown tail are barred light and dark. Below, it is slightly paler than above, more dull yellowish, with long dark shaft-streaks and fine brown vermiculations. The warm reddish or pale yellowish-brown facial disc is bordered whitish in a narrow band

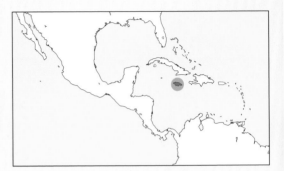

near the dark ruff, so that the face appears double-ruffed. The large eyes are hazel-brown and the bill and cere bluish-grey. Reddish-brown legs are feathered to the base of the greyish-brown toes, which have dark brown claws with blackish tips. *Juvenile* Downy chick is white. Juvenile is lighter above, with pale greyish-brown back; rest of plumage is dull cinnamon-buff.

Call Probable song is a repetition of *to-whooo* notes at varying intervals; in addition, a rough, frog-like guttural growl is heard fairly often.

Food and hunting Feeds mainly on large insects and spiders, but also takes lizards, frogs, small birds and rodents.

Habitat Inhabits lowland and mountain forests, and also semi-open country with some trees, as well as gardens with trees. Mainly at lower elevations, and only rarely on high mountains.

Status and distribution Endemic to Jamaica, in the Greater Antilles, where it is thought to be fairly common, although little studied. It could be put at risk through the continuing destruction of forests.

Geographical variation Monotypic. Molecular and biological studies are still needed in order to clarify the taxonomic relationships between this genus and the following *Asio* owls.

Similar species Clearly larger Short-eared Owl (247) has very short ear-tufts and is only a straggler and irregular breeder in the Caribbean; it has thus far not been recorded in Jamaica.

◄ Jamaican Owl resembles Striped Owl *Asio clamator* a little, but their possible relationship has not been proven with molecular studies. It is buffish-yellow to yellow-ochre below with dusky brown streaks. Eyes are hazel; in this photograph the ear-tufts are fully erected. Jamaica, November (*Yves-Jacques Rey-Millet*).

▲ Jamaican Owl is a small, tawny-brown owl with dark brown and black shaft-streaks, bars and flecks. Face amber to pale cinnamon. The owl erects or flattens the ear-tufts at will. Jamaica, November (*Yves-Jacques Rey-Millet*).

► This juvenile Jamaican Owl still has down remnants on its head. Jamaica, July (*Paul Noakes*).

242. STYGIAN OWL
Asio stygius

L 38–46cm, Wt 591–675g; W 291–380mm

Identification A medium-sized owl with long, erectile ear-tufts. Females are larger than males. The plumage is dark sooty-brown above, pale-spotted on mantle and back, the forehead appearing rather pale. The very long wings are almost plain dark brown, the primaries with indistinct rows of paler spots, the secondaries barred light and dark; the dark sooty-brown tail is relatively short, with some paler bars. It is dull yellowish below, with heavy dark markings on upper breast, the rest of underparts streaked dark and with distinct cross-bars. The dusky-brown facial disc has a finely white-speckled lateral ruff, and rather prominent but short whitish eyebrows. The eyes are yellow to orange-yellow, the bill blackish and the cere greyish-brown. The legs are feathered, and the brownish-fleshy toes partly feathered with short plumes, the claws dark horn-coloured with blackish tips. *Juvenile* Downy chick is whitish. Mesoptile is dull yellow, diffusely barred with grey, and with a sooty-black facial disc and wings. *In flight* Flies with rather slow wingbeats and frequent glides.

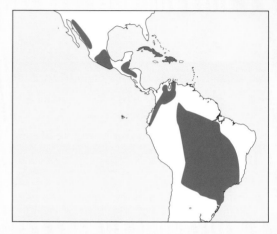

Call One song is a deep emphatic, downslurred *whuof*, repeated at intervals of six to ten seconds.

Food and hunting Eats small vertebrates and insects. It hawks bats on the wing, but normally hunts other prey from a perch.

Habitat Inhabits humid to semi-arid mountain forests and chaco, but occurs also at lower elevations and in semi-open landscapes with trees. Found from sea level to 3900m in Mexico.

Status and distribution From US–Mexico border to SE Brazil and N. Argentina; a vagrant was recorded in the lower Rio Grande Valley, in Texas. Its status is uncertain, but it seems locally to be partly overlooked as it is not very vocal.

Geographical variation Four subspecies are described: nominate *stygius* is found from Colombia to C. Brazil; *A. s. barberoi*, from Paraguay and N. Argentina to SE Brazil, is larger than the nominate; *A. s. robustus*, from Mexico to Belize, is paler below; *A. s. siguapa*, from Cuba (including Isle of Pines) and Hispaniola (including Gonâve Island), is small, with whiter markings. Distribution and geographical variation, however, are poorly known and need further study.

Similar species Allopatric Long-eared Owl (243) is smaller, and has golden-rufous facial disc distinctly dark-rimmed. Sympatric Striped Owl (246) is boldly streaked below and has dark brown eyes. Partly overlapping in range, Short-eared Owl (247) is smaller and is streaked below, with tiny ear-tufts and blackish area around yellow eyes. Sympatric screech owls are all smaller.

▲ The Caribbean race *siguapa* is said to be paler than the nominate, but this difference is nominal. Cuba, April (*Hennie Lammers*).

▲▶ An adult Stygian Owl feasting on a fledgling West Indian Woodpecker *Melanerpes superciliaris*. Cuba, April (*Hennie Lammers*).

◀ Race *robustus* has a very dark blackish back with some whitish mottling. Belize, July (*Yeray Seminario*).

▶ Race *siguapa* has whiter markings than the nominate and the face may also be a little paler. Zapata, Cuba, December (*Christian Artuso*).

243. LONG-EARED OWL
Asio otus

L 35–40cm, Wt 200–435g; W 252–320mm, WS 90–100cm

Other name Northern Long-eared Owl

Identification A medium-sized owl with conspicuous ear-tufts. Darker females are between 50g and 80g heavier than males. The crown and upperparts are darkish ochre-brown with a grey cast, the crown with fine dark mottling, nape and hindneck with blackish shaft-streaks, and mantle and back with small dark spots and blackish streaks; whitish outer webs of scapulars form a row across the shoulder. Flight feathers are barred light and dark, the primaries with yellowish-buff bases, and the tail is greyish-washed ochraceous brown with six to eight thin dark bars. Underparts are pale dull yellowish-buff with blackish streaks, belly and ventral region also with very thin cross-bars (appearing uniform at distance); underwing appears mostly off-white or whitish (barred remiges) and with dark comma-like carpal patch. Facial disc is buff with white-edged blackish rim, and with short whitish eyebrows and pale area running down centre, this bordered by rather diffuse dusky stripes extending to prominent ear-tufts, making face appear elongated. The eyes are deep orange, the bill blackish and the cere grey. Whitish-yellow feathered legs and toes, with black claws. *Juvenile* Downy chick is whitish with pink skin. Fluffy mesoptile is greyish-white or brownish-white, diffusely dusky-barred, the face almost black, ear-tufts not fully developed, but flight and tail feathers similar to those of adult. *In flight* Flies with a few wingbeats often followed by a long glide, the wings held straight in line and level with body, less often in shallow 'V'.

Call Gives a deep *whooh*, repeated at regular or varying intervals of two to four seconds, these hoots repeated in series of 10–200 notes.

Food and hunting Feeds mainly on voles and other small mammals, and also takes birds, bats and frogs; occasionally eats insects. Hunts normally by flying low along hedges and forest edges, but sometimes from a perch.

Habitat Inhabits coniferous and mixed forests, often with clearings or near open countryside; it occurs also in gardens, cemeteries and wooded areas in towns, from near sea level to 2750m.

Status and distribution In North America occurs widely from S. Canada south to N. Mexico, and in Eurasia from Scandinavia, Britain and much of mainland Europe east to N. & C. China, N. Korea and Japan and south to W Himalayas, but absent between Caspian and Aral Seas. Breeds also in N. Africa from Morocco east to Tunisia. In many regions this species is rather common, especially during years with high vole numbers. It has been estimated recently that an average of 82,000 pairs live in C. Europe; in Finland there were 1,130 nests in 2009, and only 107 in 2010 after the crash in vole populations. It winters south to C. & S. Mexico (to Oaxaca), Gulf Coast, Florida, Cuba, N. Africa, NW India and S. China.

Geographical variation Four subspecies are accepted: nominate *otus* occurs in Eurasia and N. Africa; *A. o.*

▶ Nominate Long-eared Owl has blackish streaks on the underparts with only narrow cross-barring on the belly. Speyside, Scotland (*Phil McLean*).

◀ A male brings a field vole to the female, in an old Magpie's nest in Saratov, Russia. May (*Evgeny Kotelevsky*).

canariensis, from the Canary Islands, is smaller than the nominate, with upperparts and breast more heavily mottled and vermiculated, and the belly has more pronounced barring and slightly broader dark shaft-streaks; *A. o. wilsonianus*, from Canada and USA (except western parts), is also darker, with more prominent barring below, rufous facial disc and yellow eyes; *A. o. tuftsi*, from British Columbia to NW Mexico, is paler and greyer than the previous race. It has been questioned whether this owl comprises more than one species.

Similar species Allopatric Stygian Owl (242) is larger and darker and has less prominent ruff around facial disc. Also geographically separated, Striped Owl (246) is boldly streaked below and has dark brown eyes. Sympatric Short-eared Owl (247) has very short ear-tufts, black-surrounded yellow eyes, and boldly streaked underparts. Brown-eyed Marsh Owl (249), which is narrowly sympatric only in Morocco, has tiny ear-tufts and is generally brown, with fine mottling or barring below. Sympatric scops and screech owls are much smaller and have quite short ear-tufts.

▲ Nominate Long-eared Owl from the rear. This bird has an extremely pale face, typical of older individuals. Hungary, January (*Lee Mott*).

▲▶ Nominate Long-eared Owl is heavily streaked and blotched on the chest, more finely streaked but less barred on the belly. Karlovac, Croatia, February (*Marko Matesic*).

◀ Race *wilsonianus* has a rufous face and more prominent barring below than the nominate. Quebec, Canada (*Scott Linstead*).

▶ Nominate Long-eared Owl in flight showing well the blackish comma-like carpal patches. (*Dieter Hopf*).

244. AFRICAN LONG-EARED OWL
Asio abyssinicus

L 40–44cm, Wt 245–400g; W 309–360mm

Other name Abyssinian Long-eared Owl
Identification A medium-sized owl with prominent dark ear-tufts and very long wings. Females are only slightly larger than males. Crown and upperparts are dark golden-brown, tawny-mottled, and without distinct whitish areas on scapulars. Flight feathers are conspicuously barred light and dark, and the greyish-brown tail has relatively broad dark bars. Underparts are brownish-yellow with dark brown mottling on breast, and with irregular tawny and whitish barring on belly. Rich tawny-brown facial disc has blackish-brown ruff, with inconspicuous short pale eyebrows. The eyes are yellow-orange, the bill blackish and the cere greyish-brown. Legs and toes are fully feathered,

with blackish-horn claws. *Juvenile* Undescribed, but most likely similar to that of Long-eared Owl (243).
Call Emits a deep, soft and disyllabic hoot, *ooo-oooomm*, drawn out and rising slightly in pitch.
Food and hunting Reported to take larger small mammals (up to 90g) than those eaten by Long-eared Owl. Often hunts on the wing, hovering over potential prey.
Habitat Occupies various types of mountain forest, especially juniper and *Podocarpus* forests and eucalyptus plantations, as well as moorlands with giant heath. Occurs from 1800m to 3900m in Ethiopia.
Status and distribution Ethiopian and Eritrean highlands, C. Kenya (Mt Kenya), W. Uganda, Rwanda and E. DR Congo. The status of this owl is unknown, but it is likely to be rare and locally endangered by forest and moorland destruction.
Geographical variation Two subspecies are described: nominate *abyssinicus* is found in Ethiopia and Eritrea; *A. a. graueri*, from C. Kenya and CW Uganda to Rwanda and E. DR Congo, is smaller and greyer than the nominate. Formerly treated as conspecific with Long-eared Owl (243).
Similar species Allopatric Long-eared Owl (243) is similar in size, but with shorter wings and longer ear-tufts, legs and bill less powerful, and facial disc paler; it also differs vocally. Partly sympatric Marsh Owl (249) is almost equal in size, but has a blackish-brown area surrounding dark brown eyes, and very tiny, barely visible ear-tufts.

◄ African Long-eared Owl; a little larger than Long-eared Owl *Asio otus,* and fairly dark above and strongly mottled below. This bird is of the nominate race. Dinsho, Ethiopia, January (*Paul Noakes*).

245. MADAGASCAR LONG-EARED OWL
Asio madagascariensis

L 36–51cm; W 260–340mm

Identification A medium-sized owl with long, graduated ear-tufts set wider apart than on other long-eared owls. Females are distinctly larger than males. Has blackish-brown forehead and crown with yellowish-brown flecks, and upperparts mottled, streaked and barred dark brown and golden-tawny. Flight feathers are distinctly barred pale and dusky, and the light brown tail has relatively broad dusky bars. Tan-coloured below with prominent dusky shaft-streaks and some cross-bars. Tawny-brown facial disc shades into dark brown around orange eyes, with narrow blackish rim and inconspicuous short pale eyebrows. Bill sooty-blackish. Powerful legs and toes are feathered yellowish-brown, with dusky claws. *Juvenile* Downy chick is white. Whitish mesoptile has blackish, mask-like facial disc and well visible ear-tufts, with wings and tail similar to adults.

Call Utters a long series of barking notes, *han-kan han-kan*, accelerating and increasing in volume, and dropping and fading at end. Also has a loud, lilting *ulooh*, repeated at intervals of several seconds.

Food and hunting Preys on mice, rats and small

lemurs, but also eats bats, birds, reptiles and insects. Hunts in forests and adjacent open areas.

Habitat Found in gallery forests and rainforests, but also accepts degraded forest. From lowlands to 1800m.

Status and distribution Endemic to Madagascar, where it is apparently rare and threatened.

Geographical variation Monotypic.

Similar species Partly sympatric Marsh Owl (249) has tiny ear-tufts and dark brown eyes.

▼ Madagascar Long-eared Owl's belly is very pale, with only a few fine, vertical dark lines. Perinet, Madagascar, September (*Paul Noakes*).

▼ Madagascar Long-eared Owl has a tan face, with dark around the eyes, pale 'eyebrows' and lores, and a distinct rim. Perinet, Madagascar, July (*Niall Perrins*).

246. STRIPED OWL
Asio clamator

L 30–38cm, Wt 335–556g; W 228–294mm

Identification A small to medium-sized owl with somewhat tousled but conspicuous, blackish ear-tufts. Females are up to as much as 150g heavier than males. Above, the plumage is tawny-buff, the forehead, crown and nape heavily dusky-streaked, the mantle and back with dark mottling and streaking. Scapulars have whitish areas on outer webs forming a clear row across the shoulder. Dark brown flight feathers and tail are broadly paler-barred. White throat and pale tawny to buffish-white underparts are prominently streaked dark brown or blackish. The brownish-white facial disc has a distinct blackish ruff, short whitish eyebrows extending from bill base to above the eyes, and whitish lores. The eyes are brown to cinnamon and the bill blackish. Feathered tarsi and toes are cream-coloured, with blackish claws. *Juvenile* Downy chick is whitish. Whitish-buff mesoptile is diffusely barred and washed greyish-brown above, has facial disc pale buffish and dusky-rimmed, dirty-white underparts with a slight greyish-brown wash, and on foreneck and upper breast some diffuse dark spots. *In flight* Flies over open areas with rather shallow and rapid wingbeats.
Call Emits a single nasal hoot lasting about one second; also gives series of well-spaced hoots at

intervals of several seconds, loudest and highest in middle. Hoots are higher in pitch than those of Long-eared Owl (243).
Food and hunting Feeds mainly on small mammals, bats, birds and other small vertebrates; also takes some insects, such as large grasshoppers. Hunts in flight or swoops down from fence posts or bare branches.
Habitat Prefers open or semi-open savannas and grass-lands, but also lives near humid forest edges, rice fields and airports. Absent from dense forests.
Status and distribution Ranges from S. Mexico south to Uruguay. Deforestation could locally benefit this open-country owl.
Geographical variation Four subspecies have been described: nominate *clamator* occurs from Colombia to C. Brazil; *A. c. oberi* is confined to island of Tobago, where last seen in 1971; *A. c. forbesi*, from S. Mexico to Costa Rica and Panama, is smaller and paler than the nominate; *A. c. midas*, from Bolivia to Uruguay, is the largest and palest race. A further, as yet unnamed subspecies has been found on the western slope in Ecuador and Peru, where confirmed records exist from more than ten localities; it appears to differ in voice, eye colour and morphology. This species has been placed variously in genus *Pseudoscops* and in genus *Rhinoptynx*, but is now moved back to *Asio*.
Similar species Sympatric Stygian Owl (242) is much darker, and has yellow eyes and dusky facial disc.

▲ Striped Owl of the nominate race *clamator*; this is the darkest and most heavily streaked subspecies. It is a long-eared, robust and pale-faced owl with brown eyes, and clear white 'eyebrows' and lores. Ochre-buff above and below. North-east Brazil, November (*Lee Dingain*).

▲▶ Race *midas* is the largest and palest of the subspecies. Buenos Aires, Argentina, September (*James Lowen*).

▶ A juvenile Striped Owl, spreading its wings to make itself look bigger. Its face has a distinct ruff and clear white 'commas' between the brown eyes, but otherwise this bird is still in its yellowish-white juvenile plumage, apart from the wings. San José, Costa Rica, May (*Daniel Martínez-A*).

◀ Race *forbesi* is smaller and paler than the nominate. Nicoya, Costa Rica, July (*Daniel Martínez-A*).

247. SHORT-EARED OWL
Asio flammeus

L 34–42cm, **Wt** 206–500g; **W** 281–335mm, **WS** 95–110cm

Identification A medium-sized owl with tiny, often hardly visible ear-tufts. Females are some 70g heavier than males. Has the crown and nape yellowish-brown with prominent dark streaks, and upperparts with similar ground colour marked with heavy dark streaking, mottling or blotches. Upperwing has a prominent dark carpal patch and, at base of primaries, a pale buff to whitish patch, rest of flight feathers being barred pale and dark. Shortish, slightly wedge-shaped tail has four or five dark bars in shallow arrow-shape. Underparts are pale dull yellow with clear dark streaks, these most prominent on upper breast and neck, becoming much fainter below belly; underwing appears whitish with dark carpal patch and tips. Facial disc is pale buffish-yellow to whitish with narrow dark-spotted white rim, and with dark areas around relatively small eyes; very short white eyebrows and whitish bristles around bill base. The eyes are yellow, the bill blackish-horn and the cere greyish-brown. Feathered tarsi and toes are light tawny or whitish-creamy, with greyish-horn claws. *Juvenile* Downy chick is pale ochre. Mesoptile has blackish face contrasting with yellow eyes and white eyebrows, otherwise entirely pale ochre-buff

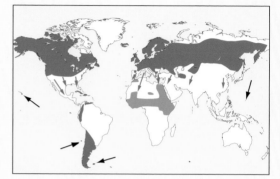

with dusky barring. *In flight* Flies with a few wingbeats followed by glide that appears somewhat wavering, with wings held forward and often in shallow 'V', occasionally also level.

Call Utters series of rather rapid hoots, *hoo-hoo-hoo-hoo….*, up to 20 at rate of about two to four per second. Song is repeated at variable intervals, in flight or from a perch.

Food and hunting Prefers voles, but also takes birds, other small vertebrates and, rarely, insects. Hunts by flying low over ground, quite often hovering, before swooping down on to prey; commonly perches on posts, on the lookout for potential prey.

Habitat Almost always found in open country, including tundra, savanna, pasturelands, swampy areas and large clearings at forest edges. Occurs from lowlands to mountain forests, up to 4300m.

Status and distribution Widely distributed in the world, but taxonomic status of many isolated island populations has been questioned. Locally common. Suffers from habitat loss caused by intensification of agriculture, which has led to worldwide decrease in populations. In C. Europe the average population has fallen to 400 pairs; in Finland 17–224 nests have been found annually between 2005 and 2010. Northern populations winter from the British Isles to the African Sahel zone, S. Asia and parts of SE Asia, N. Philippines and W. Hawaiian Islands, and from USA to S. & C. Mexico, Caribbean islands and S. Florida. The Hawaiian subspecies, *A. f. sandwichensis*, is now listed as Endangered.

◀ This Short-eared Owl has captured a fat Common Vole *Microtus arvalis*. Note that the ear-tufts are not visible. The Netherlands, April (*Lesley van Loo*).

Geographical variation Seven subspecies are listed: nominate *flammeus* has a wide distribution across the northern parts of the world; *A. f. bogotensis*, from Colombia to Peru and from Venezuela and Trinidad & Tobago to Suriname, is smaller and darker, with a more rusty wash; *A. f. suinda*, from S. Peru and Brazil to Tierra del Fuego, is similar to the nominate; *A. f. sanfordi*, from the Falkland Islands, is smaller and paler than the previous race; *A. f. sandwichensis*, from Hawaii, is generally more yellowish-grey; *A. f. ponapensis*, from Ponapé Island, in S. Pacific, is short-winged; *A. f. domingensis*, from Cuba, Hispaniola and Puerto Rico, is smaller than the nominate, and has very fine streaks on the belly and a darker ruff. This last, West Indian, subspecies is sometimes considered a full species in its own right on the grounds of plumage and vocal characters. Migrants from Asia have turned up as vagrants in a number of Pacific archipelagos, such as the Marshall Islands and Northern Marianas, and even Kure Atoll (NW Hawaii) and the Midway Islands, as well as Yap Island, in Micronesia. On all main Hawaiian Islands the subspecies *A. f. sandwichensis* is believed to be non-migratory, and thus isolated from continental taxa, and it has therefore been proposed that it be treated as a separate species, for which DNA confirmation is awaited. It is interesting that *A. f. ponapensis* has not been proposed for species status, despite its geographical isolation in the Caroline Islands; the fact that it is reportedly indistinguishable from *A. f. sandwichensis* makes the question of species status of the latter more complicated.

Similar species Sympatric Long-eared Owl (243) and almost fully allopatric Striped Owl (246) have prominent ear-tufts, and geographically overlapping barn owls have heart-shaped facial discs and dark brown eyes.

▼ Nominate Short-eared Owl. Outer Hebrides, Scotland (*Roger Wilmshurst*).

▲ The Hawaiian Short-eared Owl or Pueo *A. f. sandwichensis* has often been proposed as a separate species on biogeographical grounds. It looks very similar to continental forms; DNA results are awaited. Hawaii, November (*Kathleen Deuel*).

▲ ◄ Race *suinda* is darker rufous in colour, with a stronger bill and feet. Buenos Aires, Argentina, August (*Roberto Güller*).

◄ The nominate race of Short-eared Owl occurs in North America, Eurasia and northwest Africa. North American birds have stronger bills and feet. Washington State, USA, February (*Bonnie Block*).

▼ This Japanese Short-eared Owl (again the nominate) has a very white face. Japan, February (*Tadao Shimba*).

▲ Adult *sanfordi* in the short tussock grass of the Falklands; this island endemic race appears to be darker than mainland forms (*Martin B. Withers*).

► Short-eared Owl scanning field for prey. Belgium, February (*Lesley van Loo*).

248. GALAPAGOS SHORT-EARED OWL
Asio galapagoensis

L *c.* 35cm; **W** 278–288mm

Identification A medium-sized owl with small ear-tufts set near centre of forehead, but often concealed. Females are said to be larger than males, although no weights are known. The crown, nape and foreneck are buffish-rufous with broad sooty-brown streaks, and the upperparts are the colour of lava rock and mottled with buffish-rufous and ochre. The primaries are barred light yellowish-brown to dull yellow and secondaries are sooty brown with paler bars. Relatively short dark brown tail has rufous-buffish bars. Underparts are reddish-brown, always with distinct blackish-brown streaks. The buffish-rufous facial disc, with narrow dusky radial streaks, becomes paler towards the ruff, and has a large blackish mask-like area around the eyes. The eyes are sulphur-yellow, rimmed by blackish edges of eyelids, the bill is black and the cere dusky greyish. Tarsi are feathered to the distal phalanx of the greyish-brown toes, which have blackish claws. *Juvenile* Downy chick is brownish-white. Tawny-buff mesoptile has conspicuous black mask and pale yellow eyes. *In flight* Shows prominent dark area at carpal joint.

Call Utters a rapid *hoo-hoo-hoo-hoo-hoo*, five or six notes per second, thus being somewhat faster than that of Short-eared Owl (247).

Food and hunting Preys mainly on birds such as petrels, but also takes rats and mice. Often hunts by flying low over open areas. Able to kill birds considerably larger than itself. When dealing with large prey, such as boobies, the owl strikes with its talons at the back of the victim's neck; it is surprising that a large booby is unable to defend itself with its formidable bill, but it is caught unawares by the owl.

Habitat Found in open grasslands or areas of lava rock on larger islands.

Status and distribution Endemic to the Galapagos Archipelago, *c.* 1000km off Ecuador. It is considered vulnerable owing to human persecution and introduced cats, dogs and pigs. Study is required in order better to understand its conservation needs.

Geographical variation Monotypic. Formerly regarded as an aberrant subspecies of Short-eared Owl, but separated through long genetic isolation. This owl, which has adapted to harsh ocean-island conditions, merits further research to clarify its taxonomic relationships.

Similar species The only other owl in the Galapagos Islands is Galapagos Barn Owl (5), which is spotted below, with a heart-shaped facial disc, dark brown eyes and no blackish mask. Allopatric Short-eared Owl (247) is a little larger, paler and only streaked below, with ear-tufts less conspicuous and blackish mask reduced, and with prominent whitish eyebrows.

▶ From behind, this Galapagos Short-eared Owl shows well its heavily marked upperparts and dusky face. Genovesa, Galapagos Islands, September (*Lee Dingain*).

◀ Galapagos Short-eared Owl is smaller and much darker than the darkest Short-eared Owl *Asio flammeus* races in mainland South America. Genovesa, Galapagos Islands, September (*Fred van Olphen*).

249. MARSH OWL
Asio capensis

L 29–38cm, Wt 225–485g; W 284–380mm, WS 82–99cm

Other name African Marsh Owl

Identification A medium-sized owl with very tiny but erectile ear-tufts, often not visible. Females are some 50g heavier than males. Male is paler than female. The plumage is plain earth-brown above, crown and nape with fine buff vermiculations. Wings and tail are barred tawny and dark brown, buffish bases of primaries contrasting with darker primary coverts. Brown underparts are finely vermiculated buff, and become more uniformly light buffish on lower regions; underwing mostly buffish, with dark-barred tips and dark carpal patch. Light dull yellow facial disc has dark brown area around eyes, and narrow light-speckled dark rim. The eyes are dark brown, the bill blackish-horn and the cere grey-brown. Tarsi and most of toes are feathered, only dark brown tips of toes bare, with claws blackish. *Juvenile* Chick has buffish down, with pink skin and toes, and blackish bill. Mesoptile is buff, brown-barred above, with much darker facial disc than adult, with marked blackish ruff. *In flight* Flies with slow but powerful wingbeats, the very conspicuous dark wrist-patch visible from above and below.

Call Utters a hoarse *kerrrrrr*, repeated at variable intervals. Gives faster and louder croaks when flushed.

Food and hunting Feeds on small rodents and birds, bats, frogs and lizards; also takes scorpions and grasshoppers. Hunts by flying close to the ground.

Habitat Prefers open country, from coastal and inland

marshes to savannas, but avoids extensive long grass and forested areas. Occurs from near sea level to the hills; up to 1800m in Madagascar.

Status and distribution Found patchily from NW Morocco south to the Cape in South Africa, and in Madagascar. It is locally common in 20 countries south of the Sahara, but little studied. Migrants and vagrants seen in a further six African countries from Mauritania to Congo. Population in Morocco is estimated to be between 40 and 150 pairs, and is expected to decline further in the near future.

Geographical variation Three subspecies are described: nominate *capensis* occurs south of the Sahara; *A. c. tingitanus*, from Morocco, is darker with rufous markings, especially on the underparts, contrasting with white spotting; *A. c. hova*, from Madagascar, is much larger than other subspecies and has more distinct bars overall. Taxonomy requires further study, including molecular-genetic analyses.

Similar species Sympatric African Grass Owl (16) is larger, dark brown above and pale below, and has a heart-shaped facial disc, relatively small blackish eyes, and longer legs with bristled toes. Short-eared Owl (247), partly sympatric in winter or on migration, has yellow eyes and pale yellowish-brown plumage with distinct dusky streaking below.

▲ Nominate Marsh Owl has a whiter face, white tail barring and more yellow in the wings. Nigeria (*Tasso Leventis*).

▶ Marsh Owl of race *tingitanus* showing the uniformly very dark back when closed wings cover the yellow of the primaries – in this photo only one small ear-tuft is erected. Merja Zerga, Morocco, February (*Daniele Occhiato*).

◀ Marsh Owl has long wings, flying with powerful wingbeats. The primaries have a large yellowish area and brown bars. Dark wrist-patch visible from above and below. Race *tingitanus*, Merja Zerga, Morocco, February (*Daniele Occhiato*).

FURTHER READING

Aebischer, A. 2008. *Eulen und Käuze.* Haupt Verlag, Bern.

BirdLife International. 2000. *Threatened Birds of the World.* Lynx Edicions and BirdLife International, Barcelona and Cambridge, UK.

Baudvin, H., Génot, J. C., & Muller, Y. 1991. *Les rapaces nocturnes.* Éditions Sang de la Terre, Paris.

Burton, J. A. (ed) 1992. *Owls of the World.* Revised Edition. Eurobook, London.

Burton, J. A., & Schwarz, J. (eds) 1986. *Eulen der Welt.* Neumann-Neudamm, Melsungen.

Chiavetta, M. 1992. *Guida ai rapaci notturni.* Zanichelli Editore, Bologna.

del Hoyo, J., Elliott, A. & Sargatal, J. (eds) 1999. *Handbook of the Birds of the World. Vol. 5. Barn-owls to Hummingbirds.* Lynx Edicions, Barcelona.

Duncan, J. R. 2003. *Owls of the World , Their Lives, Behavior and Survival.* Firefly Books, New York.

Freethy, R. 1992. *Owls. A Guide for Ornithologists.* Bishopsgate Press, Hildenborough.

Fry, C. H., Keith, S. & Urban, E. K. (eds) 1988. *The Birds of Africa. Vol. III.* Academic Press, London.

Hollands, D. 1991. *Birds of the Night.* Reed Books, Balgowlah.

Hume, R. & Boyer, T. 1991. *Owls of the World.* Dragon's World, Limpsfield, Surrey.

Johnsgard, P. A. 2002. *North American Owls.* Second Edition. Smithsonian Institution Press, Washington and London.

Kemp, A. & Calburn, S. 1987. *The Owls of Southern Africa.* Struik, Cape Town.

König, C., Weick, F. & Becking, J.-H. 2008. *Owls of the World.* Second Edition. Christopher Helm, London.

Martínez , J. A., Zuberogoitia, Í. & Alonso, R. 2002. *Rapaces Nocturnas Guia para la determinación de la edad y el sexo en las Estrigiformes ibéricas.* Monticola Ediciones, Madrid.

Mebs, T. & Scherzinger, W. 2008. *Die Eulen Europas.* 2. Auflage. Kosmos Verlag, Stuttgart.

Mikkola, H. 1983. *Owls of Europe.* T. & A.D. Poyser, Calton.

Mikkola, H. 1995. *Rapaces Nocturnas de Europa.* Editorial Perfils, Lleida.

Mikkola, H. 1995. *Der Bartkauz.* Die Neue Brehm-Bücherei Bd. 538. Westarp Wissenschaften, Magdeburg.

Morris, D. 2009. *Owl.* Reaktion Books, London.

Newton, I., Kavanagh, R., Olsen, J. & Taylor, I. 2002. *Ecology and Conservation of Owls.* CSIRO, Collingwood.

Sparks, J. & Soper, T. 1989. *Owls: Their natural and unnatural history.* Revised Edition. David & Charles, Newton Abbot.

Steyn, P. 1984. *A Delight of Owls. African Owls Observed.* Tanager Books, Dover, New Hampshire.

Tarboton, W. & Erasmus, R. 1998. *Owls and Owling.* Struik Publishers, Cape Town.

Tyler, H. A. & Phillips, D. 1978. *Owls by Day and Night.* Naturegraph, Happy Camp, California.

Voous, K. H. 1988. *Owls of the Northern Hemisphere.* Collins, London.

Weick, F. 2006. *Owls (Strigiformes): Annotated and Illustrated Checklist.* Springer, Berlin & Heidelberg.

PHOTOGRAPHIC CREDITS

Copyright information for the photographs in this book; t = top, b = bottom, c = centre, l = left, r = right.

Abhishek Das 293t: **Adam Riley/Rockjumper Bird Tours** 135, 245bl, 248l, 249, 295b, 299, 376, 452: **Adam Scott Kennedy/rawnaturephoto.com** 274l, 361tr, 404l: **Adrian Boyle/www.wildlifeimages.com.au** 107, 108r, 445t, 450t, 450b: **Agami** 120: **Alain Pascua** 170, 171br, 319tl, 457: **Alex Vargas**, 380l, 380r: **Ali Sadr/www.birdsofiran.com** 265tl, 419: **Amano Samarpan** 155bl, 269r, 396tl, 396b, 431t: **Amir Ben Dov** 312, 313, 423bl: **András Mazula** 121: **Andrés Manuel Domínguez** 28r, 38c, 43t, 122, 123, 311tr: **Anne Elliott** 43b, 347: **April Conway** 301l, 301rt, 301rb: **Arpit Deomurari** 150, 318, 319tr: **Arthur Grosset** 215tr, 245br, 260, 272, 377tl, 381l: **Arto Juvonen/Birdphoto.fi** 127b: **Atle Ivar Olsen** 49b, 136: **Augusto Faustino** 311bl: **Aurélien Audevard** 365br: **Bence Mate/Agami** 62: **Bill Baston/FLPA** 296: **Bonnie Block** 498bl: **Bram Demuelemeester**, 12, 59, 101, 112l, 182, 188, 288, 464, 465l, 465r, 466t, 466bl, 471, 476: **Carmabi Institute/Peter van der Wolf** 86l, 86r: **Charles Marsh** 141l, 141r: **Ch'ien C. Lee/wildborneo.com.my** 162: **Choy Wai Mun** 286, 294, 316, 320: **Chris Brignell** 17r: **Christian Artuso** 43c, 48b, 49t, 52b, 63, 153tr, 203b, 209t, 210, 215br, 217tl, 217tr, 217b, 220, 221b, 224l, 226t, 226b, 231b, 245t, 258b, 295tl, 333, 353b, 353tr, 391b, 417r, 420rb, 437b, 487b: **Christian Fosserat** 36b, 46t, 54t: **Chris Townend** 310: **Chris van Rijswijk** 37t, 345l, 345r: **Chris van Rijswijk/Agami** 23c, 34t, 77, 81bl, 81br, 341tl, 341b: **Daniel Martínez-A** 215tl, 304, 328r, 373l, 494, 495br: **Dan Lockshaw** 235t: **Daniele Occhiato - www.pbase.com/dophoto** 267b, 267t, 309, 354, 420l, 422, 502, 503b: **Danny Laredo** 36t: **Dario Lins** 233: **Dave Watts** 110, 111l, 455: **David Ascanio** 236: **David Behrens** 287: **David Hollands** 111r, 113, 478: **David Hosking/FLPA** 443tl: **David Jirovsky** 423t: **David Monticelli**, 127tr: **David Shackleford/Rockjumper Bird Tours** 277b, 300l: **David W. Nelson** 235b: **Deborah Allen** 341tr: **Desmond Allen** 161bl, 161br: **Dick Forsman** 277t: **Dieter Hopf/Imagebroker/FLPA** 491b: **Dominic Mitchell** 378: **Donald M. Jones/Minden Pictures/FLPA** 48t, 418tl: **Doug Wechsler/VIREO** 181: **Dubi Shapiro** 139l, 139r, 145, 335t, 390, 409: **Dušan M. Brinkhuizen** 219b, 226: **Edwin Collaerts** 103tl, 103tr, 482l, 482r: **Eric Didner** 89l, 89r: **Eric Sohn Joo Tan** 479: **Eric VanderWerf** 298, 337: **Erica Olsen** 41: **Esko Rajala** 52t, 254, 359bl: **Evgeny Kotelevsky** 47c, 488: **Eyal Bartov** 48c, 79, 104, 125tr, 246: **Fabio Olmos** 90, 205, 261b: **Filip Verbelen** 165, 166, 167t, 167b, 187, 191b, 192, 447, 448, 449t, 449br, 451t, 458b, 475: **Fotonatura** 279: **Frank Lambert** 326r: **Franz Steinhauser/wedaresort.com** 185b: **Fred van Olphen** 67b, 68t, 500: **Friedhelm Adam/Imagebroker/FLPA** 435b: **Gaku Tozuka** 38b, 291t: **G. Armistead/VIREO** 394: **Gary Thoburn** 311br, 379l, 379r, 398l, 398r: **Gehan de Silva Wijeyeratne** 283: **Glenn Bartley** 223, 231t: **Guilherme Gallo-Ortiz** 331, 381r: **Guy Dutson** 468, 483: **Han Bouwmeester/Agami** 359tl: **Hanne & Jens Eriksen/birdsoman.com** 128, 129b, 266, 423br: **Hans Germeraad/Agami** 39b: **Harri Taavetti** 21b, 34b, 44b, 66, 151t, 297t, 297b, 346tl, 348l, 356, 357r, 359tr, 401tr, 431b, 432, 433, 456: **Harri Taavetti/FLPA** 349, 350t: **Heimo Mikkola** 60: **Hennie Lammers** 487tl, 487tr: **Hira Punjabi** 395: **Holly Kuchera** 4: **Hugh Harrop** 23b, 157, 346b, 430b: **HY Cheng** 19cr, 284, 295tr, 317bl, 317br, 428: **Ian Fisher** 47t, 80, 426, 427bl, 427t: **Ian Merrill** 117, 144bl, 146t, 172, 193, 194, 221t, 278, 319b, 403, 415b, 463t: **Imagebroker/FLPA** 265tr, 290: **James Eaton/Birdtour Asia** 17l, 106l, 106r, 119, 153b, 161tr, 164, 173, 175, 177, 178, 180, 184l, 186, 195, 251, 321, 323, 325l, 342, 343t, 343b, 399, 446, 473, 474t, 477: **Jacques Erard** 211l: **James Lowen - www.pbase.com/james_lowen** 257r, 261l, 327l, 389t, 389bl, 418tr, 495tr: **János Oláh** 308l, 353tl, 439: **Jan Vermeer/Minden Pictures/FLPA** 346tr: **Jari Peltomäki/Birdphoto.fi** 22t, 68t, 350b: **Jayesh Joshi** 413, 414, 415tl, 415tr: **Jérôme Micheletta** 190: **Jesus Contreras** 54bl: **Jim & Deva Burns** 24, 35b, 37b, 84l, 84r, 198, 199r, 206l, 207, 258tl, 340, 369l, 369r, 370, 371tr, 371br: **John Carlyon** 405br: **John Hicks** 453tr: **John Mittermeier** 87: **John & Jemi Holmes** 154, 396tr, 435t: **John Eveson/FLPA** 270: **Jonathan Martinez** 28tl, 324: **Jonathan Newman** 219t, 368: **Jon Groves** 357t: **Jon Hornbuckle** 224r, 240b, 361bl, 364, 383, 480: **José Carlos Motta-Junior** 31, 46b, 49c, 54br, 82, 416: **Juan Matute/vultour.es** 420rt: **Jussi Sihvo** 19tr: **Kai Gauger** 125b: **Karen Hargreave** 351, 391tl: **Kathleen Deuel** 498tr: **Kazuyasu Kisaichi** 158, 160, 344, 458tr: **K.-D. B. Dijkstra** 133l, 133r: **Keith Valentine/Rockjumper Bird Tours** 144t: **Kenji Takehara** 159t, 159b: **Kevin Lin** 149: **Kleber de Burgos** 386: **Knut Eisermann** 206r, 208, 209b, 328l, 329, 339l, 339r, 371bl, 374, 375tr, 375tl, 375b, 438: **Lars Soerink/Agami** 21tl: **Lee Dingain** 232, 307, 311tl, 495tl, 501: **Lee Mott** 45br, 491tl: **Leif Gabrielsen** 242b: **Lesley van Loo** 44t, 46c, 124, 311cr, 359br, 418b, 421, 496, 499b: **Lev Frid** 242t: **Lucas Limonta** 238: **Lucian Coman** 105b: **Luiz Gabriel Mazzoni** 326r: **Mandy Etpison** 243: **Marco Mastrorilli** 20: **Mario Dávalos P.** 95, 96, 97: **Marcel Holyoak** 109l, 109r, 127tl: **Mark Bridger** 16: **Marko Matesic** 491tr: **Markus Lagerqvist** 213, 214t, 214bl, 461r, 469: **Markus Lilje/Rockjumper Bird Tours** 105t: **Martin B. Withers/FLPA** 404, 440, 199t: **Martin Goodey** 314: **Martin Gottschling** 56, 430tl: **Martin Hale** 325r, 401tl: **Martjan Lammertink** 118, 385tr: **Mathias Putze** 265b, 427br: **Matt Bango** 38t, 436: **Matt Knoth** 29bl: **Matthias Dehling** 230, 388, 389br: **Matti Suopajärvi** 19bl, 22b, 47b, 51, 358: **Menno van Duijn/Agami** 21c, 359cr: **Michael & Patricia Fogden** 212, 234, 275: **Michael R. Anton** 102, 289tl, 289tr, 289b: **Michelle & Peter Wong**, 156l: **Mike Danzenbaker/Agami** 146b, 147, 161tl, 199l, 200t, 204, 211r, 306, 334, 335b, 366, 367tr, 373r, 462, 463b: **Miorenz** 29tr: **Murray Cooper** 385tl: **Neil Bowman/FLPA** 222, 397, 400, 401b: **Niall Perrins** 361br, 406, 493br: **Nick Athanas** 85t, 229: **Nick Gardner/**

INDEX